A HISTORY OF WESTERN SCIENCE

A HISTORY
OF
WESTERN SCIENCE

Anthony M. Alioto

PRENTICE-HALL, INC., ENGLEWOOD CLIFFS, NEW JERSEY   07632

Library of Congress Cataloging-in-Publication Data

Alioto, Anthony M. (date)
 A history of western science.

 Bibliography: p.
 Includes index.
 1. Science—History. I. Title.
 Q125.A475 1987   509.4   86-15158
 ISBN 0-13-392390-8 (pbk.)

Editorial/production supervision and
 interior design: *Hilda Tauber*
Cover design: *Wanda Lubelska*
Manufacturing buyer: *Ray Keating*

© 1987 by Prentice-Hall, Inc.
A Division of Simon & Schuster
Englewood Cliffs, New Jersey 07632

Printed in the United States of America

10 9 8 7 6 5 4 3 2 1

ISBN 0-13-392390-8   01

PRENTICE-HALL INTERNATIONAL (UK) LIMITED, *London*
PRENTICE-HALL OF AUSTRALIA PTY. LIMITED, *Sydney*
PRENTICE-HALL OF CANADA INC., *Toronto*
PRENTICE-HALL HISPANOAMERICA, S.A., *Mexico*
PRENTICE-HALL OF INDIA PRIVATE LIMITED, *New Delhi*
PRENTICE-HALL OF JAPAN, INC. *Tokyo*
PRENTICE-HALL OF SOUTHEAST ASIA PTE. LTD., *Singapore*
EDITORA PRENTICE-HALL DO BRASIL, LTDA., *Rio de Janeiro*

Contents

PART V   **THE SECOND REVOLUTION AND BEYOND**

Preface

Over four thousand years ago the Giza plateau was a vast desert of shifting sands dotted by a few small tombs. On this boundless surface an Egyptian architect staked out the ground plan of a huge pyramid for the Pharaoh, Cheops, of the Fourth Dynasty. Surveyors laid a base 755 feet on each side, and here was eventually erected a structure that reached a height of 481 feet, covered some thirteen square acres, and contained 2.5 million blocks of stone, each weighing two and a half tons. The Great Pyramid stands today, an imposing testimony to the power of the human imagination and the possibilities of human achievement.

Western science, like the construction of the Great Pyramid, is a bringing of order into what was once a desert. Enclosed by an ever-expanding perimeter, first staked out by the human mind, the once-untamed natural phenomena of science are made comprehensible for civilized humanity.

Yet the Great Pyramid is more than a metaphor for scientific achievement. Its massive blocks reveal an entire cultural complex: the religious beliefs and aspirations, the social hierarchy, the politics of the people whose arduous labors caused it to be. Likewise the history of science reveals not only a vast complex of religious, social, political, cultural, and other external factors, but also the mentality, the psyche of the peoples who engaged in it. Any study must include the psychological components. The rational and irrational motivations which drive human beings are part of the story of science.

My purpose is to tell briefly the amazing and unfinished history of Western science from its origins in the ancient world to its most modern achievements. In doing so I make the following assumptions. First, human beings think about nature as individuals and as groups in a society, and ponder esoteric problems inherited from tradition or immediate concerns posed by society, by technology, by experiment, or by observation. In some cases, science is simply the free play of "idle" imagination. Hence, the foundations of any history of science must be ideational—the story of human attempts to rise above brute instinctual concourse with the world and to bring order where apparent order is concealed. Thus an important goal of this book is to bring to life the individuals who have been instrumental in the advancement of science. I wish to convey the excitement and sense of mystery which inspire human beings to undertake the daunting task of trying to make the universe comprehensible. If the reader is left with a strong desire to learn more, then I will have succeeded in this goal.

The second assumption, of equal importance, is that science does not develop in a vacuum: All scientists are part of their time, society, and culture. The history of Western science presupposes, and is interconnected with, the history of Western civilization. For example, medieval science spoke in terms of essence, qualities, divine and final causes, thereby reflecting the influence of the Church and of ancient Greece. In contrast, eighteenth-century science used words like force, mass, inertia, and mechanism, reflecting instead the worldly preoccupations of the times, Pythagoreanism, possibly the revival of Hermetic magic, and a new level of abstraction begun by Galileo and brought to fruition by Newton. In order to communicate knowledge, scientists must speak in the language of their age; at times the language must be altered. Changes in the scientific mode of discourse may influence the meaning of knowledge itself or may be influenced by the external cultural milieu. Often the two are interrelated: The history of science blends almost imperceptibly with social and cultural history.

My third and final assumption is that the historian's task is to present information in a manner and language the reader understands. To accomplish this, I have chosen the narrative form—the style of a storyteller. Footnotes are kept to a minimum and used generally to draw attention to sources of further enlightenment. I have tried not to presume prior knowledge on the reader's part, nor to preconceive a special audience. Curiosity and a willingness to become involved in a very human adventure—these are the only prerequisites. To keep the book a manageable size, some omissions were necessary and some rather complicated scientific and cultural developments had to be reduced to a mere paragraph or sentence. But brevity need not foster confusion, and I hope readers will find the text both understandable and interesting.

In seeking to know the world, we inevitably try to humanize it. All scientific endeavors are an attempt to bring order and meaning into a fundamentally chaotic universe. Despite the ever-increasing specialization of the various sciences today, the human soul continues to harbor a profound desire for some unified view, some answer to the question: *Why all this?* A general history of science cannot and should not presume to give the answer. In telling the story, I hope to illuminate the path to understanding. The rest is up to you.

The history of science has been studied and documented at length. In addition to the suggestions for further reading at the end of each chapter, the following guides to bibliographic sources may provide useful background. The works of the great thinkers and scientists themselves are usually the most rewarding readings.

COHEN, I. BERNARD, ed. *The Development of Science: Sources for the History of Science.* 60 vols. New York, Arno Press, 1981. Reprints of important papers and books on various issues and subjects.

COHEN, ROBERT S., HIEBERT, ERWIN N., MENDELSOHN, EVERETT I., eds. *Studies in the History of Modern Science.* 14 vols. Boston: D. Reidel Publishing Co., 1977–84. Monographs on various subjects.

CORSI, PIETRO, and WEINDLING, PAUL. *Information Sources in the History of Science and Medicine.* 1983. London: Butterworth Scientific, 1983.

GILLISPIE, CHARLES C. *The Dictionary of Scientific Biography.* 16 vols. New York: Charles Scribner's Sons, 1970–80.

GUERLAC, HENRY. *Science in Western Civilization: A Syllabus.* New York: The Ronald Press, 1952.

SARTON, GEORGE. *Horus, A Guide to the History of Science.* New York: The Ronald Press, 1952.

Introduction

We live in a scientific civilization. Daily we hear of new inventions and new advances made by scientists in many fields. The pace of such advances today is greater than in any other historical epoch; indeed, many believe that science is the greatest achievement of the modern age. A wide gulf does separate us from earlier cultures; many believe, too, that the difference is science.

We are often amused to discover how earlier peoples explained natural events. The notions that the planets were deities, that the earth was the center of the universe, or that fever was an evil spirit which stalked the desert appear to us childish and naive. But much scientific perception is attitudinal. Behind our manipulation of nature—as behind earlier outlooks—there exists a certain way of seeing the world, a particular perspective which molds nature into a specific form. We tend to pride ourselves on this so-called rational or scientific outlook. Reason helps us correctly correlate our experiences in a consistent way, telling us what to expect from certain natural events and demonstrating what to rule out. Self-consistency, either/or logic, predictability, experiment, proof—such terms are all associated with scientific attitude. The technological marvels which spring from these attributes tend to solidify in our minds the "truth" of modern science.

Most people probably equate the very idea of accumulating reliable knowledge with science. Indeed, the history of science is basically a celebration of progress—of assembling an ever-increasing volume of facts to support progressively superior theories and concepts. Further, the advance of technology and its apparent success in reshaping the world have led many to believe that, beneath it all, a true and real picture, a scientific mode of understanding, actually does exist. Earlier attempts to understand the world, now repudiated or superseded by modern developments, were "unscientific" or simply false.

The first step in unraveling the history of science is to drop this attitude and dispel this parochial view of the past. Different ages have had different goals, have held different values, have asked different questions; their answers, their science, have assumed forms conditioned by, indeed dependent upon, these goals, values, and questions. The word *science* comes from the Latin *scire*, meaning to know or to understand, but knowledge can itself be understood and appreciated only *within* the culture which gives it birth. Science is a uniquely human creation tied to the emergence of culture. This is especially true of modern science: our current picture of the universe has evolved from our changing views of it over time. Given this process, no historical system can be labeled "unscientific." Because observational language itself may change with theory, and because a fact today may not be a fact tomorrow, every attempt to comprehend reality (the notion of "reality" may itself be a very human assumption) is worthy of our appreciation; deserving of our respect, and entitled to the name *science*. And so a history of science that begins in antiquity and continues into the modern world must give equal weight to every era, judging each by its own standards, aspirations, values, and presuppositions.

Science is very old, then, perhaps as old as human beings. It arose from the need to solve problems, the greatest of which was simple survival. Human beings have had "to get along" in nature in order to survive; such adaptation is called technology. But often technology results from observing and learning from experience, and not from actually understanding. For example, human beings were able to practice metallurgy, predict the seasons, and heal the sick, before they understood how or why they were able to do these things. The interrelatedness of science and technology does not mean, however, that the two are tautologous. Modern science is often highly abstract. When viewed from the perspective of mathematical physics, for instance, the universe itself becomes abstract—just as it becomes full of occult and supernatural forces when considered from the perspective of myth. Both views, however, are essentially attempts by human beings to understand and explain natural events.

The magical rites of the ancient world can thus be compared, in this sense of participation, with the technology of today: both are based on historically developed attitudes. The history of science is the story of those developments. The methods implied are important only insofar as they affect the mode of bringing ideational order to nature; the evolution and change of the presuppositions and models constitute the actual structure of what different ages have labeled "science."

The hominid family may date back some 15 million years; some branches of the hominid tree died out, but one led eventually to *homo sapiens*. The most important period of human evolution lies somewhere between 1 million and 10,000 years ago, when *homo sapiens* evolved from *homo erectus*. Accompanying this long development was a change in stone tools. We know that our distant ancestors in East Africa began using stone tools for specific tasks at least 2.5 million years ago or even earlier. We also know that the early hominids learned to shape stones by striking one against the other, such that the forms actually became more or less standardized. Then, about 100,000 years ago, more specialized tools appeared. Perhaps most important of all, Paleolithic (from the Old Stone Age, 1 million to 8000 B.C.) humans discovered the percussion method for producing fire. Hominids had become *homo sapiens*, "man the thinker."

Many animals modify their environments and many animals communicate. Only, however, when technology merges with culture, when a formal thinking pattern mirrors a purposeful thinking process, does science exist. The development of stone tools points to something more than simple survival, and that something is the beginning of science.

Somewhere in those dim ages human beings began to think symbolically. Their thoughts about nature were pure inventions of their human minds. They began to grasp their daily experiences through symbols, through patterns of thought that reflected patterns in nature. Here was a remarkable leap, for in the communication of their shared symbols the early hominids discovered a means to share experience, to order it, perhaps even to understand it. This was truly a momentous change; science since then has been one or another variation in the *means* of representing nature, both in thought and in action. In a general sense, human beings have always been scientific.

If this unique blend of imposing our symbols upon nature and transforming what we perceive accordingly is the hallmark of *homo sapiens*, we can thus assert that science *has* been around from the very beginning of the species. Of course, it was not *our* science, the science of the present. Nonetheless it greatly facilitated survival, because this ability to deal with thoughts and experiences symbolically has a tremendous liberating power. It breaks through the limitations of experience and forges a means of transcending time and passing down to future generations both the successes and failures of the present. Hence science itself has a unique evolutionary character, and it is no surprise that the survival (and possible destruction) of *homo sapiens* is intimately connected with the development of a transferable culture.

What gave rise to this sense of order, this discovery of patterns which could be represented and communicated through symbolic thought? Perhaps it was nature itself, or better, human confrontation with natural events. Simple observation reveals repetitious sequences of events. The sun rises in the east and sets in the west; the moon goes through an entire pattern of phases never separated by more than thirty days or less than twenty-nine; the seasons change; humans themselves are born, follow a regular process of growth, and die. The world is full of repetition.

Yet there are irregularities too: eclipses, storms, earthquakes, accidents, floods, deviations much more difficult to explain or represent by ideas or symbols of order. Modern rational science assumes an order nonetheless, and when it is not apparent, still expects that it will reemerge shortly, once the problem is studied. Although nature may have given the initial impetus to the search for order, at times it also defies humans to discover where the regularity lies. This is as true today as it was in ancient times. One pioneer of the quantum theory, Werner Heisenberg, recalled the early days of his work in the 1920s and his discussions with Niels Bohr that lasted until late at night. Heisenberg and Bohr often asked themselves if nature was as absurd as it seemed to be in their atomic experiments.[1] So even in modern times the symbolic rationality of science may not be the rationality of the world. The great challenge has *always* been to grasp the order of nature in thought and so to rise above the unpredictable. The history of science is the story of how human beings have met this challenge. Some past attempts to meet this challenge seem to us crude; our challenge remains to forge the historical link between ourselves and our distant relatives—and to do so less crudely.

[1]Werner Heisenberg, *Physics and Philosophy* (New York: Harper & Row, 1958), p. 42.

1

~~~~~~

Man, Gods, and the Cosmos

How did ancient people understand the world? What were the first attempts to grasp both the order and disorder in nature? Today we understand phenomena through reason, and hence we define our science by rational methods. In ancient times people also based their science on how they made things intelligible. The earliest science was myth, usually communicated in poetry—mythopoetical thought.

Modern science abstracts from experience and so is able to relate its objects and events and form them into universal laws which make their behavior predictable under given circumstances. However, strictly speaking, experience itself is never the same as the concept of it, and the kind of science which deals with such objectively conceptualized events is a recent development when viewed from the total history of human beings.

It took over four thousand years for humans to develop what we label "objective science," and, as we shall see in later chapters, the possibility of strict objectivity is being questioned even today, especially in physics. Although it is natural to want to trace the origins of modern science back into early history, we must avoid misreading the past. Ancient people, lacking our historical heritage, did not view nature in orderly, systematic patterns; rather they viewed each event as a unique experience that called forth both emotional and intellectual responses. Since every phenomenon was seen as having an identity all its own, an-

cient people comprehended life experiences in terms of their own re-sponses to them. Whereas modern science is impersonal and abstract, ancient science was a personal confrontation between the individual and nature. Thus the early peoples in the Near East did not separate their subjective responses from their representations of nature.

We tend to make a distinction between the experience of nature and the scientific concept of it. Ancient people did not do this, nor did they limit any event to an abstraction—an "it."[1] Why should they? The expe-rience of an eclipse is personal to every individual, full of significance, and not limited by any abstract concept. Thus nature and the human experience of nature do not stand in opposition, and the emotional re-sponse is just as real as the intellectual. Given this highly individual view of experience, nature possessed the same qualities as human be-ings: *it was animate from end to end*. Observation of nature was a participa-tion with it as a living thing, and every encounter was unique.

How did these people communicate such personal experiences? How did they symbolize something which resisted abstraction—in this case all of nature? Because of the personal uniqueness of events, every experience was by necessity a story. In telling the story a man included everything of himself; it spoke to the emotions and imagination as much as to the reason. Nature was seen totally from a point of view that un-derstood order and disorder in terms of human life. It did not occur to ancient people, until the Greeks, to separate the imagery of the story from the events it was meant to communicate. Mythical symbols and names were intrinsic parts of the object, just as were emotional re-sponses.

There is a certain aesthetic pleasure in the exactness of mathematics which appeals to the desire for order and certainty. Yet, the modern mathematician would probably not include this subjective feeling when using a proof to represent something in nature. But people would do so who see nature in the total human frame of reference. Duality, for exam-ple, is one of the universal experiences of human life. There are two sexes, two eyes, ears, arms and legs; there is night and day, the sun and the moon. The number two refers to more than an abstract number con-cept: it refers to fundamental experience. The number three also emerges from such experience: the trinity of the family; the sky, earth, and humanity. Experience also presents people with four directions; a fifth direction is the center, a sixth and seventh are the sky and the earth. The number one is the primal creator, the origin of all.

Of course, ancient people discovered the practical use of numbers— for keeping accounts of their resources, for grouping their tribes, and so

---

[1]H. and H. A. Frankfort, J. Wilson, T. Jacobsen, and W. Irwin, *The Intellectual Adven-ture of Ancient Man* (Chicago: University of Chicago Press, 1946), p. 5.

on—and thus probably used arithmetical operations and some concept of a number base. However, these practical uses did not preclude attributing mythical significance to numbers; we still see remnants of such subjectivity regarding lucky numbers and unlucky days.

The same subjectivity applies to space and time. The ancients' initial experiences of direction in space were personal and hence communicated mythologically. Some areas were friendly—the Nile valley in Egypt—and others hostile—the desert. The flooding of the Tigris and Euphrates rivers in Mesopotamia was erratic and brought destruction to people, crops, and animals. These people, therefore, made qualitative distinctions between earth and the waters of chaos. The Babylonian epic of creation, the *Enuma Elish*, personified the chaos as the primal goddess Tiamat, and the world's ordering as the victory of the god Marduk over her. The life-giving sun sinks every dusk into the western horizon; thus the west is associated in Egypt with death, and the east with life. The building of a pyramid repeats the mythic creation of the primal "hill" out of the chaos, and the very ground upon which the pyramid or temple stands is sacred. Like the primal creative act, the building itself gives reality to the world.

The heavens, too, were ordered by the human mind into regular patterns of celestial movement. In the spring the sun climbed high into the sky and the crops ripened; in the fall it dropped and plants died. Such a cyclical pattern had human significance, especially in agricultural societies. Nor would there have been any reason to separate meteorological and astronomical events—a storm could destroy the crops as surely as could the onset of winter. People naturally reflected upon the periodicity of such events and noticed the similarities between their own biological time and the succession of natural phenomena. Time's passage is thus rendered in human terms, symbolized in periodic ceremonies and rites. The entire outlook is not only geocentric (earth-centered) but homocentric—cosmic time is understood through human experience.

Predictability of such meteorological events as floods, seasons, and other changes was thus of great concern, given that human survival depended upon anticipating them. This led to the first astronomy and the use of numbers to calculate the positions of heavenly bodies, especially in Mesopotamia. Still, the need to predict such events carried an emotional and religious significance which cannot be separated from the practical. The harmony of celestial and earthly events represented a unity which resisted fragmentation. The planets were deities, and the heavens were sacred. Observation of the heavens was thus a religious occupation, and the first astronomers were priests.

The integration of modern science into society involves relating technology and commonly held beliefs and presuppositions to the nature of reality. For the ancients, who viewed the world through the spectacles of myth, the idea of participation in the rhythms of nature required them to form images and activities to ally themselves with the community. Here, myth passes into ritual and ceremony. A basic human experience is the trauma of birth; the earliest mythical symbols of creation dealt with a primal struggle, like the battle between Marduk and the monster Tiamat. Ritually, the cosmic struggle was repeated in rites of passage, the coming of maturity, and in ceremonies involving the harvesting of the crops. In short, the fundamental experiences of existence were sanctified by their identification with the cosmic myth. The sense of participation was objectified in the ritual.

The ceremony itself became the primal event, for by repeating the mythic gestures and rites, individuals overcame time itself and were transported back to the original moment of creation. Ritual and ceremony are timeless in the same way that Newton's law of gravity or Maxwell's equations are—they may be applied as they were in the seventeenth and nineteenth centuries. In pursuit of scientific education today we participate as ancient people participated in the cosmic rhythms of myth through which they understood nature. Ritual is mythology come alive. Joseph Campbell puts it this way:

> [Rites] are physical formulae; written not in the black on white, of, say, an $E = mc^2$, but in human flesh . . . .[M]ythology is a primitive prelude to science.[2]

So far we have been speaking in general terms, using the very abstraction—"ancient people"—that mythopoetical thought rejects. In fact, evidence of science as an entity remains scanty until we come to Egypt and Mesopotamia. Agriculture developed sometime during the Neolithic or New Stone Age (roughly 10,000–3,000 B.C.). Even so, the picture of ancient science did not become clearer until the advent of writing. Consider Stonehenge in southern England. Giant stone monoliths set into the ground form a series of awesome circles. Guides describe how the stone was brought from hundreds of miles away and how we think the Neolithic farmers set the massive slabs into place. Stonehenge is a truly remarkable feat of engineering, and there are others like it. Certainly their creators knew some physics and mathematics; perhaps Stonehenge had something to do with astronomy.[3] So we ask:

---

[2]Joseph Campbell, *The Masks of God: Primitive Mythology* (New York: Viking Penguin, 1959), p. 179.

[3]The great monument we now see is believed to be Stonehenge III, built around 2100 B.C. upon a much larger and older site. Some historians have attempted to reconstruct the alignments of its stones and trace the plans of the older structures; they believe that Stonehenge was originally a gigantic observatory whose stones formed sightlines which tracked the various risings and settings of the sun and moon through the seasons. Perhaps the sun and moon (male and female) were part of a fertility cult in which astronomy played an important role.

"What was it for?" And the answer comes back: "We don't know, probably something to do with religion."

Or, consider a cave painting in France dating from the Ice Age (roughly 60,000–10,000 B.C.). It shows the outline of a great mammoth with a blotch in the middle where one would imagine the heart to be. In fact, the patch can be nothing other than a heart. Was the hunter telling us that the best way to kill the animal was through the heart? Did he have anatomical knowledge of other animals? Of himself? Nearby were found trephined skulls—that is, skulls into which a small hole had been drilled to relieve pressure on the brain—at least we think this was the reason. In interpreting those remains, we can only guess—until we come to Egypt and Mesopotamia.

The big change that came about was writing—its earliest forms being the clay tablets of Mesopotamia and the papyrus scrolls of Egypt. Although there are still gaps in the records, by piecing the evidence together we begin to get a rough picture of these ancient civilizations.

On a map, the Nile valley resembles a narrow green strip arbitrarily painted through the desert sand. The contrast is striking, and it is no wonder that the Greek historian Herodotus referred to Egypt as the gift of the Nile. From June to September, heavy rains in the Ethiopian highlands cause the river to rise and flood the Egyptian valley, depositing fertile soil all the way to the delta where the river empties into the Mediterranean. The earliest technology consisted in raising the river banks, cutting dikes for irrigation, and using devices to bring water up to the flood plain during the seasons of low water. The delta region became the Kingdom of Lower Egypt. As irrigation methods spread south, the Kingdom of Upper Egypt was born.

Unlike Mesopotamia, in which a succession of peoples contributed to the civilization, we find in Egypt a relatively stable cultural evolution from 3100 B.C., the date of the unification of the two kingdoms, to 332 B.C. when Egypt was conquered by Alexander of Macedonia. Egyptologists label this long period Pharaonic and conveniently divide it into the Old Kingdom, the Middle Kingdom, and the New Kingdom, with periods of disorder and foreign invasion between the Old and the Middle, and the Middle and the New. Indeed, the stability of Egypt was only relative, and undue emphasis has been given to the static character of Egyptian culture. Yet, by and large, Egypt was spared the tremendous upheavals which swept over Mesopotamia and even Greece, owing mainly to mountain barriers east and west and the sea to the north.

Considering the great duration of Egyptian civilization, we must be cautious when speaking of Egyptian science per se. Nonetheless, it can be said that the basic experience of Egyptian life from the outset was the Nile valley itself, for one can stand with one foot upon the green carpet of life and the other upon the burning sands of the desert. There is a clear line between life and death—between the black, rich soil of the flood plain and the red sands of the desert. The flooding of the Nile is highly predictable, and as the source of life it resembles the cycle of birth and death. From these fundamental features of their country, it is probable that the Egyptians gained their strong sense of symmetry, balance, and geometry, perhaps best objectified by the great pyramids.

How would such a people explain their world? The earth is a flat platter with a corrugated rim, floating upon the primordial waters personified as Nūn who gave birth to the Nile. Nūn was also the waters encircling the earth, like the Greek *Okeanos*, or Great Circuit. Inside the platter is the flat plain of Egypt, and above the earth is the inverted pan of the sky, the goddess Nūt crouching over the earth with fingers and toes touching the ground. Supporting her is Shu, the air-god.

If Egypt was the gift of the Nile, it was also the creation of the sun which gives life to the crops and to man. Unlike the desert in which the sun deals death, the Nile valley, blessed with moisture, comes to life beneath its rays and flourishes in its heat. The sun is creator, bringing life from the mud of the flood plain. And so the Egyptians perceived it, naming the sun Rē-Atum and later identifying it with Horus, the hawk-god of the Osiris cycle who became the god of the living. Murdered by his brother Seth, Osiris was reanimated by his son Horus and became judge of the dead. From the chaos of the primeval slime, Rē-Atum arises and brings order out of disorder. The life-principle, called *ka* by the Egyptians, is given to the universe by Rē-Atum; the same life-principle exists in man—his *ka* or soul. Rē-Atum also created Shu and Tefmut, the goddess of moisture, and their union gave birth to the earth, Geb, and the sky-goddess Nūt. Surprisingly, there was no dividing line between the creation of gods and humans, and one text mentions the creation of man in the image of the god's own body, blessed with *ka*, the breath of life.

In another text, creation begins with a thought in the heart of the god, and once he gives utterance to this idea, order is brought to the chaos. It is as if the idea precedes the act, and activity arises from cognition. Thus creation is continuous, for where there is thought and action the principle is present. Here we begin to see the unique character of mythopoetical science. The creator-god can be the sun, a hawk, or an abstract principle, the word of command; justice can be the goddess Maat, or *maat* can be the command of justice which issues from the mouth of the Pharaoh. Nevertheless, there is a creative principle throughout the universe, giving it order and form, found in the organic, inorganic, or abstract, yet not confined to any single category. There are many gods but one nature, and no matter what form it takes, nature is seen through the whole experience of human life and in its terms. Cosmogony is a living process; the gods and the principles they represent are inseparable, just as the life experience of the Egyptian people is part of the Nile and the sun.

But is this science? In our terms, no. Yet, for the ancient Egyptians it was a way of understanding the world, the basic presupposition of their lives intimately connected to their daily existence. It was their reality, their picture of the universe. For them it *was* science.

The first great scientific achievement was the art of writing. Writing, like stone-age cave drawing, is communication. But while we can only guess what the hunter was trying to communicate, communication becomes surer when the picture becomes conceptualized (a word) and when signs are added which compel translation into the sounds of language. Most experts believe that writing originated in Mesopotamia, but Egypt took the first step toward conceptualization, when certain pictograms commonly referred to specific things or ideas. This style of writing was called hieroglyphic or "sacred carving." Most hieroglyphics contained two signs, a picture and a phonetic sign. As early as the Old Kingdom there were twenty-four such phonetic signs—indeed, alphabetic symbols. Yet the Egyptians continued to use these twenty-four symbols along with hieroglyphics.

The Egyptians wrote on papyrus, which was made from the pith of the stem of a tall sedge plant abundant in the delta. The pith was cut into longitudinal strips, and the strips were arranged crossways in two or three layers, soaked, and pressed. Writing upon papyrus was done in ink, and about 1900 B.C. there evolved an abbreviated hieroglyphic script called hieratic. Finally, a third kind of writing evolved, a shorthand of hieratic called demotic, used mainly for the ordinary writing of everyday life.

The use of symbols to represent numbers probably began very early in Egypt, for the pyramid construction of the Old Kingdom necessitated the activity of clerks in keeping accounts, solving problems, and keeping tables. The earliest Egyptian symbols for the numbers 1 to 9 were simply vertical lines: ı (1), ıı (2), ııı (3), . . . , ııııı (9). Large numbers were also represented with symbols for tens, hundreds, and thousands. For ordinary addition and subtraction, the Egyptians listed the appropriate collection of symbols in tables, since they had no sign for zero nor did they develop the concept of decimal place, as in Mesopotamia. Therefore, multiplication and division were performed by the so-called additive process. To multiply 8 by 7, for example, the scribe would first choose a multiplicand and repeatedly multiply it by 2, adding up the multipliers until they equalled the original multiplier:

1	8	Checking off the multiplier $(1 + 2 + 4 = 7)$,
2	16	he would add $8 + 16 + 32 = 56$.
4	32	
7	56	

Division was accomplished using the same method, since one simply asks for the factor required for one given number in order to obtain the second given number. Say the scribe wished to divide 24 by 4; his notation would be:

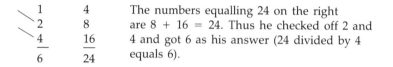

⟍ 1	4	The numbers equalling 24 on the right
⟍ 2	8	are 8 + 16 = 24. Thus he checked off 2 and
⟍ 4	16	4 and got 6 as his answer (24 divided by 4
6	24	equals 6).

Of course, the problem might require the use of fractions, and Egyptian fractions were always "unit fractions," in our notation $1/n$, denoted in Egypt by a special hieroglyph meaning "part," the nth part. Historians of mathematics use the notation $\bar{5}$ for 1/5. The exception was 2/3, represented by a separate symbol meaning 2 parts, for us 2 parts of 3. To divide 16 by 3, the scribe would again list:

⟍ 1	3	Adding 3 + 12 = 15 is one unit less than
⟍ 2	6	the required number. Hence the scribe
⟍ 4	12	would compute the fraction:
5	15	

$$\bar{\bar{3}}(2/3) \qquad 2$$
⟍ 3   1 which means that
2/3 of 3 is 2 and 1/3 of 3 is 1, and
thus he would find the answer 5 $\bar{3}$,
or 5 1/3.

Higher fractions were simply listed in tables: ¾ = ½ + ¼ for all fractions p/q. Since $\bar{\bar{3}}$ was the exception, it is believed that a two-thirds table was constructed. If the scribe was required to find ⅓ of a number, he would first find the $\bar{\bar{3}}$ of it from the table and then halve his answer. The tables for expressing higher units were intricate, and in some the expressions were written in black and red to facilitate their use.

Today when a new concept arises in mathematics we often invent new symbols for dealing with it. The Egyptians, on the other hand, invented new techniques for handling the new concept in the old notation. To us this seems cumbersome, and Egyptian mathematics appears hardly more than complicated arithmetic. Many historians believe that the Egyptians did not theorize upon the logical relationship of numbers. In the sense that they did not use mathematics to describe nature, this is probably true. But recent scholarship has shown that they did deduce mathematical relationships, and many of the operations they described in words can be transcribed into modern algebra, dealing with linear equations, second-degree equations, the sum of n terms of an arithmetical progression, and the like. Tied as we are to the logical proof of a problem expressed in symbols, we fail to see how one or two specific examples can constitute both a method and a proof. Our demand for proof is the result of a tradition arising from Greek geometry. For the Egyptians, on the other hand, the rigor was implied by the method.

The construction of the pyramids surely denotes some elementary knowledge of geometry. We know from their records that the scribes were able to compute the areas of triangles, trapezoids, and rectangles, as well as the area of a circle. They also may have known the volume of a square pyramid ($V = \frac{1}{3}ha^2$), and most certainly the volume of a truncated pyramid: $V = h/3(a^2 + ab + b^2)$. They may have known the Pythagorean theorem for a right-angled triangle ($a^2 = b^2 + c^2$), but this has been contested. However, they used knotted cords in surveying and might have known the relationship pragmatically. We must conclude that, given the restrictions of their notation, the Egyptians achieved a relatively high level of sophistication in mathematics. Yet as a method for investigating nature, mathematics did not mean to them what it has meant to us since the days of Galileo. Their natural science was mythopoetical.

The Egyptian calendar was a simple solar calendar. Observation of the heavens was useful in terms of prediction (divination) and keeping time. But the Nile floods regularly, and so their first calendar was simply an agricultural one, dividing the year into three seasons of four months each: Inundation, Coming Forth (from the flood), and Harvest. This was a practical civil calendar with a year of twelve months of thirty days each, and five days at the end devoted to the gods of the Osiris cycle. Now, the true length of a solar year is 365 ¼ days, which meant that the civil calendar slipped back one day every four years. With this slight correction, Hellenistic astronomers found the Egyptian civil calendar extremely valuable. Its basic form is still with us today.

The day was divided into 24 hours, originally of unequal length. This division was connected to the heliacal rising of Sirius, the Dog Star, which coincided with the inundation of the Nile. For ten days Sirius would herald the dawn, slipping each day, so that after the tenth day another star would serve this purpose. This gave rise to a system of *decans* in which one particular star noted the dawn every ten days, so that there would be 36 *decans* distributed through the mornings of the year. In the summer, during the inundation, 12 *decans* (in Hellenistic times, 10 degrees of the ecliptic, or the annual path of the sun on the celestial sphere) would rise during the night and determine the hours of a summer night.[4] In winter more *decans* would be visible, and thus the

---

[4]Each night the stars traverse the sky from east to west maintaining a "fixed" position relative to each other. Their nightly journey appears circular, for an observer gazing north finds a stationary pole whose elevation depends upon the observer's position on the earth. The stars nearest this pole seem to revolve around it—circumpolar stars. Traveling south, the observer finds the pole closer to the horizon and finds the stars nearest it rising and

length of the hours varied. The two calendars, civil and astronomical, existed side by side throughout Egypt and proved immensely practical.

Egypt was famous in the ancient world for its medicine, and we find evidence of practicing physicians in the Old Kingdom. As one might expect, the causes of disease were seen in terms of gods and demons, and treatment contained a great deal of ritual purification, magic, and prayer. About 2900 B.C., we hear of King Zoser's vizier, a certain Imhotep, who was an architect, astrologer, and magician, not to mention a physician of renown. He is the first physician on record and was later deified. The Greeks identified him with Asclepius, their god of the healing arts, and like Asclepius, he treated his patients through the agency of dreams. Imhotep was called the "blameless," a title a physician in any age would welcome.

Beside the mythical elements of Egyptian medicine there existed a great deal of sound empirical practice and observation. A treatise on surgery, for example, instructs the physician how to treat injuries, wounds, fractures, and tumors; treatments consist mainly in the application of dressings, general nursing, and diet. There was also an extensive use of drugs, many of which worked, and which were probably identified through repeated observation. Yet as a rule, the symptom of the disease *was* the disease, and where the cause was not apparent the physician looked to the gods.

In a society which developed mummification to a fine art, anatomy was nevertheless mainly speculative, partly because the embalmers were craftsmen and not physicians. Egyptian physiology viewed the human body as a system of vessels, like irrigation canals, carrying the various bodily fluids which were believed to originate in the heart. Seeing the heart as the central organ did make them aware of the pulse rate, yet the major organs and vessels were believed to have lives of their own, and every limb was identified with a god.

Turning to Mesopotamia, we discover a general lack of the continuity which characterized the Egyptian population and civilization. Early Mesopotamian history is a succession of struggles among various peoples for possession of the fertile valley cut in the desert by the Tigris and

---

setting. Further south, new stars appear that may not be seen in the northern skies at all. Thus the entire star field appears from the earth to be a giant rotating sphere, completing one rotation in roughly 24 hours (invisible during the day). Plotting the position of the sun through consecutive evenings on this "celestial map," it was seen that the sun moves about 1 degree each day relative to the fixed stars. The sun rises and sets like a star, yet it also moves slowly eastward, following a path on the celestial sphere. This path is called the ecliptic and it is also a great circle. The ecliptic intersects the celestial equator at the equinoxes and lies furthest north and south at the solstices. Thus the ecliptic is imagined as a great circle on the stellar sphere, tilted 23 ½ degrees to the celestial equator. For details, see Thomas S. Kuhn, *The Copernican Revolution* (New York: Random House, 1959), Chapter 1.

Euphrates rivers. The earliest civilization was a collection of city-states in the south called Sumer, representing a long tradition of agriculture and irrigation dating from before 3000 B.C. About 2500 B.C., the Semitic King Sargon of Akkad subjugated Sumer, and after a resumption of Sumerian independence under Ur, the area fell to the northern city of Babylon. After another period of Semitic invasions, most of Mesopotamia fell to the Assyrian empire of Sargon II in 722 B.C. His armies were equipped with iron weapons and siege machinery. Babylon regained preeminence under the famous Chaldean dynasty, to be followed by the Persians and eventually by the Greeks of Alexander.

Unlike the Nile, the Tigris and Euphrates rise unpredictably and threaten to destroy irrigation canals and to flood crops. For the ancient peoples of Mesopotamia, man seemed to be at the mercy of nature as well as the beneficiary of its bounties. The cosmos might exhibit an inherent order, as seen in the celestial rhythms, yet it also contained awe-inspiring destructive forces–the chaotic waters of the flood or the fierce desert storms—in which the individual human will counted for little. Of course, every natural disaster had a cause, a will, operating behind it. Nature, then, was a community of wills, the interaction of various natural forces symbolized by individual personalities—the gods. More than in Egypt, nature in Mesopotamia seemed to be full of strife and struggle in which order was based upon authority and strength which harmonize the anarchical tendencies of natural forces. The cosmos was similar to the state.

It is nearly impossible to give a comprehensive picture of Mesopotamian mythopoetical thought. Each Sumerian city had its own god, the temple or ziggurat being the god's home, and with the march of empires the standings of the various gods changed just as did the fortunes of the state. The most powerful Sumerian god seems to have been Anu, the sky-god, who symbolized the force lifting the world from chaos. Enlil, the god of Nippur, was Anu's son and the god of the storm. Taken together, they also embodied the principles of the state: authority and force. The local deity of Eridu, Enki (later Ea) was the god of the waters surrounding the earth, master of the arts, known for his cunning and wisdom.

There were many other deities, each symbolizing a natural power. Yet, the cosmos as a whole functioned as an assembly of the gods, a consensus of wills, like any state. Over the assembly ruled Anu, the sky, and Enlil was the archetype of the human king.

Mesopotamian cosmogony is best represented by the Babylonian creation epic, the *Enuma Elish* (When Above), composed during the first Babylonian dynasty. The epic is written in Akkadian, and the central figure is the old solar deity of Babylon, Marduk, later replaced in Assyria by Assur. It is recorded upon seven clay tablets found in the Assyrian library of Nineveh and probably represents an older Sumerian tradition, and therefore, in general, the cosmogony of Mesopotamia.

The earliest stage of the universe is chaos, disorder, represented by the intermingling of fresh water, salt water, and mist, all personified by deities. From the fresh and salt waters, Aspu and Tiamat respectively, comes the silt of the primal hill (as in Egypt). The sky and earth are two enormous disks built up from the silt deposited along the inside rim of the horizon, later forced apart by the winds into a great oyster. The origin of the world order is a prolonged conflict between the primal chaos, Tiamat, and the principle of power and authority, Marduk.

The gods who were given birth by the primal couple, Aspu and Tiamat, symbolize the drive to order and activity. First Aspu is killed and his body, the fresh water, is locked in the earth. But Tiamat is aroused, and gathering an army of monsters, she attacks the gods who killed her mate. We can well imagine a sudden flood, a desert storm, or a nomadic invasion reflected in this scene, for the destructive forces attack the whole person, reason and emotion, understanding and awe.

Today we would generally seek a scientific explanation in the face of such forces, and perhaps technological control; we search for an intellectual and physical ordering of phenomena. The Mesopotamians, too, sought to order the chaos, only for them the physical aspect of order was the state, Babylon, and its intellectual concept was myth, the god Marduk. The young god Marduk, given authority by the older gods, kills Tiamat and cuts up her body to form the world. His creative act of ordering the chaos reflects the power and authority of his rule and therefore the state he represents. The *Enuma Elish* presents a sanctification of the earthly state by the original act of cosmic creation. Participation in the life of the earthly state and obedience to its laws recalls the victory of Marduk. In Mesopotamia, mankind was created for the service of the gods; from the blood of Kingu, Tiamat's vizier, the victorious gods form mankind to serve them. Thus order is maintained through human obedience to the state and hence the cosmos.

Turning from cosmogony to the exact sciences, we find ourselves in the same position as with Egypt. What seem to us the proper achievements of science were for the Mesopotamians closer to the realm of practical life. Fortunately, we have a rich record of more than two thousand years of civilization, for the Mesopotamians wrote on clay in a script composed of wedge-shaped signs called cuneiform (from the Latin *cuneus* meaning wedge). Scattered throughout the world's museums are forty to fifty thousand of these tablets from all areas of Mesopotamian life. Besides the vast record-keeping of everyday life, we find a great philological tradition among the scribes, refining and standardizing the Sumerian and Akkadian languages. This alone must have taken a great deal of energy, for the Sumerian script, based upon roughly 350 syllabic signs, never reached a true alphabetic stage. Further, since clay dries quickly, each tablet had to be small and completed in one session.

An important category of the cuneiform tablets is the omen texts. In a society subject to the virulence of both nature and man, divination played an important role. In fact, a great deal of the astronomy and medicine in Mesopotamia arose from the desire to divine the intentions of the gods—nature—and thus counteract them. Disease, for example, was the mark of a god's displeasure, and the chief method of diagnosis and prognosis was the interpretation of omens. Treatment might be a combination of magic, sacrifices, and religious rites to appease the cause of the disease—a god, evil spirit, or demon. As in Egypt, the disease was the symptom and the physician was a priest. Giving drugs and performing surgery were manual parts of the spell. The emerging separation between the empirical and the magical that we saw in the Egyptian surgical papyrus has no counterpart in Mesopotamia. Yet many remedies which simply worked and required no explanations were used. Whenever explanation was sought, however, it was mythical—demons, witchcraft, sorcery, the evil eye, the evil tongue. Ea, the lord of wisdom, was the god of the physicians.

The Mesopotamians did not dissect, though they did speculate about anatomy. The heart was the seat of the intellect, the ears and eyes of attention, the stomach of cunning, the liver of activity, and the uterus of compassion. Dreams, of course, were significant—a view common to the ancient world but largely forgotten in Western science until Sigmund Freud.

From the Old Babylonian period, roughly 1600 B.C., we find tablets which contain a more highly developed mathematical system than in Egypt, although it is important to be careful when making such value judgments. The tablets are basically "problem texts" and "table texts" in which specific problems are used to illustrate what modern mathematics states in generalized, symbolic procedures. The system was basically sexagesimal—that is, based on 60 and powers of 60—but it also contained features of our decimal system. What made the Babylonian system easy to use was that it expressed numbers by means of only two symbols: a vertical wedge Γ meaning 1 or 60, and the symbol $\langle$ for 10. Combinations of these signs are used up to 59, and then the 1 becomes 60. What is more, the system used place-value, like our own, inasmuch as a symbol changed value as its position shifted. The wedge symbol Γ in end position has the value 1; $\Gamma\Gamma = 2$, $\langle\langle\Gamma\Gamma = 22$, and so on; after 59 the wedge stands for 60 (and is sometimes written larger). Thus $\Gamma\langle$ stands for 70 (60 + 10 = 70); $\Gamma\Gamma\langle$ for 130 (2 × 60 + 10 = 130); and $\Gamma\Gamma\Gamma\langle\langle$ for 200 (60 × 3 + 10 × 2 = 200). Historians of mathematics generally transcribe the notation as 1,10 (for 70); 1,20 (for 80); . . .; 2,10 (for 130). To express zero quantity the texts simply left a space.

The use of a decimal system made fractions much easier to work with than in Egypt, since a fraction may be treated in the same manner as an integer. Usually the specific nature of the problem text, or the table being computed, determined the integer or fraction represented. Obviously the system facilitates calculation with integers and fractions alike.

There has been a great deal of speculation about why the Babylonians made use of the sexagesimal system, which we continue in our division of hours and minutes into units of sixty, and in the 360 degrees of our circle. It should be noted, however, that they applied their system consistently only in the context of strictly mathematical and astronomical texts. In other calculations such as weights and measures, and even some geometrical problems, use was made of mixed systems similar to our divisions of feet, inches, and yards.

In one tablet we find a remarkable approximation for $\sqrt{2}$ as 1; 24,51,10 or 1.414213 instead of 1.414214. This, in effect, is also the determination of the square from its side representing the Pythagorean theorem which says that the sum of the squares of the sides of a right triangle equals the square of the length of the hypotenuse. It is possible that the Babylonians knew the relation a thousand years before Pythagoras but never generalized it.

Using their straightforward notation, the Babylonians prepared many kinds of tables (squares, square roots, cubes, reciprocals, and others), and their problem texts indicate solutions involving unknowns. The problems were solved verbally by describing the steps required to calculate the solutions and were not given in general symbolic terms. Yet, these mathematicians were apparently able to solve problems of two unknowns, and they reached the solution of a problem in which the unknown is squared, arising from the problem of asking for a number which added to its reciprocal yields a given number. In our notation they asked for x and $\bar{x}$ (reciprocal x) such that $x\bar{x} = 1$, $x + \bar{x} = b$. The two equations yield a quadratic equation, $x^2 - bx + 1 = 0$, which they solved by making the left side a perfect square and obtaining the expressions $b/2 + \sqrt{(b/2)^2 - 1}$ and $b/2 - \sqrt{(b/2)^2 - 1}$ through a number of steps stated in concrete examples. Certain words were used for unknowns, and while only concrete examples were worked, some might have been intended to illustrate general procedures, although this can be debated.

Babylonian astronomy took the extraordinary step of applying mathematics to the data of celestial observation. We, of course, see this activity as characteristic of modern science, yet the procedure itself arose from the Mesopotamian practice of using the lunar calendar and the need to make adjustments with the seasons determined by the solar year. Since the lunar month began with the first visibility of the new crescent shortly after sunset, the time between successive visibilities is never more than thirty days or less than twenty-nine. The problem arises: when is a specific month twenty-nine or thirty days? The answer is based upon a number of complicated periodic phenomena such as the length of daylight, the variable velocity of the sun and moon (viewed geocentrically), and the moon's movement in latitude.

It appears that regular observations were practiced by the Assyrian period, and the "intercalculation" of these variable phenomena was fully accomplished by 300 B.C. Since a sequence of twelve lunar months is about eleven days short of a solar year, an extra month is added at intervals (intercalculated), giving a solar nineteen-year cycle of 235 lunar months. In order to do this, the Babylonians recognized that the complicated phenomena were the result of a series of independent variables. By listing the separate variations as arithmetical progressions in tables called ephemerides, combining both lunar and planetary movements, they noticed that the terms of the tables uniformly increased to a maximum and decreased to a minimum. The rules for such computations were written in the procedure texts, and the regular progressions gave rise to linear functions enabling them to predict lunar-solar conjunctions, variations in planetary movements, and even make rough approximations of lunar eclipses. More information would be required to predict a solar eclipse—the actual distances of the sun and moon from the earth as well as the sizes of these bodies. Yet, the Babylonians could make a rough estimate of when a solar eclipse was possible.

The zodiac was invented about the fourth century B.C. However, it should be noted that it was for mathematical reasons alone that the Babylonians divided the sky into a great circle of 360 degrees, conveniently marked off by certain constellations in thirty-degree sections (12 x 30 = 360). The zodiac was simply an ancient coordinate system arbitrarily imposed to measure the progressions of the celestial phenomena in question. Originally it had nothing to do with astrology, although in later times the Chaldeans were frequently referred to as a nation of astrologers. Paradoxically, their zodiac was a mathematical idealization used exclusively for calculations, a method which strikes us as very modern indeed.

Here we have come full circle. What seems to us the beginning of scientific astronomy was simply a handy way of making phenomena predictable for the Mesopotamians. It was for practical purposes alone: the keeping of time and the lunar measurement of the agricultural seasons. That it became more highly developed in Mesopotamia than in Egypt may well be due to the unpredictability of the life-giving rivers and the basic insecurity of the civilization.

Predicting the phenomena added nothing to understanding them, making them intelligible. The application of rigorous methods to an understanding of nature was yet many centuries away. This is our method, and although we see the rudiments of it in the ancient Near East, we must realize that these peoples' picture of nature, their "science," came from other sources. Speculation was confined to the realm of myth, and though this strikes us as totally "unscientific," it is still speculation based upon the need to explain experience. And this it did quite well. It is only when man begins to desanctify nature, speculate upon the "it," that the use of reason comes into play. This we owe to the Greeks.

SUGGESTIONS FOR FURTHER READING

ELIADE, MIRCEA. *The Sacred and the Profane: The Nature of Religion*, Willard R. Trask. New York: Harcourt Brace Jovanovich, 1959.

———. *Cosmos and History: The Myth of the Eternal Return*, trans. Willard R. Trask. New York: Harper & Row, 1959.

FRANKFORT, H., and H. A.; WILSON, JOHN A.; JACOBSEN, THORKILD; and IRWIN, WILLIAM A. *The Intellectual Adventure of Ancient Man.* Chicago: University of Chicago Press, 1946.

GILLINGS, RICHARD J. *Mathematics in the Time of the Pharaohs.* New York: Dover, 1982.

KLINE, MORRIS. *Mathematical Thought from Ancient to Modern Times.* New York: Oxford University Press, 1972.

KUHN, THOMAS S. *The Copernican Revolution: Planetary Astronomy in the Development of Western Thought.* New York: Random House, 1959.

NEUGEBAUER, O. *The Exact Sciences in Antiquity*, 2d ed. New York: Dover, 1969.

OPPENHEIM, A. LEO. *Ancient Mesopotamia.* Chicago: University of Chicago Press, 1977.

SARTON, GEORGE. *A History of Science*, 2 vols. Cambridge, Mass.: Harvard University Press, 1952–1959.

SIGERIST, HENRY E. *A History of Medicine: Primitive and Archaic Medicine*, Vol. I. New York: Oxford University Press, 1951.

2

~~~~~~

Lighting the Lamp

Early Greece

In the sixth century B.C., Thales of Miletus stated that water was the origin of all things. Perhaps this marks the birth of Western science. Well, nothing in history is ever so simple. The Greeks of Ionia, on the shores of Asia Minor, did not get up one morning and say: "Today we are going to speculate about nature in a rational way. Tomorrow we will generalize our observations and produce theories." But the Greeks did become dissatisfied with mythopoetical science, without at first rejecting it outright. And they began asking questions which the old tradition found difficult to answer: What is the underlying ingredient of nature? What causes change? Why do some changes appear to exhibit order and others not? Such questions, coupled with proposed answers, would eventually change attitudes toward nature. Yet this takes time, and those who ask the questions are seldom conscious of themselves as radical innovators. They only know that they are dissatisfied with previous explanations.

What factors led to this change of attitude among the Greeks? After all, mythopoetical thought did explain nature in its own way: people were able to predict events, deal with numbers, heal the sick. In fact, the Greeks themselves were acutely aware of their debt to the older tradition. Thales, Pythagoras, and others visited Egypt, bringing back to the Greeks its wealth of geometrical and mathematical knowledge. Greek astronomers used the painstaking recorded observations of the Babylonians. Even Thales' principle of water reminds us of Marduk's ordering of the world from the waters of chaos. Yet, they asked questions which went beyond the realm of the older tradition. Why?

The Greeks developed a critical, intellectual curiosity which myth was unable to satisfy. Now, curiosity is the basis of all science. At the dawn of civilization in the ancient Near East mankind had little in the way of a tradition to fall back upon in order to satisfy this curiosity. So one was created. For them it was good; it worked. Nonetheless, curiosity becomes stifled by satisfaction. If you feel that all the important questions have been answered and your science is foolproof, it is extremely doubtful that you will wish to change it. New problems will be translated into the language of the tradition and dealt with accordingly. Sometimes the answers will satisfy, and sometimes the problem itself will be deemed insolvable. Throughout the history of science, we shall see this process happening again and again. But someone will become dissatisfied. Perhaps it will be for rational reasons, or, in the majority of cases, it will simply be a vague feeling that something is wrong with tradition. It is also quite possible that the stimulus will come from factors external to pure intellectual pursuits.

The early history of Greece was as restless as the curiosity of its peoples. During the centuries from the early Minoan civilization of Crete to the creation of the magnificent acropolis of Mycenae and the glorious city-states of mainland Greece, Italy, and Asia Minor, the Greeks and their ancestors navigated the sparkling blue waters of the Mediterranean and Aegean Seas. It is true that the Egyptians traveled and explored, but the Greeks settled—colonized—and came into extended contact with other peoples and traditions. It is no wonder, then, that the trading cities of Ionia provided the stimulus for a new outlook. Here on the shores of Asia Minor, the overland trading routes terminated in Greek ports where the sea-borne trade began. Along with oil, figs, flax, wool, and cedar came ideas, the varying traditions of the Babylonians, Phoenicians, Egyptians, and the Greeks themselves. How to reconcile them?

The Mycenaean age ended about 900 B.C. as waves of Dorian invaders swept down from the north into Greece, bringing iron which probably originated with the Hittites in Asia Minor. These northern invaders merged with the older populations, creating the historic divisions of the Greek people into Ionians, Dorians, Aeolians, and Arcadians. On the mainland of Greece, the Ionians held the peninsula of Attica. Slowly they spread east, along with settlers from Crete, colonizing the coast of what is today Turkey. There they founded the great trading cities, of which Miletus was one. It was probably in Ionia, too, that the Greeks first adopted the Phoenician alphabet, improving it by adding short vowels. The Phoenicians were also traders, navigating across the Mediterranean from cities such as Tyre, and Herodotus said that Thales himself was descended from Phoenicians. If true, this provides further evidence of the admixture of cultures in Ionia.

Historians have also noted the social and political innovations in Greek life which emerged from the collapse of the Mycenaean age. Like the nearly omnipotent god-kings of the Near East, the Mycenaean kings ruled by religious authority; politics was ritual anchored by mythology and social life was centered upon the palace. Society itself was a complex hierarchy with the king, fusing all sovereignty and power, at its apex. Political discourse was the ritual word, the precise formula of a mythologically based culture, and the human social order was an analogue of the mythical cosmos.

From the ashes of the Mycenaean age arose a new political and social entity, the *polis*, the city-state. Politics as it developed in the *polis* became discussion among citizens in the broad daylight of the marketplace; sovereignty was desanctified and submitted to debate. Political debate, in turn, became competition, the goal of which was persuasion, not by divine authority, but by logical argument. The survival and vitality of the *polis* rested upon its social universe—various competing groups submitting to a balance of interests, a kind of social symmetry or equilibrium. This problem of balancing the many in the one brought the question of human order into the province of rational discussion.

It is quite likely that other areas besides politics, other claims to wisdom and authority, were submitted to this rational argument and competition. Thus it has been suggested that this quest for a demythologized geometrical balance among social groups, the *polis*, expanded its borders to include physical nature itself, the cosmos. Indeed, the desire for certainty in an axiomatic system—which was a distinct characteristic of Greek science and often resulted in a certain lack of empirical content—may have been the result of a reaction to the merely plausible arguments of the competition. Hence, evidence would be used to support rather than to test theories, and self-criticism would be on the whole lacking since the goal was to emerge victorious in the contest. If the rational secularization of politics in the city-state became the rationalization of the elements in the cosmos, hence a stimulus for rational science, the style of such inquiry would be a kind of dogmatic dispute between the contenders. Indeed, we see this style among the early Greek philosophers.

Still, the Greeks of Ionia had their own ancient mythological tradition, and here too some historians have looked for the seeds of Greek rationalism. For the end of the Mycenaean age produced the poet Homer, and later the poet Hesiod. In fact, it is probably best to view Homer himself as a tradition, the greatest of a long line of traveling minstrels who sang of the past glories of Mycenaean civilization, and whose songs were finally recorded in the *Iliad* and the *Odyssey*.

At first glance the two poets portray a world-view not much different from the myths of Egypt and Mesopotamia. To Homer the sky is a solid hemisphere; as in the Babylonian oyster universe, it covers a round, flat earth like an upper shell. The upper air is fiery and below the earth is Tartarus, the gloomy abode of the dead. Around the rim of the world flows the primordial river *Okeanos*, the origin of all things. The same is true for Hesiod, who clearly says in his *Theogony*: "Verily first of all did Chaos come into being, and then broad-bosomed Gaia (earth) . . . " The production of the world resembles the birthing process of human beings; things are brought forth like offspring, including the gods. The primal forces of nature are often violent, titanic; the gods must battle them in order to establish their rule. Thus Zeus, perhaps the ancient Minoan sky-god, leads the gods against the violent Titan Cronos much as Marduk fought Tiamat. In the end, the forces of nature are subdued, ordered according to the law of the gods, and the world is divided into departments under their rule.

The laws of nature are thus legislated by the gods, and again it seems that we are in the presence of mythopoetical thought. Yet, in the *Iliad* and the *Odyssey* there is a subtle difference. We find the gods and goddesses acting much like humans; the Trojan War originates from a quarrel between three goddesses over who is the most fair. The gods visit man unrecognized, they fight in battles, they are wounded, they display the entire range of human characteristics. While natural law seems to be an act of legislation by the will of Zeus, the god himself has become a personality, drifting away from the sky he represents—as do the other gods. They live upon Mt. Olympus in Thessaly, separated from the elements they govern. The implication could be that nature itself, the living and self-changing world the Greeks called *physis*, is something *apart* from the gods, like a common stage upon which both men and gods act out their roles.

It almost seems inevitable that mythopoetical thought would have the tendency to put too much of the human personality into nature. Eventually nature would be seen to be at the mercy of a remoter power, primary and more ancient than the gods. In the later Greek tragedies which built upon the myths, especially Sophoclean tragedy, it is evident that even the gods are subject to such a power, for Zeus recognizes that the gods cannot escape *Moira*, fate or destiny. It is possible for gods and men to struggle against fate, but in the end disaster strikes, for fate is that which limits individual power, be it of gods or men. *Moira* is as much a natural moral law as it is a predetermined destiny. No one escapes fate, neither Greeks, Trojans, nor gods. It stands behind everything, an eternal principle governing the motion and growth of *physis*.

For rational science to begin, there must exist a faith that there are unchanging principles in nature, principles to which nature conforms and which underlie the constant flux of sensate experience. *Moira* implies necessity, a pattern giving rise to change and governing it. Fate is the bridge between myth and rationalism.

Was there a natural order which reflected the *polis*? Could it be demonstrated in logical argument? What was left in nature once the gods had become distinct personalities, apart from natural forces, and not much different from humans? How could one regularize and make sense out of these myths? What was the principle or order once the gods were made distinct from *physis*? What was nature made of? From such questions came a change of attitude, the rejection of *mythos* for other means of harmonizing experience. The ground is prepared for Thales and the Ionian science of *logos*—reason. Yet, it should be recognized that the totally rationalist human being is an abstraction which nowhere exists in the history of science. What changed among the early Greek philosophers was the emphasis of the intellect, and this emphasis did not preclude the magical, nor the cults and mysteries dominant in Greek society.

Nonetheless, with Thales a new tradition began, based upon a theoretical, generalized science of nature. Unfortunately, our only knowledge of this important step rests upon fragments from pre-Socratic philosophers (before Socrates and Plato) preserved as quotations in other ancient authors from Plato, to Aristotle and his school, to the sixth century. Often, as in the case of Aristotle, these fragments are interpreted in the light of later Greek philosophy, or presented by way of critique. The mere fact that they were preserved, however, indicates their importance for later Greek science. In general, the questions asked and even some of the answers given began theoretical discussions which even today form the precincts of science.

As far as we can tell, Thales was a man of many parts: engineer, traveler, astronomer, mathematician, businessman. He was, perhaps, a bit absent-minded—Plato says that he fell into a well while observing the stars. Herodotus claims that he predicted the eclipse of May 28, 585 B.C., and others reported that he predicted the solstices—the highest point of the sun over the horizon for summer and the lowest point for winter. These predictions would have taken considerable astronomical knowledge which it is probable Thales did not possess. Indeed, he assumed that the earth was a disk, or short cylinder, which floated upon the primal water like a cork. Yet, he probably was familiar with the Babylonian tables. Using them, he would have known approximately when an eclipse was possible. He may have also been familiar with the gnomon, the column on a sundial to measure the sun's height by its shadow cast at noon. It was said that he estimated the size of the sun and moon by the ratios of their diameters to their orbits, but this is hard to believe, since in his system they did not pass beneath the earth.

Thales visited Egypt, measured the pyramids by their shadows, and brought geometry to Greece. It is claimed that he introduced some geometrical generalizations: a circle is bisected by its diameter; the angles of an isosceles triangle are equal; if two straight lines intersect, the opposite angles are equal; the angle inscribed in a semicircle is a right angle; and a triangle is determined if its base and the angles relative to the base are given. Unlike Euclid, he gave no proofs. Nevertheless, it seems that he did generalize, speculating upon the nature of triangularity (not a specific triangle) and upon the relationship of angles (not a certain measurement). His geometry is abstract, a new step in mathematical thinking.

Yet, the old tradition was still alive in Thales. For him the world was full of gods—a lodestone (magnet) had a soul since it moved iron. Mechanical motion and energy were still equated with vital activity. The early concept of *physis* referred to growth, change, and life. Motion was a function of some animate principle. Throughout change, however, the basic substratum, water, remained the same. Here in a rudimentary form is stated one of the basic problems which was to haunt all Greek science: the problem of change and permanence. It is the same problem which haunted man's first attempts to explain nature.

Anaximander, son of the Praxiades of Miletus, was a contemporary of Thales and probably his student. With him, we suddenly enter the realm of pure physical speculation and encounter an idea which, through various forms, reappears again and again in the history of science: the idea of the nonperceptibility of the world's ultimate stuff. Anaximander said that the principle of the world was *apeiron*, the indefinite or infinite, and from this principle comes the eternal motion of coming-to-be and passing-away. Moreover, all the heavens and all the elements of the cosmos are separated from the *apeiron* through eternal motion and will be reabsorbed, the entire process beginning anew. Thus creation and destruction occur through cycles of an eternal dance in which opposed natural substances appear at intervals and are eventually reconciled into the infinite.

It is remarkable that Anaximander chose to label his primal stuff the indefinite or boundless, yet in the search for rational explanations his step almost seems a natural one. In a world of multifarious change, where all definite things are only temporary, what can be everlasting except the infinite? To give the eternal principle solidity, definitiveness, is to subject it to change and hence alteration. But what is the source of change, of qualities? Can the source itself be subject to change? With these questions, Anaximander plunges into the abstract.

Anaximander also maintains that the earth is suspended without support in the center of the universe. It is like a cylindrical drum, its depth a third of its width. The sun, moon, and stars are set between the layers of a sphere of flame surrounding the blanket of air covering the earth, and these bodies are encased by air, with pipelike passages through which they show themselves. Eclipses occur when the passages are blocked. The heavenly bodies are carried by circles and spheres, like the solid rungs of a cartwheel, around the earth. Celestial motion is thus accounted for by the rungs of this spinning cartwheel which are concentric to the earth. It is a purely physical and rational theory.

Finally, as if this were not all, Anaximander proposes that the first living creatures were born in moisture, encased in thorny bark like seeds, and as their age increased they came forth from the moisture to a different life. Man was originally an embryo in a fishlike creature until he came forth to the land. What are we to make of this? Does Anaximander anticipate Darwin by saying that living things evolve? There is much debate among scholars over this; nevertheless, such possible similarities exist only in retrospect. The *apeiron* has all the qualities of an animate primal being, a god above the world of coming-to-be and passing-away. It is mystical and rational at once; Nietzsche called it a "metaphysical fortress" from which Anaximander gazed out over the world of change. In such a world humans too must be subject to becoming.

Our final Milesian is Anaximenes, the pupil of Anaximander. To him, the principle of *physis* is air, and material things arise from the condensation and rarefaction of air. Perhaps it is a come-down from the lofty thought of Anaximander, this principle of air, yet it may also be seen as the protest of an empiricist. Anaximenes has returned to natural causation—that is, the cause of change which can be verified by observation. The process of the condensation and rarefaction of air is such a cause; that which is condensed and compressed is cold, ponderable, heavy, and that which is rarefied is hot, light, flexible. The heavenly bodies come from the airy exhalation of the earth, rising and becoming fiery. They are supported by condensed air, and the world is encased in a crystalline sphere in which the stars are implanted like nails. The life-soul of man is equated with the breath, air.

We must take especial note of this closed crystalline sphere. Until Copernicus, the idea of a rigid sphere surrounding the universe remained the basic postulate of all astronomy. It is even possible that Anaximenes was the first to distinguish clearly the planets from the fixed stars, for in his system the planets float freely in the air while the stars are fastened to the heavenly sphere.

The Milesians presented Greek science with a basic problem, that of motion and matter—or, what amounted to the same thing for them, the problem of accounting for both stability and change. The story of Greek speculation until Plato's time is the gradual pulling apart of *physis*, the separation of its vital spirit from its material base. By rights we ought to consider Pythagoras next, who by defining things as number—numbers are the principle of nature—set aside matter in favor of form. But this leads naturally to Plato and must wait for the next chapter. For now, let us follow the problem and see where it leads. One road ends with Parmenides and the completely logical denial of change; the other ends in atomism, a purely mechanical picture of nature, more dead than alive.

But Heraclitus, from Ephesus, a city north of Miletus in Ionia, called "the riddler" by the ancient authors, had the courage to draw some frightening conclusions from the problem. What if stability is illusion? What if the principle of being is change itself, the eternal war of opposites in constant tension?

From what we know of Heraclitus, he was an isolated man, disdainful, misanthropic—a mystic. "Learning of many things does not teach intelligence," he supposedly said; also that, although the *logos*, divine reason, permeates all, few men listen to its *inner* promptings. In other words, Heraclitus asserted the role of intuition in the study of nature, an important factor in the history of science we often overlook.

His intuition told him that change is all; that there is no harmony of opposites but only their conflict. War is the ruling and creative force, symbolized by fire. The material aspect of the intelligent principle *logos* is fire. There is no permanent substratum of existence, only the eternal flux, the eternal fire. "Observe the world," Heraclitus might have said to the Milesians or the followers of Pythagoras. "Nothing remains the same, there is no mathematical harmony of things, only eternal strife. You can never step into the same river twice. Reality is fluid, pure activity, and any given quality battles with itself, separates into opposites, and seeks to reunite—eternal change!" To some it was a liberating thought, but others recoiled in horror.

All along the Milesians had been speculating about nature, searching for the stable principle which anchored the world. It was most important that they gave a *logos*—reason—for their answers to the question and that this *logos* did not conflict with the phenomena they observed. For this observation they relied upon their senses; the idea of experiment which guides, corrects, and proves theory was yet some time away. Now Heraclitus saw that the use of the senses alone could not lead to the discovery of the principle the Milesians sought. Nowhere did experience present a stable form of being, except, perhaps, in the heavens. Pythagoras might speak of triangularity and present an unchanging mathematical theorem to describe a triangle, yet nowhere in nature (as perceived through the senses) does such a triangle exist. Theories are one thing and actual experience is another. Be it through myth or reason, we are always in danger of becoming separated from the phenomena themselves. The problem is by no means solved, even today.

Across the Adriatic, in southern Italy, the Greeks settled and built city-states as they did in Ionia. One of these was Elea, and it was to become famous as a center of Greek philosophy. Xenophanes of Colophon settled in Elea and spent the last years of his long life there (he lived to age ninety-two), dying about 525 B.C. Xenophanes was a critic, a profound doubter, who wondered if *any* knowledge was possible. His doubt was basically over religious questions, but it would have wider implications for early Greek science as a whole. "Look at the Homeric gods," he was believed to have said. "They behave like men, with all the frailties and faults of human beings. If cattle or lions had gods, and hands to draw them, lions would draw gods like lions and cattle like cattle!" In response to his own criticism, Xenophanes posited a single, invisible god, and further, reserved all certain knowledge only for that god. Man may know nature, and even approach the truth of it, yet he can never be sure of his knowledge, for it is always clouded by his uncertain senses.

Xenophanes posited his own ideas about nature. The cosmos is a spherical body, alive and conscious, the cause of its own change. All things come from earth and water, and there are cycles of flooding in which everything is covered by mud. He pointed to fossils as proof that during certain periods long ago the earth was covered in mud. What strikes us most, however, is his theory of knowledge. For the first time we have a separation between the way things seem and the reality of the way they are. Xenophanes said that men would claim figs are the sweetest things in nature—were it not for honey. Honey only *seems* sweetest in relation to human experience. Things *seem* this way or that, yet the truth is something different. Change seems to be all, air seems to be the basic stuff. How do we know?

The problem became the dividing line of pre-Socratic philosophy. In a way, the Ionians had begun theorizing prematurely, for in their enthusiasm they neglected to clarify how scientific knowledge itself was possible. What were the rules of thinking which led to the way of truth rather than to the relativity of seeming? This was the very question asked by the next Eleatic thinker, Parmenides, and his answer had momentous consequences for the course of Western science.

Parmenides was born in 515 B.C. and Plato claims that he visited Athens and met Socrates when he was about sixty-five. Scholars tend to doubt this; nevertheless his ideas certainly influenced the course of Plato's own philosophy. Parmenides believed that the "Way of Seeming" and the "Way of Truth" were forever separate, yet truth was attainable, and the road to it was through reason. The truth of nature is logical, based upon the law of contradiction: either a thing is or it is not. It was a decisive act, this assertion that truth must conform to logic, and it led Parmenides to a position opposite that of Heraclitus. In the end, his logic told him that change was impossible.

To Heraclitus, all being was really becoming and opposites were continually at war. Yet, what is the opposite of being? It is not-being. Here Parmenides applies his iron law of contradiction. To say that not-being is the opposite of being is still to posit something which is (not-being). To say that all is in change and motion is also to say that being, as the result of change and motion, comes from something which does not exist, or its opposite. Being comes from something that was not, and through alteration, becomes something different. But you cannot even speak of that which is not, for the law of contradiction compels you to posit the not-being as something that is! The same is true of motion, void, and infinity. To think of void is to think of something, an object of thought, which is a contradiction to emptiness. To think of infinity is to mark it off in thought by a limit; plurality is really unity, and things are only opposite in appearance.

It is as if logic has suddenly run amuck in Parmenides. Faced with a world of ceaseless change in which certain knowledge appears impossible, Parmenides completely denies the evidence of his senses and thus banishes their offending relativity. Reality is motionless, finite, change-less, eternal—like the law of contradiction or a mathematical equation. Truth is the pure application of reason and that is all.

Parmenides' picture of the universe is consistent with his logic. The universe is spherical, for the sphere is continuous yet finite. Wreaths or bands of pure and mixed elements which include the heavenly bodies encircle the earth. The idea of perfection and wholeness assigned to the sphere would have lasting consequences for astronomy. A circle has no beginning or end, it simply is, like the being of Parmenides. The circular motion of the heavens expresses stability rather than change, and so ce-lestial movement must be circular.

Modern physics speaks in the timeless language of mathematics, yet these constructions are arbitrary and may be revised or changed by ob-servation and experiment. On the other hand, Parmenides deemed ex-perience itself arbitrary and in doing so condemned it as illusionary. Thus motion was banished and with it the possibility of physics.

Parmenides left it to his pupil Zeno (born in 495 B.C.) to defend his logic, and this Zeno did quite well through the use of paradoxes de-signed to illustrate the logical absurdity of motion and the infinite divisi-bility of space and time. The first paradox is what Aristotle calls the Di-chotomy. Let us take motion from point 1 to point 2; if space is infinitely divisible, one must first arrive at the halfway point, $\frac{1}{2}$; but to arrive at $\frac{1}{2}$ one must first arrive at $\frac{1}{4}$, and for $\frac{1}{4}$ first at $\frac{1}{8}$, and so on (Figure 2.1). The conclusion is that if a finite length contains an infinite number of

Figure 2–1

points, it is logically impossible to cover even a finite length in a finite time.

The same kind of paradox holds for time. Say an arrow takes 2 seconds to reach its target; it must take 1 second to reach the halfway mark; but then it must take ½ second to reach the halfway mark of the next half, and ¼ to reach the halfway mark of this remainder, and so on (1 + ½ + ¼ + ⋯ = 2). The arrow can never really reach the target no matter how closely it approaches, because we are forced to posit still more steps ahead—an infinity of them. Also, consider the "moving" arrow itself. At every instant in time, the arrow occupies a definite position and so is at rest, and at the very next instant it occupies a new position and hence is at rest. So if time is mathematically divisible into instants, and if at each instant the arrow is at rest, the arrow never moves!

Next let us consider a race between the swift Achilles and a tortoise in which the tortoise has a head start. At each instant of time, both Achilles and the tortoise must occupy a given position, and again if length and time are infinitely divisible, the tortoise passes the same number of instances as does Achilles. So at each instant Achilles reaches a point and the tortoise also reaches a point and since infinity equals itself (we think) the number of points in both cases must be equal. But Achilles must pass through *more* points to overtake the tortoise and, alas, this he cannot logically do. Thus we are forced to conclude, contrary to our outraged common sense, that Achilles can never beat the tortoise in the race. (The paradoxes are easily solved by the differential calculus. However, in regard to the Achilles paradox, Zeno's trick lies hidden within his introduction of time: we are to consider only that length of time in which Achilles has not overtaken the tortoise. While the time intervals in this segment are infinite, the sum of their lengths is finite, which means that "can never" or "forever behind" refer to infinitely many terms but are confused with "forever" in the sense of time. Also, it is assumed that the number of points on a line segment are related to the segment's length.)

Parmenides and Zeno posed a major problem for the early Greek philosophers: how to rescue the physical world, the apparent world of the senses, from the tyranny of their logic. The key to the problem was to account for the movement and change while still retaining the old Milesian substratum of nature which could be known by reason. Up to this point in time, Greek science had been trying to account for both motion and matter in a single principle. To separate the two, motion and matter, would be the task of those who met the challenge.

One of these was Empedocles of Sicily, who died in 434 B.C. Empedocles denied the unity of Parmenides and Zeno and instead claimed that the world was made up of four distinct elements: fire, air, water, and earth. At first glance, it seems that this is not much of a step away from the materialism of the Milesians, and it involves facing the problem of knowledge, hence change. However, Empedocles gave up the tenet of an absolute *single* substratum. He also stated that all things were a mixture of the basic elements and that the agents responsible for these mixtures were not the elements themselves, but outside forces he called love and strife—attraction and opposition. Everything is distinguished by the quality of the mixture of the four elements, and they are conserved through the natural process of combination and dissolution. Even perception and cognition are reducible to this interaction.

With Empedocles, the idea of the four primary elements makes its first appearance. Most important, the cause of change is now separate from these basic principles. Empedocles went on to explain the celestial phenomena through the use of the four elements, dividing the crystal sphere into two hemispheres, one filled with fire and light, the other a mixture of air with darkness and a little fire. Further, he explained the sun as a concentration of light rays reflected upon the sphere, and the moon as borrowing its light from the sun (which he may have known from his contemporary Anaxagoras). He was aware of the true explanation of eclipses, saying that the moon shuts off sunbeams by its shadow on the earth. Remarkably, he even suggested that light takes time to pass from one point to another—that light has speed.

The definitive step toward atomism was taken by a contemporary of Empedocles, Anaxagoras, who brought Ionian science to Athens and was supposedly a friend of Pericles. Anaxagoras held that there was a portion of everything in everything; different kinds of matter were really mixtures of countless "seeds" which were infinite and imperceptible. That which we perceive is really built from that which we do not—the seeds. Nothing comes into being or perishes; rather, everything is compounded and dissolved from things (seeds) that already are. In answer to Zeno, Anaxagoras distinguished between the division of things into infinitely small magnitudes and sheer nothingness. Things can be divided, it is true, yet they still contain magnitude even if we cannot see it. They do not necessarily become points without extension.

According to Anaxagoras, the world began as a vortex of all things together and started whirling, separating its constituent parts off by the force of its motion; in other words, he conceived of centrifugal force. And what causes motion? Anaxagoras took the almost inevitable step and separated mind (*nous* in Greek) from matter. For the first time mind becomes nearly incorporeal, a ghost, and the problem for atomism ever since has been how to account for it in a purely material universe. For Anaxagoras, mind is the *deus ex machina* of movement.

In the second half of the fifth century B.C., with Leucippus of Miletus and Democritus of Abdera, Greek atomism achieved maturity. They actually used the term atom (*atomos*, uncuttable). The world is made up of imperceptible, infinite atoms, moving in a void, and their motion is simply an eternal jostling in which some atoms, upon collision, adhere to each other and produce compounds. Where space is empty of atoms, they pour in, setting up a circular eddy or vortex, in which heavy atoms tend to seek the center and lighter ones are squeezed outward. From this vortex the cosmos comes into being.

What causes motion? According to these atomists, nothing! Motion is natural to matter and was with it for all time. It is purely mechanical—the movement of atoms in the void—and thus the last traces of animism are removed from Greek *physis*. Through infinite time, worlds come and go, never the same, an infinity of worlds built from the never-ending collisions, adherences, and rebounding of the invisible atoms. There is no longer any need to account for motion and hence change. Motion is natural to matter—a necessity.

What, then, in this almost lifeless, mechanical world, is mind? Democritus may have been the first Greek to use the term microcosm as applied to man. Man and the universe are built up of the same stuff, following the same laws. Mind is only a collection of smooth, rarefied atoms, and sensation is due to the impact of atoms upon the body. For example, bitter taste is caused by smooth, rounded atoms, salty taste by heavy, jagged ones. The primary qualities of sensation are really only size, shape, and arrangement, and these produce secondary qualities such as taste, smell, and even color. Both sensation and thought derive from the same thing—the physical properties of atoms. A human being is nothing more than a miniature picture of the universe.

Thus the atomists arrive at a position nearly the opposite of Parmenides' yet just as radical. While Parmenides completely denies change and makes his reality totally mental, the atomists explain everything, including the human thinking ability, in terms of unseen atoms in motion. For both, the true nature of reality lies beyond phenomena, but whereas Parmenides begins by rejecting the phenomena as perceived by the senses, the atomists take matter as a whole as their starting point and appeal to the purely *hypothetical* atoms in order to explain it. They save the phenomena through theory instead of rejecting the evidence of their senses. Yet in the process they lose the mind, for in a purely atomistic theory, devoid of energy or any organizing activity, the integration of organic behavior and the phenomenon of thought itself remain a mystery. Theirs is a thoroughgoing materialism, and it was never really popular among the Greeks; it is a long way from the animate, half-mythical concept of *physis*.

Meanwhile the Greeks built their civilization and went on celebrating the mysteries of Eleusis, the frenzies of Dionysus, the rites and magical practices of the mythical world-view. Yet something was stirring; a new child now sat at the feet of the Great Mother Mythology. This infant held what seemed to be a lamp fueled by strange new questions. The flame of the lamp flickered and only a few noticed it. To us, looking back, it was a beacon.

SUGGESTIONS FOR FURTHER READING

COHEN, M. R., and DRABKIN, I. E., eds. *A Source Book in Greek Science.* Cambridge: Cambridge University Press, 1958.

CORNFORD, FRANCIS M. *From Religion to Philosophy: A Study in the Origins of Western Speculation.* New York: Longmans, Green, 1912.

GUTHRIE, W. K. C. *A History of Greek Philosophy,* 6 vols. Cambridge: Cambridge University Press, 1961–1981.

KIRK, G. S., and RAVEN, J. E. *The Pre-Socratic Philosophers: A Critical History with Selection of Texts.* Cambridge: Cambridge University Press, 1957.

LLOYD, G. E. R. *Early Greek Science: Thales to Aristotle.* New York: W. W. Norton, 1970.

SAMBURSKY, S. *The Physical World of the Greeks.* London: Routledge and Kegan Paul, 1956.

∽∽∽∽∽∽∽∽∽∽∽∽∽∽∽∽∽∽∽∽∽∽∽∽∽∽∽∽∽∽∽∽∽∽∽

3

The Elements
of All Things

Numbers and Ideas

Underlying the physical theories briefly outlined in the last chapter was the mathematical philosophy of the Pythagoreans. Owing to the secrecy of the school and the half-legendary life of Pythagoras, it is extremely difficult to separate the developed doctrines of the school from those which may be safely attributed to its founder. Aristotle himself seldom, if ever, spoke of Pythagoras alone, referring to his doctrines as those of the Pythagoreans or the so-called Pythagoreans. Nonetheless, a great many of the school's leading ideas were common knowledge among the Greeks and were considered important enough to be the subject of criticism or praise. Pythagorean mathematical philosophy, as well as Heraclitus' thinking, certainly influenced the Eleatic school, and it is probable that the targets of Zeno's paradoxes included the Pythagorean effort to derive the physical world from number and geometry.

Thales, it should be remembered, is credited with bringing the principles of geometry to Greece. His theorems also constituted demonstrations which could be recognized by practical constructions. In a sense, the construction could be deduced from the theory; it was not necessary to illustrate each principle by a specific problem as in Egypt. In other words, with Thales, mathematics became deductive and therefore abstract. The Pythagoreans extended this process of abstraction and in turn infused all of nature with mathematical concepts. It seems that they were the first to stress the idea of number and geometry underlying diverse natural phenomena. The result, adapted and enshrined in Plato's later philosophy along with an ethical, transcendental corollary, was the important recognition that numbers are abstractions, mental concepts, suggested by material things but independent of them. For the early Pythagoreans, however, the physical world was actually constructed from numbers.

Pythagoras himself was born on the island of Samos sometime around 569–572 B.C. It is reported that he traveled to Egypt, and even to India, finally settling in the city of Croton in southern Italy. At Croton he became famous for his esoteric teachings and gathered a community of followers. However, the mysticism and secretiveness of the Pythagorean community seems to have aroused the suspicions of Croton's population and Pythagoras was forced to flee to the nearby city of Metapontum. He died there around the age of eighty—some say that he was murdered—but the community flourished and kept his teachings alive.

The early Pythagorean community was a mystical, religious group; researches into the science of mathematics were part of a larger philosophy. It is interesting to note this connection, for even today mathematics is fundamentally a science of ideal knowledge which to the uninitiated may at times seem mystical. It is said that Pythagoras believed in the transmigration of souls, the passage of an immortal soul from one living body to another. Purification was important, and rules of abstinence were strictly adhered to. Pythagoras abstained from eating or sacrificing animal flesh, since all animal life was akin, and he also forbade the eating of beans for reasons probably known only to himself.

The Pythagorean theory of the transmigration of souls seems to indicate that events recur in cycles as well as suggesting a profound harmony in the universe. Indeed, Pythagoras himself probably believed that the universe was alive and intelligent, and he may have been the first to claim that the earth is spherical like the entire cosmos. It was this purely religious faith in the spiritual harmony of the universe which would suggest some interesting relations in the realm of numbers to the Pythagoreans.

Imagine Pythagoras sitting in the sand one day, under the hot Italian sun, musing over some pebbles. He had no symbols for numbers and so used his pebbles to represent the whole numbers. A single pebble represents the number 1, a unit-point in the chaos of the sand. He adds another pebble and notices that the two pebbles generate a line. Now he adds a third and finds an enclosed surface. Finally, with a leap of imagination, he adds a fourth and discovers that he has constructed a solid. Now here is something to think about! Four pebbles, four unit-points, have generated the dimensions of nature.

The pebbles, remember, can represent numbers too. Maybe their arrangement in the sand, their shape, has something to do with their strictly numerical properties. The numbers 3, 6, 10, and so on can be arranged to form triangles:

(3) •• (6) • •• • (10) ••••

The numbers 4, 9, 16, and so on form squares in the sand:

(4) •• •• (9) ••• ••• ••• (16) •••• •••• •••• ••••

But if all things in the universe are harmonious, there must be some relationship between the triangular and square numbers, hence between whole numbers in general. Considering the square numbers, for example 9, Pythagoras draws a diagonal in the sand:

and finds the triangular numbers 3 and 6. Going back, he discovers that the square number 4 is the sum of 1 and 3, and likewise 16 = 6 + 10, the sum of triangular numbers. In fact, the sum of two triangular numbers is always a square number.

Next, Pythagoras glances at the square numbers and sees that in order to pass from one square number to another he must add a gnomon, which in his time meant a carpenter's square. For example, given the square number 4 •• •• to obtain another square he adds a gnomon, 5:

which is also the square number plus one. In our notation the Pythagoreans discovered that $n^2 + 2n + 1 = (n + 1)^2$ for the square number n^2. Therefore $2^2 + (2 \times 2) + 1 = 3^2$ or the square 9, and the square $16 = 3^2 + (2 \times 3) + 1 = 4^2$, and the next square number would be 25, or $4^2 + (2 \times 4) + 1$. Further, if we start with 1 and add gnomon 3, then gnomon 5, then gnomon 7, we find a square. So, adding the sequence of odd numbers always results in the same figure, whereas adding the sequence of even numbers always gives an oblong figure, for example, 2 + 4 + 6 = 12 which is a ratio of length to height:

(2) •• (2+4) ••• ••• (2+4+6) •••• •••• ••••

The Pythagoreans were quick to assign special significance to this arrangement, for the odd numbers remain the same figure after each addition while the even numbers change. Thus they identified the unlimited or infinite with even, and the limited with odd, since even numbers can always be halved, while the addition of odd limits the regress by putting an end to the halving.

These fascinating relationships between numbers and the forms they generated suggested to the Pythagoreans that the very harmony they worshipped in their mystical doctrine was somehow related to the harmonious combinations of numbers. If numbers generated geometrical relationships, why not assume that this was true of cosmology as well? And so they did, saying that the numbers were actually the essence of nature, and that every object in nature consisted of material unit-points like atoms. Besides being alive, nature was rational from end to end.

The Pythagoreans looked for signs of this mathematical harmony everywhere. Perhaps Pythagoras himself discovered that the chief musical intervals were expressible as simple ratios of integers—in fact, the first four integers. Harmonious sounds were created when two equally taut strings were plucked, provided these were in definite proportions. Thus the ratio of 2 to 1 is still called an octave, 3 to 2 is known as a fifth, and 4 to 3 is a fourth. So was born the science of acoustics.

The number 4 seemed to have special significance. Addition of the first four integers gave the number ten, which was a perfect triangular number. Ten thus became for the Pythagoreans a sacred number and figure, or what they called a *tetraktys*. The triangular number ten has four members on each edge:

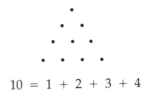

$$10 = 1 + 2 + 3 + 4$$

Ten, in fact, seemed to be an ideal number, and the Pythagoreans held it to be the very nature of number. It must, therefore, be basic in the design of nature, the sacred *tetraktys*.

If ten was the sacred number, it was logical to assume that it and its harmony should be basic in the design of the heavens, too. Pythagoras had given the planets a revolution of their own which was opposite the daily rotation of the fixed stars from east to west. Yet, the planets did share in the daily rotation of the fixed stars, following the celestial equator, while their own special movements followed the plane of the ecliptic—the motion of the sun through the constellations. Where was the harmony here? Further, since the sun, moon, and planets all had their motions derived from these compound rotations, it would follow— inharmoniously—that the proportions of velocity and distances would not hold from one instant to the next.

To meet these objections and some others (for example, that the earth was of too gross a nature to be fittingly placed in the center of the universe), the later Pythagoreans developed an amazing new theory. They took the earth out of the center of the universe and replaced it with a "central fire," also called the hearth of the universe, or the watch-tower of Zeus. Greece was situated upon the side of the earth which was always turned away from the central fire, but some Pythagoreans held that the fire might be seen in India. When the inhabited world (at least Greece) swings around in sight of the sun, we have day, and when it rotates to the other side of the central fire, we have night. Still later Pythagoreans assigned a daily rotation to the earth, abandoning the central fire, or placing it in the center of the earth, claiming it to be the origin of volcanic eruption. According to any of these Pythagorean theories, then, the earth moved!

Let us not forget the sacred number ten. The central fire may have made the heavens more orderly, but there were still only nine bodies— earth, moon, sun, the five planets, and the fixed stars in that order. The Pythagoreans added a tenth body, calling it the *antichthon*, or counter-earth, placing it between the earth and the central fire. It was invisible, for it always kept pace with the earth. Nonetheless, we now have our sacred ten. The phenomena are saved and harmony rules.

The search for unity in nature based upon numbers and geometry led the Pythagoreans to great heights. They developed a theory of application of areas which allowed them to compare the ratios of surfaces. Hippocrates of Chios (not to be confused with the physician) even came up with the idea of arranging theorems so that those presented later could be proven on the basis of earlier ones. Antiphon and Bryson, mathematicians of the sophist school, were influenced by the Pythagorean notion of comparing figures. They approached the problem of squaring the circle by inscribing or circumscribing polygons of more and more sides. This "method of exhaustion" would later be used by Plato's student Eudoxus and finally by Archimedes.

Yet for all their outstanding work, mathematical disaster struck the Pythagoreans. They experienced a crisis which shook the whole foundation of their number theory, and indeed, the whole of Greek mathematics. The problem arose from a rather innocent discovery. They found that the sum of some of their square numbers was also a square number—for example, $9 + 16 = 25$, or $3^2 + 4^2 = 5^2$. They soon found that this relationship, called Pythagorean triples, expressed a general fact about right triangles: the sum of the squares of the arms equals the square of the hypotenuse. Pythagoras sacrificed an ox—perhaps breaking a vow in doing so—to honor the discovery. Yet, their greatest triumph led to their undoing.

One day an unfortunate Pythagorean decided to examine the simplest form of the relation, a right triangle with each arm a single unit long. Now, the Pythagorean number theory expressed fractions as the ratios of two whole numbers, as we do today. However, they found that their celebrated theorem, when reduced to units of one, gave them $\sqrt{2}$ for the hypotenuse. But $\sqrt{2}$ *does not* equal a ratio fraction of whole numbers (the proof is found in Euclid); it goes on indefinitely as 1.4142135 . . . . Suddenly an "irrational" element had arisen which did not exemplify the harmony of numbers. Indeed, the Pythagoreans labeled their discovery irrational and the relationship incommensurate. Legend has it that the discovery was made while they were at sea, and they threw the man who pointed it out overboard, swearing themselves to secrecy.

Because of the problem of incommensurability, the Greeks never did develop a rigorous concept of number. Most important, the Pythagorean effort to identify number and geometry was abandoned, the two becoming separate domains. The new emphasis was upon form and quality, no longer upon the purely quantitative use of numbers. It would take the development of algebra, and the wedding of equation with line by Descartes, before the two were again reunited.

Also, the paradoxes of Zeno—involving notions of continuity, limit, and infinity, concepts fundamental to the calculus—demanded a theory of number which the Greeks never developed. Problems of motion and variability remained upon the plane of qualitative explanation and even metaphysics.

This is not to say that the Greeks stopped thinking about number. The mistake of the early Pythagoreans had been their very desire to see the relations of numbers as actually existing in their perceptions of material nature. They had failed to make the distinction between the purely rational, abstract concept and the world of sensate change. In turning to mathematics, the Pythagoreans had actually tried to settle the issue by appealing to unchanging forms and the deduction of relationships from intuitive principles of thought. The problem was how to account for the world of the senses in terms of these rational forms. One answer was the idealism of Plato.

Plato's influence has been felt in many areas of Western thought, from religion, politics, and ethics to science. Some people see his contributions to science as being mainly negative, divorcing the world of fact from the pure realm of ideas, and proclaiming the latter to be the sole repository of certain knowledge and hence reality. To others, Plato's cosmological intuitions and his geometrical theory of the world's structure come remarkably close to the general tendencies of modern phys-ics. Still, there are those who see Plato standing upon the periphery of science, a moral and political philosopher who, although not a scientist himself, did spur the scientific tendencies in his students.

Plato was born in Athens in 428 B.C. of a distinguished Athenian family descended from the old kings of Attica. The Athens of Plato, like all of Greece, existed in a time of great turmoil, and it is during such times that people tend to question most of the fundamental beliefs of their civilization.

During the early fifth century B.C., the city-states of Greece had rallied together and twice humiliated the mighty Persian Empire. The Greeks had stood their ground against the powerful, older empires of the Near East and had triumphed. Athens, which later boasted the wis-dom of Pericles and the magnificence of the Parthenon, had been one of the leading cities in this victory. It was a glorious moment in Greek his-tory, full of confidence and pride. Yet the very strength of the Greeks in their independent city-states led to conflict among them, once the Per-sian enemy was defeated. A long and disillusioning conflict, the Pelo-ponnesian War, broke out between two confederations of cities, one led by Athens and the other by Sparta. When Athens fell to Sparta in 404 B.C., something of the Greeks' confidence in their civilization had been lost.

In this maelstrom of war and disillusionment, a group of profes-sional teachers, collectively known as the sophists, began their teaching. The sophists claimed to teach wisdom, but theirs was a wisdom which rested upon superior arguments and rhetoric, not necessarily upon truth. As one might expect, the proliferation of scientific theories cou-pled with political upheaval led people to wonder how to separate the true from the false and what were the prerequisites for a virtuous life. The sophists had a simple answer: all opinions are equally true, for there is no judge of what is true other than the human being. What is true for me—what is hot for me, tastes sweet or sour, what has unity or plurality—may not be what is true for you. Therefore, scientific or moral truth is only opinion based upon the experience of one man or the agreed-upon experience of the group. The sophist Protagoras suppos-edly said that man is the measure of all things, and hence truth is a mat-ter of preference. Learning, then, is reduced to skillful argument and morality to practical expediency.

Let us imagine young Plato walking through the *agora*—the marketplace of Athens—listening to these various discussions. Sud-denly he is accosted by a rather ugly, barefoot stonecutter named Socrates.

"My, my," Socrates might have said, "but don't these sophists and philosophers use big words and argue so convincingly! What marvelous speculations they concoct to describe the nature of the universe. Yet, have all these things resulted in a better life? In virtuous conduct? I sometimes wonder if they know what virtue actually is."

Plato and other young men gather about Socrates. "Tell us, then," they ask, "which of these theories you believe to be good. Which gives us the knowledge we need?"

"Ah," answers Socrates, a glint in his eyes, "perhaps we ought to begin by asking what knowledge itself is, what virtue is. What is the nature of the good?"

"Tell us, Socrates!" they chorus.

"I?" says Socrates. "But I am an ignorant stonecutter who is only conscious of his own ignorance. I don't know. I only ask questions of those who claim they do."

"And what have you discovered?"

"I found that when I pressed them with my questions they were driven to perplexity and could only give me *examples* of, say, virtue. They could not tell me what virtue *is*. How strange, it seems to me, to expound theories about something and yet be unable to say clearly what that something is."

Here a young Athenian interrupts. "The sophists claim that everything is true according to each individual's measure of truth, and thus all theories are equally true and false."

Socrates looks puzzled. "Then I would say that they must admit that their own statement can be false too! For how can they hold that all opinions are equal and not include themselves in this? How can we even believe their principle when the principle itself is relative to the man or group which holds it?"[1]

Socrates would probably have stopped here and not troubled himself over scientific questions, for he was more interested in what makes a godly and just life. Though he seems to have subordinated science to ethics, he held that whatever men discuss, they first must be clear and logical in their definitions and classifications. For this he was condemned by the government of Athens on the charge of corrupting its youth. But his deep convictions galvanized Plato, who made Socrates his own mouthpiece in the dialogues, fictional renderings of his conversations. In the Socratic quest for universal definitions can be found the origins of Platonism—the theory of Forms. The key word was virtue. Virtue, for Plato, was knowledge, and knowledge—if it is to mean anything at all—must be knowledge of reality.

Thus Plato's contribution to the history of science orginated from an ethical basis in which knowledge was not only the foundation of tech-

---

[1]Adapted from Plato, *Theaetetus*. In *Plato's Theory of Knowledge*, trans. Francis M. Cornford (New York: The Liberal Arts Library, 1957).

nical skill, *techne*, but also the awareness of how to act virtuously, *arete*. How does one separate true knowledge from opinion and the alteration of material instances? The problem, again, was to reconcile the changeless being of Parmenides with the Heraclitean flux of sensation.

How does one recognize the principle of equality? I look at some pieces of wood or stone, for example, and say that their number is equal, or that they are equal in size. Yet, are they exactly equal? Close up, I see that they are not. The same pieces of wood (or stone) appear at one time equal and at another unequal. So, for Plato, the *idea* of perfect equality is not the same as its instance in the material world: the former is abstract and perfect equality while the latter is only a partial and defective illustration. It is true that both ideas and their instances exist. But real knowledge of anything (hence science) depends upon the recognition of the idea *apart from* sense-objects, which vary from place to place or from person to person.

Science (and knowledge in general) is a grasping of the universal Idea or Form which is changeless and eternal through its various manifestations in physical things. The world of these Forms is true reality, supersensate and divine, and the Forms are only imperfectly contained in the changing material world. A person tied exclusively to the physical world gains only opinion—at times illusion—from the senses. Knowledge, on the other hand, is of universal ideas and gained through pure reason.

Men who do not use their reasoning abilities to separate the universal Form from its material instances are like prisoners in a cave, chained with their backs to a fire, looking at the cave wall and seeing only shadows of the real things behind them. For the prisoners, the shadows or images (*eidola*) cast upon the wall by the firelight are real. But if one broke his chains and turned, he would see clearly that these shadows are only reflections. And if he should get out of the cave, he would perceive the sun and would see that it produces the seasons and controls everything in the visible world. This simile of the cave is probably the most famous passage in Plato's writings, and it is found in the dialogue *The Republic*. The gaining of knowledge is analogous to the ascent out of the cave, and the final vision of the sun is the Form of the Good.

In *The Republic*, Plato recommends that the training of the philosopher-ruler incorporate mathematics divided into five disciplines: arithmetic, plane geometry, solid geometry, astronomy and harmonics. Mathematics is a step toward the world of Forms, training the mind to think abstractly in order to lead the philosopher to the ultimate vision of reality. The heavenly movements, for example, are only illustrations of eternal geometrical truths, and so it is pure geometry which ought to occupy our investigations, not the celestial phenomena themselves—which are inferior to this ideal reality.

Plato's philosophy evolved from this point to the concept of a hierarchy of ideas demonstratively linked to one another by deductive relations similar, if not identical, to the principles of mathematics. The world of Forms in the later dialogues is linked to numerical relations in which the integers are generated by unity, one, lines generated by fluxion from the integers, surfaces by lines, solids by surfaces, upon all of which the harmony of the world is founded. In short, Plato adopted Pythagorean mathematical philosophy, freeing it from immersion in the physical world.

Legend has it that a sign was affixed to the gate of Plato's Academy: "Let no man ignorant of geometry enter." Whether this is true or not, it is certain that his later dialogues became more and more mathematical and his cosmology Pythagorean.

We can see the process unfolding in the dialogue *Meno*. Meno asks Socrates whether virtue can be taught and, as expected, the question leads to the sophist dilemma of knowledge itself. Socrates, of course, is looking for definitions, and the examples of what he wants come directly from geometry. The *Meno* is also concerned with the idea of the soul's immortality and its prior acquaintance with the world of Forms. The problem is to construct a square with its area double that of a given square, and through Socrates' questions, Meno's slave *sees* for himself that the answer is not to make the sides twice as long but to build upon the diagonal. Thus the truths of geometry are timeless, divine, and eternal, and the question refers to the universal mathematical concept of square, glimpsed by the soul in a transcendental realm.

Here, by the way, is a partial answer to the Pythagorean problem of irrationals. Although Plato himself was not a mathematician and never developed a number theory which incorporated irrationals, his Forms do provide a way out. That $\sqrt{2}$ cannot be exemplified by experience does not mean we cannot reason about it. For Plato, it can be seen as one of those divine ideas our incorporeal intelligence encountered before birth. We may analyze it, see if we can deduce any contradictions from it, and if our premises prove correct we may deduce other principles. The results we get by working from pure ideas are definite and certain. From Plato's time onward, mathematics was totally deductive and no longer *techne*, a simple skill, but a science in its own right.

That mathematics gives us a model of physical reality and is yet apart from the corporeal is the heart of Plato's contribution to science. Plato, however, did not view his mathematical constructions as arbitrary creations of the human mind. The question of the ultimate reality of scientific theories has plagued scientists to this very day. Most scientists see their pictures of nature as representations approximating the world of fact, conditioned and corrected by experiment and observation. Yet, deep inside, there may well be the conviction that these ideas are in a sense real, and it is faith in their reality which leads us onwards. At the beginning of this road stands the imposing figure of Plato.

The culmination of Plato's Pythagorean mathematical philosophy is the *Timaeus*. In earlier dialogues Plato spoke only briefly about cosmology: The universe is a sphere, super-celestial space is occupied by eternal ideas, and the earth stands in the center of the heavens. The picture is vague and presented mythologically. In the *Timaeus*, however, the purpose is to place man in the cosmic setting and give a "likely" account of that setting.

The *Timaeus* proposes to give us a rational account of the harmony of the universe. Since that account (in being rational) must be based upon the Forms, the visible world must be a *likeness* of something, in that it partakes of unchanging reality. Hence Plato's account of the visible world can be only a "likely story" which mirrors the divine intelligence. This means that there can be no *exact* science of nature. Rather, Plato's purpose is to show that the *eikon* (image) points to the Forms just as corporeal phenomena approximate changeless mathematical laws. In this way the visible world is like a work of art which the artist or craftsman (Plato calls him the Demiurge) shapes from materials already at hand, injecting reason and order into material chaos. The Demiurge is more of a technician than a creator-god, for he is limited by the materials with which he works. Any rational account of his work must also be limited—a "likely story."

The world is finite and created by a harmony of four elements—fire, earth, water, and air. The harmony of these elements is a copy of geometrical proportions, for Plato speaks in terms of planes and solids bound together by a continued proportion of square and cubed numbers. Yet while mathematics yields rigorous deductions whose validity is certain, we may expect the copy, the visible world, to match the ideal only provisionally. The generation of the universe, therefore, is a mixed result of Reason and Necessity, with Necessity (or, to use Plato's other term, the Receptacle) being the irrational and indeterminate limiting factor. Since the four elements are always changing in the corporeal world, they are transient appearances in the Receptacle which is molded into various shapes by the qualities of the Forms.

Plato's conception of the four elements is fully geometrical and based upon the five regular solids of his contemporary, Theatetus. The right-angled isosceles triangle (90°, 45°, 45°) and the half-equilateral (90°, 60°, 30°) are the two irreducible elements in the construction of four of these regular solids.

The simplest solid is the pyramid of four triangles of eight half-equilaterals. With five faces (four triangles and a square base) and the sharpest edges, it is fire. The cube, constructed with right-angled isosceles triangles, is the most immobile and hence earth. The octahedron of eight faces of sixteen half-equilaterals is air, and the icosahedron of twenty faces or forty half-equilaterals is water.

The transformation of elements occurs through the combinations of these unit-triangles. Water, for example, breaks down into two atoms of air plus one of fire, or, in terms of triangles, 40 (water) = 2 × 16 (air) + 8 (fire). Since the triangles of earth are different, solids can mix only with solids. The last of the five regular solids, the dodecahedron, has pentagonal faces and cannot be formed out of the elementary triangles. It is equated with the cosmos, since its form is nearly spherical (remember, this is a "likely story"!).

The triangles can be of different sizes, and hence there are grades of elements which account for their various manifestations. For example, water can be ice or liquid but still the same elemental shape. Wine is a mixture of various grades of water with a little fire (obviously). One merit of Plato's choice of triangles is that they can yield these grades much closer together in size. Another merit of this choice (as pointed out by Karl Popper, whom we will meet in a later chapter) is that it allows the incorporation of irrationals by substituting a fundamentally geometrical system for the Pythagorean number theory. Pythagorean mathematical harmony was saved through an emphasis upon its geometrical properties as well as the transcendental world of the Forms.

The astronomical system of the *Timaeus* is also a "likely story" which cannot account exactly for the observed motions of the celestial bodies. Plato divided the heavens into two movements. One, the Circle of the Same, accounted for the entire daily rotation of the heavens. The other, the Circle of the Different, was divided into six places, seven unequal circles, to account for the movements of the planets—moon, sun, Venus, Mercury, Mars, Jupiter, and Saturn—as seen from the earth in the center. The planetary circles are unequal, of different diameters, so that they fit inside one another, corresponding to the sequence 1, 2, 3, 4, 8, 9, 27. This is probably not based upon any serious estimate of observable distances or their ratios, but simply a piece of Plato's ideal astronomy.

All the planets share in the rotation of the Same, which is also, according to Plato, the motion of the World Soul and the fixed stars. The sun, Venus, and Mercury have a proper motion identical to the Different; the moon has an additional movement which carries her faster. These motions probably are meant to correspond to the solar year or equal angular speeds. Venus and Mercury, we are told, also have a contrary tendency, overtaking one another and being overtaken, which seems to contradict the former statement. It is possible that this contrary movement applies to the fact that, when seen from the earth, these inferior planets, having their orbits inside that of the earth, appear to move from side to side of the sun and thus may rise ahead of the sun in the morning or after sunset in the evening.

The three outer planets, Mars, Jupiter, and Saturn, share in the motion of the Different but counteract that motion by slowing down and by "opposite tendencies." Again, Plato may be referring to the varying speeds of the orbits of these planets, which are slower than the solar year, and also to the phenomenon of *retrogression*. When seen from the earth, the planets seem upon occasion to stop and retrace their steps. The reason for this retrograde movement is that the earth's orbit actually overtakes and passes the orbit of the superior planet, which, when seen against the stellar background, makes it seem that the planet stops and moves in the opposite direction.

The rational principle of the soul infuses the entire universe as well as man, whose immortal soul is evidenced by his ability to see the principle at work in nature. The principle is living and eternal and cannot be found realized in its full completeness. To make it apparent, therefore, the Demiurge resolved to have a moving image of this eternal Form which he set in motion within the visible order. Plato calls this moving image of eternity *time* and has it revolve according to number. Time, and hence change, came into being with this world, and both are thought of as the movement of the entire heavenly sphere.

Some ancient writers, like Plutarch, claimed that Plato in his old age regretted that he had given the earth a central position and followed the Pythagoreans in conceiving of a central fire as a worthier body for the center. Modern historians believe that this change was probably that of Plato's followers in the Academy, for there is no evidence of it in his writings. Many hold, however, that he taught a secret unwritten doctrine which may have been even more Pythagorean.

Although Plato's Academy included instruction in preliminary disciplines such as mathematics, it was more of a community of philosophers, seekers of knowledge, with no single orthodoxy or doctrine. It seems, however, that Plato's students were split into roughly two groups: those who favored the Forms, popularly labeled the "gods," and those who wanted to bring science back down to the visible world, the "giants." Plato did not totally reject the world of the senses, as Parmenides did, but in studying natural phenomena he desired the mind to be elevated toward the eternal Forms. For many, the end result was neglect of the material world.

Eudoxus was born in Cnidus in Asia Minor and came to the Academy at the age of twenty-three, about 368 B.C. He was probably not a true student of the Academy, and Plutarch says that Eudoxus and his pupil Menaechmus irritated Plato by trying to solve stereometric problems (construction problems like doubling the cube) with mechanical aids and not by reason alone. Archemides credited Eudoxus with the proofs of two propositions first discovered by Democritus: A cone is one-third the volume of a cylinder and the volume of a pyramid is one-third that of a prism with the same base and equal height.

However, as a mathematician Eudoxus is probably best known for his theories of magnitude and exhaustion. The theory of magnitude, designed to deal with the problem of irrationals, was concerned mainly with practical construction problems of geometry. Instead of assigning numbers to geometric figures, Eudoxus assigned magnitudes to lines, areas, angles, and volumes which had no quantitative values but were used in terms of equality and multiples of ratios. In this way both rational and irrational numbers could be treated as proportions based upon a definition of equal ratios. Thus Eudoxus completed the separation of arithmetic and geometry. Today we often refer to x^2 as "x squared" and x^3 as "x cubed," reflecting this geometrical emphasis.

His second contribution has already been touched upon. The theory of exhaustion (a term he himself did not use) aided him in his proofs. It is possible that the method had been formulated earlier by Hippocrates of Chios, and it is reproduced in Book V of Euclid's *Elements*. Nonetheless, the credit usually goes to Eudoxus.

In the computation of areas and volumes of curved surfaces, Eudoxus showed that we can "exhaust" the continuum of the curve by the continued division of a given magnitude, thereby getting an answer that approaches the required area as closely as desired. In this way, one could show that the areas of circles are to each other as the squares of their diameters or volumes as cubes of radii. Yet it should be cautioned that the Greek mathematicians never considered the process carried out to an infinite number of steps, as modern calculus does in passing to the limit. The area of volume was never truly exhausted, as defined by the limit of an infinite numerical sequence, and so there was always something left over.

It is no surprise, then, that Eudoxus applied his considerable mathematical talents to the problem of the celestial movements. Like everyone before Kepler, Eudoxus assumed that the planets moved in circular orbits—geocentrically, of course—and thus he had to account for their irregular wanderings by a combination of circular rotations. He posited a series of spheres, one inside the other, all having the earth as their common center. To explain such things as varying speeds, retrogradations, and deviations in latitude, Eudoxus assumed a combination of circular motions, which meant assigning each body a certain number of spheres. Further, the poles of the sphere carrying the planet itself were not fixed but moved in a greater sphere concentric to the carrying sphere, which moved about two different poles with a speed of its own.

Eudoxus found that he could sufficiently represent the motions of the sun and the moon with three spheres for each. In order to account for the stationary points and retrogradations of the five planets he assigned four spheres each, and the fixed stars were given a single sphere. Thus the total number of spheres in his system was twenty-seven.

Of the four spheres given to each planet, the outermost sphere pro-

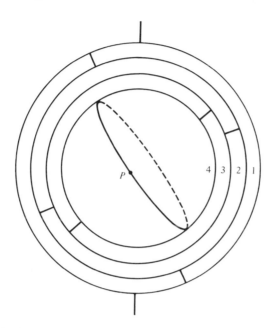

Figure 3.1 The four concentric spheres of Eudoxus (from G. E. R. Lloyd, *Early Greek Science: Thales to Aristotle*, New York: Norton, 1970). Sphere 4 carries the planet (*P*) which moves along the equator of the sphere; the axis of sphere 4 is carried by the rotation of sphere 3 about its own axis; the axis of sphere 3 by 2, and so on.

duced the daily rotation of the planet about the earth and the second gave the motion through the zodiac—the sidereal period or the yearly movements of the planet which, for the superior planets, varies in relation to the solar year. The third sphere had its poles at two opposite ends of the zodiac, carried around by the second, and its rotation equaled the synodic period of the planet—the time elapsed between two successive oppositions or conjunctions with the sun. The poles of the fourth sphere were attached to the surface of the third at a constant angle which differed for each planet. The planet itself was fixed on the equator of the fourth sphere whose rotation took place in the same period of the third but in opposite direction. The effect of the last two spheres is to give the planet a curving path, the familiar figure eight.

Figure 3.1 is a modern reconstruction of the model, and there is no ancient authority for this diagram. It is not certain that Eudoxus provided definite parameters for all the planets or if the model was fully quantitative.[2]

[2]See G. E. R. Lloyd, *Magic, Reason and Experience* (Cambridge: Cambridge University Press, 1979), pp. 175–76.

For the sun, moon, Jupiter, Saturn, and to some extent Mercury, the system was adequate in explaining principal phenomena. It ran into problems for Venus and failed completely in the case of Mars, for that planet's retrograde motion could not be made to agree with its synodic period at all. Further, the system did not take into account the full eccentricities of the planets' orbits nor the precession of the equinoxes due to the earth's own rotational spin (later discovered by Hipparchus).

Callippus of Cyzicus, who followed Eudoxus, attempted to correct some of these deficiencies by adding a fifth sphere to Mars as well as an extra sphere to Venus and Mercury. Callippus also introduced two new spheres into the solar theory to account for the unequal motion of the sun in longitude, and he increased the lunar spheres by two. Each addition was designed to approximate as closely as possible the observed motions of the planets, yet in the end the entire complicated and ingenious system still failed.

Whether the material spheres actually existed in the sky or were, in the end, only mathematical ideal constructions was another question and one Eudoxus did not answer. Nor are we certain of the extent of his own observations or his debts to Near Eastern astronomical sources. If we can say that Eudoxus represents a more empirical turn of mind than Plato, the tendency toward empiricism becomes even more evident in the development of Greek medicine.

As in the Near East, medicine among the Greeks originated from a religious, mythological system, most notably the cult of Asclepius. In Homer's time Asclepius was not a god; he is referred to in the *Iliad* as a "blameless physician." Asclepius was supposed to have been the son of Apollo and to have been instructed in the healing arts by the centaur Chiron. Hesiod gives us the story that Asclepius became so self-assured in his practice that he even resuscitated the dead—whereupon Zeus slew him. This story is worth our attention, for it shows that very early in Greek medicine the danger of the physician's interfering with the laws of nature is present.

By about 500 B.C. Asclepius had been deified, and many temples of his cult were spread across the Greek world. The most important was at Epidaurus, where the patient would sleep in the temple in order to be visited by the god in a dream. The dream would then be interpreted by a priest and the cure prescribed. Asclepius became the patron god of physicians. His symbols were the snake and dog, the former preserved in the caduceus, the symbol of the medical profession to this day.

Around these temples, schools of a purely secular and rational nature grew up, deriving medical practice from observation and the theories of Greek philosophy as well as experience in the gymnasium. The rational theories of the early Greek philosophers had a profound influence in these schools, especially the ideal of Pythagorean harmony.

Nature as a harmony of numbers could also be seen in terms of an equilibrium in which all the elements are in perfect balance. Man as a microcosm of the universe owes his health to the same balance of elements; disease is therefore a disturbance of this balance. Hence the physician's job was to *aid* nature in restoring the equilibrium, and the basic cure was regulation of the diet. By "diet" the Greeks meant not only food but a person's entire mode of living, including sleep, rest, and exercise.

The extent and significance of the natural philosophers' influence upon medicine is a complex subject. It is possible to see the pre-Socratic emphasis on natural causes reflected by the Hippocratic corpus (discussed below). However, such speculations and axiomatic systems were balanced by an empirical tradition concerned with observation, the description of diseases, and specific treatments. Here the practical value of observing the process of the disease, rather than speculating on its causes, was emphasized. The name associated with this "empirical" approach is Hippocrates, the "Father of Medicine."

Hippocrates of Cos (460–379 B.C.) came from a family of Asclepiads. He is supposed to have traveled widely in Greece, studying philosophy, politics, tragedy, and even sculpture. Some fifty to seventy books collected in Alexandria in the third century B.C. are attributed to Hippocrates. This collection represents a long tradition rather than the writings of a single man or group. Nonetheless, the stress upon observation, upon naturalistic explanations over supernatural, and upon knowledge gained from experience as a guide to treatment, justify the general label Hippocratic medicine.

Hippocrates probably had some knowledge of bone structures—one book is devoted to fractures and how to set bones—but his knowledge of internal organs was vague. Influenced by the common belief in four elements (air, fire, earth, water), Hippocratic physiology was most notable for its theory of the four humors (Figure 3.2). The four humors were bodily secretions related to the four elements and to the seasons: blood (the heart) was equated with air (spring); yellow bile (the liver) was equated with fire (summer); black bile (the spleen) with earth (autumn); and phlegm (the brain) with water (winter). Good health meant the four humors were in equilibrium. Disease occurred when the balance of the humors was upset, and treatment entailed methods to right the balance. Also related to the four humors were the four qualities: hot, dry, cold, and moist. Fever was hot, and a hot sweat might completely carry off a fever, or it could be cured by a discharge of blood. Here we can see the origins of bloodletting, or leeching, as a method to redress the imbalance of hot blood.

The idea of "critical days" is important to Hippocrates, and the physician should be able to form a prognosis of the disease and foretell the danger ahead. In *The Book of Prognostics*, the physician is said to be held

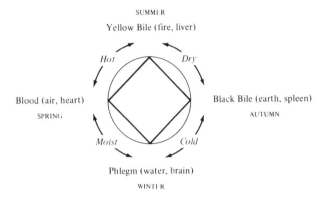

Figure 3.2 The four-humor theory. The four humors and four elements are related to the four qualities. An imbalance could be compensated by drugs associated with the opposite quality.

in esteem if he can tell beforehand which cases are incurable. This is done by examining the overall bearing of the patient and knowing the telltale signs of death. The book *On Airs, Waters, and Places* is the first to lay stress upon the climatic conditions of disease and what illnesses are peculiar to certain areas. *Of the Epidemics* is a collection of case histories, straightforward and unemotional descriptions of particular cases, about half of which end in death. The physician is told that if he cannot do any good, at least he should do no harm—which remains a basic tenet of modern medicine.

The book *On the Sacred Disease* is believed by many to have actually been written by Hippocrates, and it certainly illustrates his belief in natural causes. He says that the sacred disease (epilepsy) is called divine simply because man cannot comprehend the nature of it. But it is no more sacred than any other disease and originates from a stoppage of air by phlegm in the brain.

The famous Oath of Hippocrates was probably taken by apprentices before they were accepted as members in the physicians' guild. It stresses the professional attitude and ethical obligations of physicians. Finally, the most popular book of the entire corpus is the *Aphorisms*. One entry says: "Life is short, and Art long; the crisis fleeting; experience perilous, and decision difficult."[3]

[3]Hypocrates, *Aphorisms. Great Books of the Western World* (Chicago; Encyclopaedia Britannica, 1952), vol. 10, p. 131, Aphorism 1.

SUGGESTIONS FOR FURTHER READING

ACKERKNECHT, ERWIN H. *A Short History of Medicine*. Baltimore: The Johns Hopkins University Press, 1982.

CORNFORD, FRANCIS M. *Plato's Cosmology: The 'Timaeus' of Plato Translated with a Running Commentary*. New York: The Liberal Arts Press, 1957.

FINDLAY, J. N. *Plato and Platonism*. New York: New York Times Books, 1978.

LLOYD, G. E. R. *Magic, Reason, and Experience: Studies in the Origin and Development of Greek Science*. New York: Cambridge University Press, 1979.

SIGERIST, HENRY E. *A History of Medicine: Early Greek, Hindu, and Persian Medicine*, Vol. II. New York: Oxford University Press, 1961.

4

The First Scientist

Aristotle

The walls of the Stanza della Segnatura, oldest apartments of the Vatican, are adorned with frescoes by Raphael representing the operation of the Intellect symbolized by the four faculties: Theology, Philosophy, Jurisprudence, and Poetry. Painted between 1508 and 1511 for Pope Julius II, the fresco symbolizing philosophy has come to be known as *The School of Athens*. Two figures occupy the center of the painting. One is Plato holding his *Timaeus*, pointing directly upward; the other is Aristotle, his arm extended straight ahead, palm down, as if he is pointing to the earth. Science begins with the things of the earth, Aristotle seems to be saying, not with transcendental poetic Forms.

Up to this point, we have been unable to draw a clear distinction between religion and science, or even between the different branches of scientific endeavor. Indeed, at a very deep level, it is probable that no sharp lines can ever be drawn. However, on the level of practical research, the establishment of principles and methods, everything so far has appeared jumbled. In a single treatise like the *Timaeus*, we leap from astronomy to theology, from physics to poetic metaphors of reality, from pure mathematical operations to the constitution of elements. Certainly this kind of thinking has its rewards, for science is as much intuitive guesswork as it is painstaking cataloguing of separate disciplines. Nevertheless, when science leans too much on intuition it threatens to become mere fanciful imagination, and when it becomes too immersed in particulars it declines into superficial fact-finding. Often there is need of a balance, a reminder that science itself deals with a world of things, with corporeal nature, while remaining basically the creation of human thought. This reminder, above all, we owe to Aristotle.

Aristotle was the first Greek to construct a system which gave a theoretical account of common sense. That his authority in many areas would become dogmatic in the Middle Ages (although not above criticism) attests to the strength of his synthesis. Indeed, it did accord in a rational and consistent manner with the spontaneous experience of nature, providing a closely knit world-view which found in nature the long-sought goals of order and purpose. Nature *was* purposeful. Its basic principles were no longer removed to a realm apart from everyday experience, but were to be discovered within nature itself and grasped by the operations of abstract reasoning and empirical observation. Thus Aristotle stands shoulder to shoulder with Plato, for both saw a purpose operating in nature and both believed that human reason was equipped to discover it. To Aristotle, however, the meaning of the world was not separate but embodied in its matter. His explanation of that operation was no "likely story."

Aristotle was the first to see himself and his science in a historical setting. His writings begin with the opinions of his predecessors coupled with his criticisms of their ideas. There is also an overriding desire to clarify the question, much as in Socrates. Aristotle does nothing less than experiment, although on the whole his experiments are confined to the realm of ideas. Nonetheless, Aristotle's explanations *are* self-evident; we can actually see them for ourselves. We are no longer called upon to imagine things which are impossible to experience; no longer are we taking leaps in the dark. With Aristotle we proceed step by painful step, sometimes so slowly it seems we crawl.

The heart of Aristotle's science was qualitative. He makes a clear distinction between the celestial, eternal realm above the sphere of the moon, and the central earth where all change is finite, having beginning and end. The entire Aristotelian universe is finite, extending to the sphere of the fixed stars and qualitatively differentiated by the concept of natural place. Space is full, a plenum, and in this plenum all the elements are assigned a natural place which, when unobstructed, they automatically seek. The concept of a stable earth and fully occupied finite space in which all the basic elements have distinct potencies forming a natural hierarchy of being is the cornerstone of all Aristotelian science. Each region, the sublunar and the celestial, and every element, aether (the celestial element), fire, air, water, and earth, has by nature characteristics all its own. Much as in mythopoetical science, these regions of space exert influences determined qualitatively and not relative to the observer. Perhaps this similarity is mere coincidence, but Aristotle's entire outlook, corresponding closely to the experience of the senses while at the same time meeting the demands of reason, is far nearer to the animate world of mythopoetical thought than that of some of the earlier Greeks. This relationship should not be forgotten by those who would draw too strong a contrast between the poetic Plato and the scientific Aristotle.

Aristotle was born in the city of Stagira near Mount Athos in Thrace about 384 B.C. His father Nichomachus was a physician belonging to the guild of the Asclepiads, and perhaps a court physician to Amyntas II, king of Macedonia. The Macedonian connection would surface again, for Aristotle was destined to become the tutor of young Alexander. It is probable that his interest in biology began as he assisted his father and learned to dissect, for the medical profession was usually inherited, passing from father to son. Medicine, as we have seen from the Hippocratic corpus, represented a true empirical science among the Greeks. When disease strikes, it is a particular disease afflicting a particular person, and treatment begins with the individual. Only after repeated observation will general statements be made about this certain disease which may be applied to other persons who exhibit its symptoms.

After the death of his parents, his guardian sent him to Athens at the age of seventeen to study in Plato's Academy. Aristotle stayed in the Academy for about twenty years, until the death of Plato in 347 and the succession of Plato's nephew Speusippus, who was said to have seprated the Forms even further from the physical world. Aristotle left Athens and traveled in Asia, where he probably did a great deal of biological research, especially upon marine animals found around the island of Lesbos and in its lagoon. Many of his observations and recordings were later incorporated into his massive works on biology, which mark its beginnings as a true scientific discipline. In 343 B.C. he became the tutor of Alexander the Great at the Macedonian court, returning to Athens after Alexander became king.

Upon his return to Athens, he established his own school east of the walls on the road to Marathon, signifying, perhaps, his break with Platonism and the Academy. The grove in which his school was located was sacred to the wolf-god, Apollon-Lycerius, and hence named the Lyceum. The grove had also been a favorite haunt of Socrates. Aristotle offered two types of lessons at the Lyceum, in the morning for his regular students and in the evening for the general public. The bulk of his surviving writings belongs to the former category—lectures and tracts meant for private study rather than publication. In 323, Alexander the Great died in Babylon, and Athenian reaction against the Macedonian subjugation forced Aristotle to leave, for the charge of impiety was raised against him as it had been against Socrates, and he would not give Athens the opportunity to sin twice against philosophy. He went to Chalcis, where he died in 322 at the age of sixty-three. It is probable that he was never in sympathy with his former student's vision of a universal empire of mankind, since he held to the unquestioned superiority of the Hellene over the barbarian. Yet, he may have benefited from Alexander's conquests, for his nephew accompanied the Macedonians and may have sent him specimens from Asia. The Lyceum, then, contained these collections as well as a library; it was a place of research and learning. More than the Academy, it was the ancestor of our modern universities.

It is difficult to say just when Aristotle began to entertain doubts concerning Plato's doctrine of the Forms, and the evolution of his thought has been a subject of debate among scholars. Certainly Aristotle's basic character, his empirical tendency, was quite different from that of Plato even when he came to the Academy at an early age. Yet, he stayed for twenty years, and it is likely that the young man would have found it exceedingly difficult to oppose outwardly his famous teacher while Plato was still alive. He may have sided with Plato against Eudoxus and the giants, although in the end he would become the greatest giant of them all. On the other hand, there was a tradition among later Platonists which had Plato himself complaining about Aristotle's rebelliousness.

Aristotle agreed with Plato that the senses per se do not give us unchangeable knowledge and that nature's most characteristic mark is change and motion. But the fact that nature does not cooperate with our search for universal principles does not mean that we are to run away from change, leaving it unexplained or rejected for some supersensible form of reality. Aristotle's starting point is the exact opposite of Plato's. He will not deny the evidence of his senses; rather, he begins with what is before him and seeks to explain it. When I am burned by fire it is not the Form which burns me, but this particular fire. In this way my senses are infallible—they tell me that the fire is hot and burns. However, they do not tell me *why* the fire is hot; this is the object of knowledge.

How, then, do unmoved Forms, static and eternal, explain movement and change, growth and decline, in the physical world? For example, if Forms are eternal, there must exist the Form for something which has not yet come into being. The Form of Socrates exists whether the physical Socrates exists or not, and the same is true for other things which have yet to be. On the other hand, nothing comes into being without an originating movement. If Forms are the cause of being and becoming, how is it that they exist while the physical things which are their copies do not? What moves or grows? The Forms? But they are unmoved, and yet we see their physical copies come-to-be. One might say that a unicorn does not exist while, at the same time, one has a Form of it in the mind. If there is a Form of the unicorn and one of Socrates, why should one exist and the other not?

Let us consider negation. A horse is not a man. Therefore, horse partakes of the Form "not-man." Yet both are animals. Thus, in the case of man, there will be several patterns of the same thing. Animal, two-footed, and even man himself are all Forms of man as animal; four-footed and horse are all Forms of horse. Thus Forms exist not only as negations, but as patterns of themselves. How can the same Form serve both as a pattern and a copy, the Form of a species copied from the genus yet different? Is there a "third man," the Form of the species man, the Form of the genus animal, as well as the Form of a particular man, say, Socrates?

Further, the later Platonic identification of Forms and numbers led, according to Aristotle, to manifest absurdities. Mathematics deals with universal propositions which may or may not be sensible. A number or ratio of numbers can represent anything—men, horses, unicorns, or simply purely abstract quantities. When the mathematician deals with numbers, he concerns himself not with the physical things they may stand for but with numbers as pure objects of thought. And even if the subjects of mathematics are sensible things, the mathematician does not treat them as sensibles but will abstract them from things. The ideas of line, plane, and solid are abstracted from real bodies to be used in the realm of abstract magnitude. Now, let us say that the original sensible body moves or is destroyed. The universal mathematical theorem still exists as the Form of something which has actually perished or changed. This problem led Aristotle to the conclusion that mathematics could only treat of primary movement, that is, movement in its simplest, eternal form—the rotation of a circle. Therefore, mathematics could be used to describe heavenly rotation but not the variable rectilinear movements of earthly physics.

These and other objections led Aristotle to reject the supersensible world of eternal Forms. However, like Plato, he held that true knowledge could only be knowledge of some unchanging definition, the form or essence of what the thing is. Aristotle did not reject the fundamental tenet of Socratic wisdom; rather, he found Plato's solution untenable in the everyday common-sense world of perception. How to reconcile the two?

Aristotle was really facing a complex of problems with the extra requirement that change itself must now be accounted for—one must be able to identify the *cause* of change. The first step was to recognize that the essence of something is never separate from the particular thing. The scientist must begin with perception—the perception of something real and concrete. From the sensation of many particular things, the scientist, using logical analysis, is able to abstract certain common features which exist in the things themselves, only to be separated in the mind of the perceiver. We never experience Form without matter. Form and matter are distinguishable logically but not in fact, at least in the sublunary world.

To know the essence is to know what is necessary for definition. For example, a musical man *is not* the prerequisite of all men, but a two-legged animal is predicated of man universally. As in biology, the individual is the embodiment of the species—all men are two-legged—and the species is contained by the genus—man is a warm-blooded, air-breathing animal. The emphasis, therefore, is upon classification, and forms are abstract categories which aid the scientist in defining the individual thing. At the same time form is fully real, immanent in matter.

The same is true for change and motion. Looking to nature, Aristotle saw an answer to his problem in the process of birth and growth. An acorn is not a tree, yet, through growth, it realizes a *potential* already there. Movement, growth, is always movement toward something, toward an end or *telos*. The unqualified matter of the acorn has the potential inhering within to become an *actual* tree, which in turn can be known by the essence of tree. Pure form is equated by Aristotle with *actuality*, yet nature in the sublunary world always presents us with a combination of matter and form, and matter always has the potential to become something else. Thus, change is a passing from potential to actual, and unqualified matter has the potential to be informed (that is, to be given form), while informed matter has the potential to assume a contrary form. When a man is healthy, this particular form exists actually in a given person. At the same time, a man is also potentially ill, the contrary of health. The two cannot exist simultaneously in *actuality*, the actual state of a man at a given time. However, health may withdraw to be replaced by disease. In many of his writings Aristotle equates matter with potential and form with actuality.

Nonethless, essence is, for Aristotle, an existent thing and not merely a logical abstraction. It is a nonmaterial entity which exists, but in the sublunary realm it exists in particular manifestations of nature. The actuality of form is still prior to the potentiality of matter, and thus the potential changeability of matter is not evolutionary; it is change to a goal, the form or essence. The doctrine of substance, matter and form, culminates in teleology, and this teleology pervades the entire Aristotelian edifice.

Now that Aristotle has established the reality of nature and the goal of knowledge, his next task is to describe the methods which govern thought. The necessary preliminary to all science is logic, called analytics by Aristotle, which is the tool (*Organon*) of science. The individual substance is the starting point of science, so Aristotle begins by listing the widest possible predicates which are the primary classes of everything else. In the *Categories* he lists those terms which distinguish things without combination. For example, Socrates is white, white is a color,

color is a quality, and quality is the virtue of something, that which defines a thing as being such and such. Likewise, quantity, relation, place, date, action, and so on are the ultimate classes into whatever is real falls. Locomotion, for example, is change of place; two is quantity; roughness and smoothness are relations.

On Interpretation deals with propositions and the rules which govern them. Words, says Aristotle, are symbols of mental experience, and propositions must use these symbols in a noncontradictory way. Everything must either exist or not exist, a thing must be this or that, but it is not always possible to say determinately which of these alternatives should necessarily come about. The same thing can both be and not be *potentially*, not in actuality. Actuality, therefore, is primary to potentiality.

The most important treatises of the *Organon* are the *Prior Analytics* and the *Posterior Analytics*, which lay down the rules of the syllogism and how these rules operate in scientific reasoning. The word *syllogismos*, occurring in Plato, means reckoning up, working out, or understanding. It is, for Aristotle, the method of drawing conclusions from premises, or deductive reasoning. In science, knowledge and demonstration are concerned with necessity, the eternal and universal connections between things. The syllogism, by the necessity of the connections between its terms, demonstrates the universal. What happens by chance or by accident cannot be shown by necessary demonstration and hence cannot be an object of science. The gamblers had not yet entered the scientific fold!

In *Prior Analytics* the key to the syllogism is necessity. If A belongs to all B, and B belongs to C, A *must* be predicated to all of C. In the syllogism the major term belongs to the third term by the middle, and the middle must not be indefinite (B may or may not belong to C), since science is concerned with things that can only be demonstrated universally. Aristotle goes on to illustrate the various forms of the syllogism, but we need not follow him here. His most important contribution to formal logic was the use of variables as illustrative symbols. Not until the introduction of whole propositions as units instead of single terms did the fundamental science of logic surpass what was done by Aristotle.

The *Posterior Analytics* introduces the application of logic to science. Science must depend ultimately upon the premises chosen to demonstrate truths about nature. Without primary, undemonstrable truths, however, we are faced either with infinite regress or a vicious circle. The laws of contradiction, the excluded middle, and such assertions as "equals taken from equals give equal results," cannot be demonstrated. They are the starting point of science, the faith that nature conforms to reason.

Sensation is another indispensable precondition of science. From experience we recognize the common characteristics of things, establishing them as universals *inducted* from their concrete instances. Though we are able to supply many instances of man, count many men, and therefore establish the universal premise that man is a two-footed animal, we can never draw this premise from *all* the particular instances of man. We must assume that our major premise is composed of all particular cases. It is simply not possible to examine all the individuals of a certain species, both past and present. In the end, the entire edifice of rational science is built upon a faith that there is a constancy in nature which conforms to the use of logical analysis. Aristotle recognized this, and his system was not as dogmatic as later philosophers made it.

Aristotle was too much of the biologist to surrender the complexity and vitality of nature for completely abstract principles. To explain movement and change, one had to begin with these phenomena as they presented themselves to perception. Aristotle begins his physics with the recognition that changes—coming-to-be and perishing—were basic facts of earthly phenomena which had to be accounted for. Following the dictates of his own logic, Aristotle's starting point is with those things which are nearest to perception, earthly changes. Having explained these, he then applies himself to the heavens. In effect, such a method was exactly opposite that of the Newtonian world-view.

The first decisive step was to banish mathematics from earthly physics. For Aristotle, Plato's building up the physical world from triangles was untenable. The material substratum, pure potentiality, is that "out of which" things come into being, not Plato's receptacle "in which" mathematical forms arise. Simple matter is privation of form and without qualification. Yet, perception everywhere presents us with qualities which we generally recognize by their opposition. Change among the four basic elements (fire, air, water, earth) is change between qualities (moist, dry, hot, and cold). These qualities are tied to the substratum, for only physical experience gives them to us. Hot is always *something* hot and, at the same time, has the potential to become *something* cold.

Each element has within itself a tendency to occupy its natural place, for Aristotle's universe is a highly stratified, ordered, and finite system built upon the qualitative distinction of place. Again, this distinction comes from sensible experience. Fire has a natural movement upward; it is light, and its natural place is between air and the celestial element aether. Earth is heavy; it has a natural downward movement which seeks the center of the universe. Water and air fall between the two, water rising above earth and below air, air existing between fire and water. The elements seek their natural places in the cosmos; left to themselves, they will go directly to their places, which are the goals of natural movement. They also seek their natural state, which, in the case of the four elements below the sphere of the moon, is *rest*. In short, the *telos* of terrestial movement is rest and natural place; locomotion (or change of place) is violent and must have a cause when it disturbs that natural place.

Early in his treatise *Physics*, Aristotle divided the causes of change into four classes: the substratum or material cause; the essence of a thing or formal cause; the source of movement or efficient cause; and the goal of movement or the sake for which change occurs, the final cause. The material cause of a bronze statue is the bronze; its formal cause is the pattern or figure the statue is moulded to portray; its efficient cause is the sculptor; its final cause, the most important, is the end for which the statue is made. Cause, then, includes all the factors which are necessary for a thing to come into being, whether naturally or artificially, and although the different types of causation may be separated in thought, they are collectively gathered in change.

That an ordered universe should come about from the chance collisions of atoms, as Democritus held, would mean that blind forces were responsible for the apparent order. Chance, according to Aristotle, is an incidental cause, the name for connections between things which have no intrinsic or logical relationship. I may go to the marketplace and incidently meet someone who owes me money, yet the final cause, the purpose for which I went to the market, has no relation to my chance of collecting a debt. Spontaneity, on the other hand, is something occurring in nature which is contrary to natural purpose. Monstrous births, for example, may happen spontaneously and serve no purpose, or they may actually hinder the individual from achieving the form of the species which is the final cause of growth. Aristotle does recognize the failure of purpose in nature, which is basically due to the limitations of matter. These chance and spontaneous events are relegated to an inferior position. And as we have seen from his logic, chance cannot be the object of scientific knowledge. Thus he can neither be accused of strict determinism nor of complete adherence to the idea of Heraclitean flux.

We may now turn to the types of movement or change observed on the earth. For Aristotle, motion is divided into three classes: qualitative change or alteration; quantitative change or increase and decrease; and change of place or locomotion. A thing may be moved by its own nature or something cojoined with it. The most perfect movement, as we have seen, is rotation. This is natural to the heavens, the actuality of the fifth element (aether) which constitutes celestial phenomena. Since for Aristotle, this motion belongs to the science of astronomy, physics must treat of earthbound motion, which, because the four elements naturally seek their place of rest, is the result of some efficient cause when that place is disturbed.

Thus, all motion on the earth, when it is not natural to the body, must have an efficient cause, a mover which moves the thing by contact. The same is true for qualitative change or alteration. The transformation of elements—the alteration of elements in a compound body—produces change. For example, if something which is predominantly earth is exposed to heat which changes it to a mixture in which fire predominates, its motion wll be upward. In his treatise on meteorology, Aristotle described various phenomena such as rain, clouds, hail, snow (all moist), and shooting stars, the aurora borealis, comets, the Milky Way (all dry) as exhalations of the earth which have been transformed by some influence and transferred to higher cosmic regions. Water, when heated by the sun, becomes mist and rises; when it is in the air, it seeks its natural place and so falls, becoming rain.

In the study of movement, the concepts of weight and medium assume an important role. Aristotle's study of the movement of bodies by impressed force was of profound importance for later scientists, and was very early subjected to criticism, especially at the hands of the Byzantine, John Philoponus, in the sixth century. Yet, these theories were actually peripheral to Aristotle's major concern in the *Physics*, which was to deny the existence of the void and infinity. We must first turn to these arguments to grasp the significance of his theory of impressed force.

First, Aristotle says that an infinite body would be by definition unknowable, for to know the infinite would be to give it a limit and make it into a sensible thing with a specific form—which it cannot have. Second, since Aristotle's universe is finite, based upon the qualitative distinction of place, the finite cannot contain the infinite, for it would have no natural place or natural movement. Moreover, any infinite element in a finite universe would overpower the other elements, interfering with their natural movements. For Aristotle, the infinite is only potential when applied to numbers or extended magnitude. There is always the possibility of a greater magnitude, but the infinite itself can never exist actually, only potentially in the realm of abstract mathematics. Hence the infinite is banished from physics, for no sensible magnitude can exceed the cosmos—which is finite. In the same way, a magnitude can be divided potentially to infinity, but not actually, for there is in actuality an end to division—which would be a limit. This concept of potential infinity which can never be actual is also Aristotle's solution to Zeno's first two paradoxes.

The argument against the void is probably the most important for the physics of motion. Place, says Aristotle, is that which contains a body and so is separate. Yet, place is neither form nor matter, for it admits of various bodies. Here the Aristotle of common sense steps in. Place exerts an influence upon a body, for by their very nature, the elements seek their natural places. This means that place carries a qualitative distinction—up or down—which reduces to the intrinsic qualities of bodies—heavy and light. Further, place is understood by motion. The place of heavy bodies and hence rectilinear motion is the sublunary world; the heavens are a place by virtue of continual rotary motion and the aether. But in a void there can be no natural places, just as an infinitely extended body has no natural form. Unfilled place, therefore, cannot exist in actuality, for it would admit of no natural movement, having no places which make movement possible. Thus, the cosmos is a plenum, and place is determined qualitatively by the kinds of material which fill it.

In regard to kinematic movement, Aristotle holds that the motion of a body is based upon two things: the weight of the body and the density of the medium through which it moves. In the same medium two bodies move at speeds proportional to their weights. Now, the medium offers resistance to movement, and so the speed of, say, a falling body will be inversely proportional to the density of the medium through which the body moves. With forced motion, the speed of the body is proportional to its motive force, the force impressed upon it, and the resistance of the medium. Movement occurs when the force is sufficiently great to overcome resistance.

Almost as an aside, Aristotle notes that in order to explain the movement of projectiles, one must account for continued impulse once the motive force has been removed. For example, once a thrown stone leaves a person's hand, one would assume that it should immediately seek its natural place and fall straight down. But since experience shows that this does not happen, Aristotle says that the air is pushed aside by the projectile, and, with a circular thrust, fills the void left behind. This circular thrusting of air pushes the projectile faster than its natural locomotion, which seeks its natural place. A variation of this theory says that the air receives the force of motion and simply carries the projectile. This is certainly the weakest point of Aristotle's theory of motion, and it would be replaced in the Middle Ages by the theory of impetus.

If the void existed, none of these motions, natural or forced, would be possible. Falling bodies, without resistance, would fall instantaneously, for the density of the medium would be zero. Although Aristotle does not use modern formulas, it might be helpful to see what he means in our notations. The movement of one body to another is said to be "quicker" by the ratio $V = S/T$, where V is velocity or speed, T is time of the motion, and S is the distance covered. $V = S/T$ is also proportional to F/R, where in forced motion F is an external force and R is the weight of the body, while in natural motion (such as falling bodies) F is weight and R is density of the medium. Natural motion in a void would make the density zero, and therefore in the modern notation V would go to infinity. Forced motion would be impossible in the void for the same physical reasons noted above: once the body was out of contact with the mover, nothing would be left to carry on the movement and the body would fall. Therefore, since instantaneous movement is unthinkable, and forced movement would be impossible, the void (or vacuum) does not exist.

From such concrete sense-perceptions, Aristotle was able to encompass the entire universe in a closed, finite, and rational system. In *De Caelo (On the Heavens)* his cosmology springs directly from his physics—which, as we shall see later, was exactly the opposite of what happened in the scientific revolution. For both Galileo and Newton to account for the moving earth that was necessitated by the Copernican system, they had to formulate a new theory of motion. This meant rejecting the qualitative theory of natural place, which, in turn, finally demolished the logical objections to a vacuum and infinite space, as well as the qualitative distinction between earthly change and heavenly perfection. Aristotle's theory of motion, in the words of Thomas Kuhn, is "inextricably bound to the conception of a finite and fully occupied space. The two stand or fall together."[1]

The heavens must have an element which makes celestial phenomena unique and explains what Aristotle considers to be a primary fact of experience: the ceaseless circular motion of the celestial spheres. This fifth element is the aether, filling the space between the sphere of the moon and the fixed stars and having circular motion as an intrinsic quality—just as the earthly elements contain in themselves the motion of up and down. Thus the concentric spheres of Eudoxus and Callippus are no longer simply abstract geometrical constructions meant to represent the phenomena but are fully real, making up a vast mechanical system. Yet here Aristotle discovers a problem the others did not. If the universe is full, a plenum, how, in the concentric system, is the motion of one sphere prevented from interfering with the movement of another if the two are in contact?

In order to counteract this "heavenly friction" and yet maintain the constant circular motion of the spheres, Aristotle found it necessary to add *more* spheres, calling them "unrolling." The system becomes exceedingly cumbersome, for Aristotle is forced to insert unrolling spheres to neutralize other spheres whenever their motion interferes with a lower sphere. Because we are now dealing with a truly physical system, not a simple geometrical representation, the innermost sphere of each planet and the outermost sphere of the next planet below had to be separated by these unrolling spheres; only the moon (with no spheres below) required no new additions. In all, Aristotle added twenty-two new spheres to the thirty-three of Callippus, thus giving a total of fifty-five

[1]Thomas S. Kuhn, *The Copernican Revolution* (New York: Random House, 1959), p. 89.

spheres in the heavens. The movements of the final spheres, those which carry the moon, are communicated to the four elements in the sublunary realm, jostling them about and thus accounting for their combinations and incessant change (see Figure 4.1). Nonetheless, the universe as a whole is eternal, for time's most basic measurement is the motion of the heavens, circular motion, and all other motions having beginning and end are measured by it. Thus time itself, having no beginning and end, is tied to the perfect circular motions of the heavens. The universe is thus finite and eternal, and any questions of origins or of what lies outside the sphere of the fixed stars are irrelevant.

Aristotle also brings forward arguments for the sphericity of the earth: heavy elements tend to the center of the universe and obtain spherical equilibrium; eclipses of the moon show a curved (convex) surface which is actually the earth's shadow; some stars seen above the horizon in Egypt and Cyprus are not seen further north, while some stars set in these regions which in the northern latitudes always remain above the horizon. Thus, not only is the earth spherical, but it is not a very large sphere at that. Aristotle gives an estimate of its circumference as 400,000 *stades* (a stade is about 607 English feet), which, surprisingly, is nearly twice that of Eratosthenes' more accurate estimate of 252,000 *stades*. Aristotle adds that it is not too incredible to suppose that the sea from the Pillars of Hercules (Gibraltar) to India is continuous—maybe providing Columbus with the clue to sail west!

Simple observation shows us that the earth does not move. Not only would a moving earth inhibit natural motion—all heavy things seek the

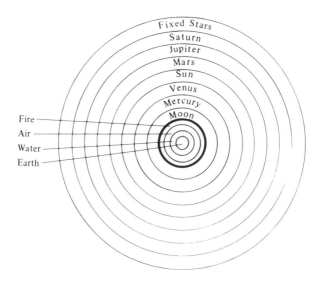

Figure 4.1 Aristotle's universe (simplified).

center of the universe and rest—but all one need do to prove the earth's stability is look at everyday experience. If I throw a stone vertically into the air, the stone moves upward and then falls in a vertical line. That line is always at the same angle—a right angle—with the earth's surface, showing that objects fall in the direction of the center. Not only would a moving earth contradict this natural fall to the center, but the stone would always fall to the earth at a different point from the point it was thrown from, left behind by the earth's motion. Even if the air were carried along by the earth's rotation, it would not push the stone hard enough to keep in step. In the same way, clouds would be left behind if the earth moved, and the stars would not rise and set in the same place. Now, a rotation of twenty-four hours would meet this last objection, yet if this were the case, Aristotle claims that the motion would be so fast that *everything* on the earth would be left behind!

A word must be said about Aristotle's Prime Mover, although this falls into metaphysics more than physical science. The motion of the heavens is eternal and indestructible, without beginning or end. But the cause of eternal movement cannot itself be moved, otherwise the causal chain goes on *ad infinitum*, which is impossible in a finite universe. Therefore, in the *Metaphysics*, Aristotle posits an unmoved mover or a *primun movens* which itself is not moved but imparts movement to everything in the universe. Being eternal and of pure actuality, the prime mover is self-sufficient, perfect, pure form, and perfect goodness. Since it is pure form, unencumbered by matter, it is pure thought—which is God.

God as pure form does not move by physical contact with the material universe. Rather, he moves things as a Final Cause—the object of love and desire as all things strive to realize as closely as possible pure form. Because God is pure thought, existing at the extremity of the heavenly sphere where movement becomes the most perfect, the only object of his thought is himself, the most perfect form. Pure actuality admits of nothing potential, which, in the case of God, would be outside himself. Yet, he is the fulfillment of all potential and hence unmoved, for any change must be change to something, and the only potential change of the most perfect is change to the worse. Thus, throughout eternity, God does nothing more than contemplate himself. Except for being the object of desire, the Final Cause, he is totally removed from the world and, in effect, ignorant of it. In the humorous words of Will Durant: "No wonder the British like Aristotle; his God is obviously copied from their king."[2]

It should be noted, however, that Aristotle's God as pure actualized form stands at the very pinnacle of nature's hierarchy. This idea, which

[2]Will Durant, *The Story of Philosophy* (New York: Simon and Schuster, 1961), p. 82.

can be labeled the scale of perfection, connects Aristotle's biological, physiological, and even psychological researches to his overall cosmology. Descending from the pure actuality of God, everything in nature has within itself a degree of privation (potentialities), which by the very nature of each species means a certain lack of perfect actuality. Below the orbit of the moon, we find nature in a constant state of generation and corruption in which all species, while having their own forms and actualities, are yet inferior to the changeless heavenly realm. Since form is not transcendent but embodied in matter, each species can be classified by the form to which the species strives. All living creatures, as a combination of form and matter (actuality and potentiality), can be roughly arranged upon a scale of being—or, as it was later called, the *scale naturae.*

Beneath the apparent change and variation of nature could be found evidence of this hierarchy of being, later called the Great Chain of Being, which exposed the structure of the entire cosmos, descending from God to the inanimate world. Although the lines of classification, the taxonomy of living things, were difficult to draw, the idea of a static species and the Great Chain of Being were accepted by most biologists without question until the late eighteenth century. Aristotle himself did not assert that there was *conscious* purpose in nature, yet he still spoke of the final cause for which change occurred, the teleology in which nature fulfills its ends. On the whole, therefore, chance is ruled out of the works of nature, and mutations are simply failures within a species to realize its ends.

Beginning with plants, Aristotle sees a progressive ascent to animals. In the sea, some forms of life have both plant and animal characteristics—especially sponges and zoophytes. His classification of animals is based upon the presence or absence of red blood and is roughly equivalent to our invertebrates and vertebrates. He also knew of the mammalian placenta and the general tendency of mammals to be viviparous (live-bearing) with a placenta nourishing the embryo in the womb. Some animals which were not mammals were also viviparous, even though they had no placenta, so that the egg develops within the parent. Aristotle also gives a description of the dogfish (smooth-skinned shark), which is externally viviparous, having its embryo attached by a navel-string to the placenta, much like the structure in the mammalian womb. This curious phenomenon was disbelieved by zoologists until 1842, when it was described again by Johannes Muller. Aristotle also recognized the sexual arm of the octopus and stated correctly that it was more pointed than the others, having two very large suckers.

One of Aristotle's outstanding observations was that whales and dolphins are not fish. They have hair and lungs, they breathe like terrestrial animals, and they suckle their young by means of mammae. He also knew of the compound stomach in sheep, oxen, deer, and goats, and he must have dissected the organ. He held to the theory of spontaneous generation—as in the case of some insects which, he maintained, came from putrefying earth or vegetable matter, or even from inside animals. On the other hand, he recognized that in an embryo the heart is the first organ to be formed. He was convinced, again from pure observation, that hands and talons, nails and hooves, and feathers, scales, and scutes are all analogous structures; but he went on to claim that the arms of humans, the wings of birds, and the pectoral fins of fish are comparable organs, too.

Aristotle's physiology was dominated by the search for final causes in organs. Thus, respiration was to cool the heat of the animal, especially around the heart, which in higher animals is hotter. Fish cool their hearts by water drawn through the gills, while mammalian lungs are described as bellows, passing air through their vessels into the pulmonary artery and vein. He had no knowledge of the circulation nor the pulsation of the blood in relation to the heart, yet he gives a detailed account of the vena cava and its branches as well as of blood vessels deriving from the aorta. However, he failed to distinguish between veins and arteries, indicating only that the blood issues from the heart and is dissipated in the tissues. For him, the heart, as we shall shortly see, had other uses.

In reproduction, we are told that the semen of the male constitutes only the form and efficient cause of generation while the female contributes the matter or material cause. Digestion is compared to boiling or cooking food—concoction—which nourishes the body.

Man is the highest of the animals because of his reasoning ability, his rational soul. Here we pass to Aristotle's psychology. Aristotle uses the term soul in a wide sense to include nutrition and reproduction, sensation, imagination, appetite or will, and, of course, reason. The distinction among these various aspects of the soul also leads to further classification of living things. Both plants and animals have a Nutritive Soul (nutrition and reproduction), animals have a Sensitive Soul (sensation and desire), but man alone has all of these plus a Rational Soul. Thus each step up the "scale of Soul" includes the previous stages. Further, soul and body make up a composite whole in which the body supplies matter and the soul form. Since the soul is in this way the actuality of the body, it clearly cannot exist apart.

Sensation, according to Aristotle, is the reception by a sense organ of a perceptible form minus its accompanying matter. Yet, the sense organs have a certain limiting range of perception (perception of color falls between the limits of white and black) which, if exceeded, causes damage to the organ. The central organ of perception and the source of life, connected to the others by blood vessels, is the heart. Since some creatures that have sensation appeared to lack a brain (Aristotle was unaware of the importance of the central nervous system), he chose the heart as the organ of sensation. The brain is the coldest organ of the body—the eyes, ears, and tongue are found near it so that they will not be overwhelmed by the heat of the heart! Yet for the discharge of its functions the heart depends upon the brain as a kind of intermediary organ correcting excesses of heat.

Aristotle uses the word *phantasia* for imagination, seeing it basically as the ability to retain, visualize, and recover past sensation from memory. In man the imagination can be both sensitive and deliberate; that is, man has the ability to recall and compare past sensible forms at will. Now, there is a difference between sensible forms of individual objects (dependent for their reception upon some organ) and the essence or intelligible form which is grasped through reason. Individual instances implant in a rudimentary way the universal form, which is only fully perceived by a leap of abstraction. The faculty which performs this leap is none other than the rational soul. There is passive reason, the potential of apprehending intelligible forms. There is also active reason, the light which makes apprehension possible.

Here we arrive at the most difficult and debated aspect of Aristotle's psychology: the question of immortality. Like Anaxagoras, Aristotle believed that *nous*—mind—was ultimately responsible for universal order and was something totally apart from matter, unmixed and eternal. In pure thought, the *nous*, the existing essence of a thing and the idea of that essence are identical, admitting of no potential for change. That which grasps these essences, the active reason, is eternal and immutable—like God. Being pure actuality, it cannot admit of potential. Hence if the active reason is that part of the soul which is immortal, it cannot retain any impressions from this life nor any individuality. Aristotle's final prospect of immortality, the impersonal survival of active reason, is rather bleak and airless—like his God.

It is impossible to overestimate Aristotle's influence, not only upon Western science but upon Western thought as a whole. No brief summary can do justice to the range and power of his mind, for besides the subjects which we have treated in this chapter, he investigated politics, art, and ethics. It is true that his thought did not remain unaltered for two thousand years, nor did it escape criticism, correction, and the outright rejection of many particulars. But his concepts—a finite universe, a closed and essentially qualitative world, a central and immobile earth, a celestial realm of eternal and circular movement—remained fundamental to Western science until Copernicus. No other single scientific system equals his for duration.

How do we explain such longevity and power? It has been pointed out that Aristotle provided logical and common-sense explanations for everyday experience. In so doing he neither denied nor neglected the phenomena, nor did he posit a transcendental realm of reality to explain them. Yet, in the end, through a rational and self-consistent series of ascending steps, he reached to the very throne of the deity. Like Plato, he presented the sceptical with a powerful argument for the existence of reliable knowledge in reasonable propositions they could verify with their own eyes as operating in the world. Raphael might well have painted Aristotle too, with an upraised arm. Without rejecting the phenomena he yet rises above them, saving the sensate world through a system of eternal scientific laws. We whould not be surprised, then, that his natural philosophy dominated Western physical thought for so long; rather we should wonder why it was rejected at all.

SUGGESTIONS FOR FURTHER READING

CLAGETT, MARSHALL. *Greek Science in Antiquity.* New York: Collier, 1963.

COLE, F. J. *A History of Comparative Anatomy: From Aristotle to the Eighteenth Century.* New York: Dover, 1975.

LLOYD, G. E. R. *Aristotle: The Growth and Structure of His Thought.* Cambridge: Cambridge University Press, 1968.

LOVEJOY, ARTHUR O. *The Great Chain of Being.* Cambridge, Mass.: Harvard University Press, 1936.

SOLMSEN, FREDERICK. *Aristotle's System of the Physical World.* Ithaca, N.Y.: Cornell University Press, 1960.

5

The Temple
of the Muses

Science after Aristotle

In our age of specialization a universalist like Aristotle, or even Plato, awes and, at the same time, disquiets us. We distrust grand systems of thought such as theirs, perhaps doubting that an individual can maintain a high degree of intellectual integrity through such diverse fields as physics, biology, ethics, and political theory. The age which followed Aristotle—the Hellenistic or Alexandrian age of Greek science—was, in general, a retreat from sweeping philosophic and scientific systems. Its scope was narrower and its approach to problems more specialized and conservative. This is, of course, a relative statement—as we shall shortly see with the systems of Epicurus and the Stoics. Yet, taken as a whole, the aims of science after Aristotle became those of various individuals working in separate areas, usually upon specific problems.

Specialization has always been a sword with a double edge. Greek science after Aristotle can be seen as representing great achievements in some areas, especially those we label the exact sciences. However, it can also be seen as a period of beginning stagnation in which the fundamental presuppositions of earlier thinkers became cemented in men's minds. On the face of it there is still a continuity in the development of rational science, and there are thorough and rigorous investigations which sometimes make the earlier natural philosophers appear to be groping in the dark. However, new questions were seldom asked and the basic tradition remained intact.

Certainly a great many factors are involved, including the social upheavals which followed Alexander's death and the eventual triumph of Rome. It is true that radical innovations do often occur during periods of social turmoil. But on the whole Alexandrian science became involved with filling in the details of the tradition, correcting and expanding it, rather than branching out in new and creative directions. Those thinkers who did break with tradition, like Aristarchus with his heliocentric theory, were ignored or criticized on the very strength of this tradition. Indeed, diverse and creative speculation was not lacking—it hardly ever is, in any era—and the seeds of many valuable scientific theories were planted. Yet, the tree seems to require fresh soil in which to grow, and in the well-worked soil of tradition the seeds usually fail to germinate.

Alexandria's stimulus to scientific research centered around the Museum and the Library. The Alexandrian Museum was a temple to the Muses, the mythical nine daughters of Zeus and the goddess Memory. It was founded by Ptolemy Soter and expanded by his son Ptolemy II (Philadelphus) about 280 B.C. The Museum was a group of buildings equipped for scientific research in which members worked together like the fellows of a medieval college. The Library, which was a part of this research center, contained about a half million volumes—certainly the largest collection in the ancient world. The Ptolemies tried to attract scientists to Alexandria to add to the luster of their reputations, and the city itself became a cosmopolis, the first of its kind, the meeting place of east and west. On the well-planned avenues of Alexandria one could rub shoulders with people of many nationalities, including Greeks, Africans, Jews, Syrians, Babylonians, and Persians. If there ever was a realization of Alexander's dream of a universal empire and the brotherhood of man, it was Alexandria.

It is no surprise, then, that the Egyptian city would supersede the insular Greek city-states and assume the intellectual leadership of the Hellenistic world—indeed, of the Roman world as well. Greek cities like Athens were reduced to municipalities in the new social setting of the Mediterranean world. However, for some time after Aristotle, Athens itself remained preeminent in philosophy, logic, and physics, while Alexandria became the center of mathematical, biological, and astronomical research.

After the death of Aristotle in 322 B.C., Theophrastus of Lesbos became the head of the Lyceum, remaining there for the next thirty-six years. He is said to have written over two hundred treatises, mostly commentaries on the works of his great teacher. Yet, his commentaries were not without their criticisms of the master, and Theophrastus took exception to one of Aristotle's basic tenets, the final cause. Many things in nature, he said, do not occur for the sake of an end—the tides, for example, or breasts in male animals—and the classification of causes is much more difficult than Aristotle conceived it. Theophrastus had his reservations about spontaneous generation as well, wondering, perhaps, if air or rivers carry seeds (as Anaxagoras said) which are responsible for what appears to be spontaneous growth.

Theophrastus was succeeded by Strato as head of the Lyceum in 286 B.C. Like his predecessor, Strato had doubts about some of Aristotle's theories, especially in physics. Strato thought that there was no need to postulate an upward tendency for fire and air; their upward movement could be explained as displacement caused by the downward motion of heavy bodies. Strato also conceived of acceleration, saying that bodies move more swiftly as they approach their natural places. The Alexandrian Hero's experiment of blowing air into a bottle to show that vacuums between air molecules actually do exist may have been derived from an observation by Strato about the compressibility of air. If Strato did conceive of the existence of a vacuum, it was a noteworthy departure from a fundamental Aristotelian principle, yet we have no indication that he considered motion as being possible in the void.

The question of the void arose in another philosophic school—that of Epicurus. For Epicurus and his followers, as well as for the Stoics, physics and natural science were subordinated to ethics. The goal was to secure a happy life free from needless fears and imagined anxieties. Science became a program to provide a clear and comprehensible picture of reality which would dispel the harmful effects of superstition. Thus conceived, any science which denies basic sensate experience is useless, for it is no better than fantasy. The senses are thus infallible.

The universe, according to Epicurus, is constructed from atoms and the void. It is said that he smashed two boards together and quickly drew them apart, showing that air rushes into the gap, which, therefore, must be a vacuum. Compound bodies are always a mixture of atoms and void, since there is always the potential for destruction—that is, the splitting apart of compound matter. Matter itself is made up of indivisible atoms, entirely without void, having three properties: size, shape, and weight. We know from the sensate world, extended in space, that a body is composed of distinguishable parts; in thought we recognize that the least perceptible points on a plane are aggregates of smaller particles, and in thought we may carry this division out. So, by analogy, atoms are also made of parts—but since atoms are the "least possible" minima of extension, these parts are inseparable and exist *only* in thought. Yet atoms vary in size and shape by the disposition of parts. Atoms may collectively acquire new properties (qualities), but qualities are not independent. Since all thinking begins with sensation, these qualities arise from the compound itself, fired off as "idols" which impinge upon the senses.

Epicurus broke from Democritus and Leucippus with his idea of motion in the void. All atoms, heavy or light, fall at equal ratios and "as fast as thought," for the void offers no resistance to their fall, making their differences in weight meaningless in regard to their speed. Atoms moving in the void fall downward, yet since downward is meaningless in infinite space, that direction must be regarded in relation to ourselves.

If all atoms are falling uniformly in the void "as fast as thought," then how does Epicurus account for the existence of compound bodies? Atoms, he says, sometimes push a little from their vertical paths; that is, at times they "swerve." This swerving of atoms causes collisions with other atoms and also changes in direction in the movement of atoms. When atoms meet one another, they become locked in a bond of union. Thus is formed the entire physical world. Movement in the world of compound bodies is a secondary result of the swerve, for upward and sideways motion are due to "blows" by which the atoms rebound off each other and change course. Now, the compound body is an aggregate of atoms in constant motion—vibrating, as it were. While at rest, the internal vibrations of the atoms counteract each other, achieving a state of equilibrium. When bodies move, the direction and velocity of their movement is the resultant sum of the movement of the atoms caused by external blows which upset the equilibrium. A complex body, however, is more than the aggregate of simple atomic motion. In its unified form it acquires new characteristics which allow for both the bonding of atoms and their collective movement.

The swerve has been one of the most criticized theories of Epicurean physics. The swerve is actually an indeterminate mechanical motion, and the swerve of the fine atoms which compose the mind expresses volition or free will. The swerve is a conscious break with determinism on the part of Epicurus. It is meant to rescue his ethics from the consequences of his physics. Although the world is a fully mechanical structure, its very ground in the swerve of atoms and subsequent construction of matter, motion, and mind is indeterminate or uncertain. Epicurus has rescued mind and free will from the deathlike grip of the old atomism. The swerve may seem to be a gratuitous concept, yet it is a necessary one.

Opposed to this revival of atomism was the school of the Stoics, named from the porch or colonnade (stoa) under which Zeno taught in Athens. The main body of the Stoic continuum theory was developed by Zeno and Chrysippos. For them, the proper model of physics was the biological and animate world. Like Aristotle, the Stoics denied the existence of a void *within* the cosmos, although they held that the ordered universe was like an island floating in an infinite void. Most important was their concept of *pneuma*. The universe is permeated by a pervasive substratum, which is the dynamic agent holding all its parts together. This substratum, called *pneuma*, was identified with fire and air in Stoic physics. Without the active properties of *pneuma*, the other elements would break apart, since they are passive and do not possess a cohesive force. Structure, then, is the result of a dynamic quality which pervades the universe, not the basically static geometrical continuum of Plato and Aristotle.

What is interesting about this Stoic *pneuma* is its dynamic quality; its movement is ceaseless. The tension of the *pneuma*, its process of penetration and cohesion (*hexis*), gives rise to the physical, organic, and even mental states of existence. Imagine a wave in the ocean. The simple matter, water, is continuous; the tensional fluctuation of the wave gives it form, motion, and coherence. Yet, the wave cannot be separated from the water.

Among the later Stoics the idea of the universal *pneuma* led to the concept of cosmic sympathy. The cosmos is a complete, animate, unified structure in which even the celestial bodies are held together by the cohesive forces of the *hexis*. It is thus a closed structure in which each of the heavenly bodies has a center of attraction all its own, making for a plurality of such centers.

If nothing can be added to or subtracted from the cosmos, an uncaused event is strictly excluded, for in essence such an event would be creation *ex nihilo* (out of nothing). This idea also suggests an all-embracing determinism in which humans are at the mercy of arbitrary fate. The Stoics did try to soften this determinism, giving humans the free will to yield or not to yield to an impulse, within certain parameters. Yet, in a strictly determinate universe, to know the antecedent cause is to be able to predict the future. Stoicism provided the intellectual validation for divination and astrology, since these were based upon the fundamental sympathy of the cosmos.

The Stoics conceived of geometrical figures in terms of material shapes which were held together by the *penuma*. A straightened cord (line) is the extreme case, being a curved cord stretched to its utmost tension. In other words, the Stoics held that straightness is a *function* of tension and curvature a *variation* of this function. And since all material change and structure is the transmission of dynamic force (the *pneuma*), the distinct surface or boundary of a body must be replaced by a sequence of boundaries or dimensions, converging like inscribed and circumscribed figures to form the dynamic entity.

Now we return to Alexandria and the most lasting accomplishments of Greek mathematics and astronomy. The influence of Euclid of Alexandria can hardly be overstated. Euclid's *Elements*, composed c. 300 B.C., remains one of the most important mathematical treatises ever written. Until the nineteenth century, it was not even necessary to affix the name Euclidean to his geometry; the *Elements* represented THE geometry of the physical world. To the German philosopher Immanuel Kant, Euclid's proofs were unassailable *a priori* truths. It was generally held by mathematicians that Euclidean geometry was the most rigorously constructed branch of mathematics.

What is so special about the *Elements* is not so much the originality of the work—many of the theorems and demonstrations are not Euclid's own discoveries. Rather, it is the highly systematic and deductive way in which the thirteen Books are constructed. Except for Book V, the Theory of Proportions, and Book VII, the Number Theory, the later Books presuppose the earlier ones—and these build directly upon a system of axioms, postulates, and definitions. The entire treatise, then, is a massive exercise in geometrical deductive logic, a mathematical counterpart of Aristotelian formal logic.

The *Elements* was probably originally a textbook; indeed, most elementary geometry texts still follow its form. Although we have no extant copies, the traditional form of the treatise is that of the thirteen Books. Euclid's logical basis is not the hypothetical formal logic that mathematics follows today; instead, Euclid, much like Aristotle, conceived of geometry as an idealization of the real world. Euclid's geometry, like Aristotle's science, is a rigorous and logical extension of common-sense intuition, rooted in sensory spatial experience. His proofs are accompanied by clear figures which enable one to *see* the necessary deductions being carried out. It is no wonder, then, that Euclidean geometry would maintain such a hold upon the minds of generations of scholars, for it is nothing less than refined common sense.

Here we can provide only a brief summary of the treatise, with a few examples of its methods and proofs. Book I begins with twenty-three definitions, among them that a point has no part (in other words is discrete), line is breadthless length, extremities of a line are points, extremities of a surface are lines. From these definitions follow others concerning angles, circles, figures in general, and finally the assertion that parallel lines do not meet. Next come five postulates dealing entirely with geometry and finally five general notions which roughly correspond to some of Aristotle's logical axioms. In fact, Aristotle's logic is readily felt in the common notions, for they are the fundamental principles not only of geometry but of science in general. They cannot be proved or demonstrated; rather, they are clear intuitive statements which all humans recognize. The third common notion seems to be taken directly from Aristotle's *Organon*: equals subtracted from equals give equal results.

Among the postulates, the fifth, which was felt by later mathematicians to be one of the weaker parts of the entire treatise, is historically the most interesting. It is generally known as the parallel postulate. The very way in which it is presented suggests that Euclid may have had some misgivings about it. The postulate says that if two lines in one plane intersect a third in such a way as to make the sum of the interior angles on one side of the intersecting line *less* than two right angles, these two lines will eventually meet on that side. In effect, Euclid states indirectly that there is only one line through a given point on the same plane above a given line which is parallel to that line. So why all the fuss? This seems self-evident. Yet, recall that Euclidean geometry presumed to assert rigorous truths which were facts about *physical* space and which could be visualized and confirmed by experience. The parallel postulate asserts that the two parallel lines *will never meet*; it goes beyond our limited range of spatial experience and says that the lines will not meet even when extended to infinity. Who can make rigorous statements about what happens infinitely far out in space? The postulate *is not* self-evident, and all attempts to deduce it from other postulates failed.

The thirteen Books of the *Elements* contain 467 propositions and adopt various methods of proof. Some include reduction to the absurd, first assuming the contradiction of the thesis to be proved and then showing how it leads to impossible or absurd results. Most of the proofs, however, rely upon the assumed axioms and use such methods as comparison of magnitudes, areas, and the principle of exhaustion.

Book II is a contribution to geometrical algebra, although the quantities are represented by line segments and the product of two numbers is actually the area of the rectangle whose sides are represented by the segments. Thus the product of three numbers is a volume. Using this method, Euclid is able to avoid the Pythagorean curse of irrationals. Euclid does not, of course, use algebraic formulas to present his proofs; they are fully geometrical.

Book III of the *Elements* deals with the properties of circles, chords, tangents, and inscribed angles, and Book IV treats inscribed and circumscribed figures. Book V, the Theory of Proportion, is based upon the work of Eudoxus. It has been called the greatest achievement of Euclidean geometry, because here the theory of proportion is extended to incommensurable ratios without the use of irrational numbers. Euclid had already used the Eudoxian notion of magnitudes in his earlier Books for lengths and areas. In Book V, however, he begins again, since now he must treat the *ratios* of magnitudes and proportions, which may be commensurate or incommensurate. In order to do this he must introduce a general notion of magnitude. But Euclid gives no clear definition of magnitude. We simply start with the statement that "a magnitude is part of a magnitude, the less of the greater, when it measures the greater." But Euclid does not use numbers here.

Definition 5 is probably the most important of the Book. It states that magnitudes are said to be in the *same ratio*, the first to the second and the third to the fourth ($a/b = c/d$) when, if any equimultiples whatever be taken of the first and the third, and any equimultiples of the second and the fourth, the former both exceed, are equal to, or fall short of the latter equimultiples taken in order. This simply means that if we multiply a and c by any whole number and b and d by any other whole number, the relationships of greater than, equal to, or less than are implied between the terms of the ratio. Again, Euclid does not use algebra. Mathematicians who followed him believed that the rigor of magnitudes applied only to geometry, and thus geometry was the most firmly established branch of mathematics. Indeed, Book VI uses the theory of proportion to formulate proofs dealing with similar figures, which served to cement this impression.

Books VII to IX deal with the theory of numbers, the properties of numbers and ratios. As one might expect, the numbers are whole numbers and the propositions are verbal rather than symbolic. Proposition 20 of Book IX is interesting, for it uses the method of reduction to the impossible. The proposition states that the number of primes is infinite. Euclid begins by supposing the contrary, that the number of primes is finite, or $P_1, P_2, \ldots, P_n$. Yet, one may easily form the sequence $P_1, P_2, \ldots, P_n + 1$. If the new number is a prime, we have a contradiction to the proposition of finite primes, or, if $P_n + 1$ is a composite number, it must be exactly divisible by a prime. But the divisor cannot be $P_1, P_2, \ldots,$ or P_n, because there will always be a remainder of 1. Hence $P_n + 1$ is some other prime, and the original proposition is proved by the contradiction of its contradiction!

Books XI through XIII treat solid geometry and the method of exhaustion, using the latter to prove, for example, that the areas of two circles are to each other as the squares of their diameters. The areas are "exhausted" by inscribed regular polygons, and it is assumed that, since the theory is true for polygons, it is proved for circles.

It should be emphasized that no amount of exposition can do justice to the *Elements*. The very basis of the concept of proof in mathematics, as well as the logical ordering of theorems, can be learned from it. The treatise is a great masterpiece of synthetic thought, and it may be called the realization of Aristotelian logic in geometry. It served as a model for subsequent mathematical thought and was held to be the most rigorous rationalization of common spatial experience.

Before passing on to Archimedes, we should take note of Apollonius of Perga, born about 262 B.C., who studied under Euclid's successors in Alexandria. He was known as a "Great Geometer" as well as an astronomer, and his most famous work was the *Conic Sections*. Like the *Elements*, it is a well-organized and logical piece; it contains, moreover, some highly orginal material. In fact, the eight Books and 487 propositions of the *Conic Sections* practically closed the subject for later mathematicians.

As indicated by the title, Apollonius undertook the first thorough (and complete) investigation of the curved sections of solid cones. The terms he introduced to describe them are basically those used in the subject today: the ellipse, parabola, and hyperbola. The construction of conics from given data, tangents, focal properties, and intersecting cones are among the propositions Apollonius proves.

It is generally agreed that one of the greatest mathematicians of antiquity was Archimedes of Syracuse. He was born in the Greek city-state of Syracuse on the island of Sicily in 287 B.C.. As a youth he went to Alexandria, where he received a mathematical education. He did most of his work in Syracuse, but he corresponded with other mathematicians in Alexandria and elsewhere, addressing many of his treatises to them.

Archimedes followed the meticulous demonstrations of the Euclidean kind, yet he also displayed a strong imagination which served him well. He was often able to leap from mechanical examples to idealized geometrical proofs, using the former as a method of discovery and intuitive guide. His works cover a variety of subjects including mechanics, where he used a combination of mathematical and mechanical reasoning to find centers of gravity in planes and solids. He introduced theorems on the lever and actually founded the science of hydrostatics, which deals with the pressure and equilibrium of water. The invention of a water-raising device (the Archimedean screw) is attributed to him, along with such mechanisms as compound pulleys, cogwheels, and the endless screw. In his youth he constructed a planetarium, supposedly operated by water power, which produced the movements of the planets. He conceived of a scheme for representing large numbers and, using this system, calculated the number of grains of sand in the universe! Plutarch, however, called his inventions simple diversions, geometry at play!

One of his most interesting works is entitled *The Method*. In it he demonstrates the use of mechanical ideas to prove mathematical propositions and stresses the importance of having some preliminary knowledge of the problem before carrying out the deductive proof. He cites the use Eudoxus made of the unproved assertions of Democritus regarding the cone and cylinder. Pure mental insight, he seems to say, is an aid to rigor, in a way combining the methods of the gods and the giants. Yet, Archimedes tempers this intuitive method by warning that it is only a method of discovery; proof must follow the Euclidean deductive model.

The Method was discovered as late as 1906 in a library in Constantinople. One of its proofs is the proposition that a parabolic segment is $\frac{4}{3}$ the triangle having the same base and vertex (the vertex taken as a point on the segment from which a perpendicular to the base is the greatest). In order to facilitate his proof, Archimedes uses basically physical arguments taken from theories on levers and centers of gravity, and the fruitful idea that a surface is made of of lines. The fact that Archimedes supposed that surfaces may be regarded as consisting of infinite lines has led some to assert that he anticipated the integral calculus. He also discovered such things as the volumes of segments of conoids and cylindrical wedges, centers of gravity of the semicircle, centers of gravity of parabolic segments, segments of a sphere and of a paraboloid. Yet, he did not speak of the *number* of elements in each figure as infinite, nor did he define the definite integral as the limit of an infinite sequence. He simply assumed this method as a preliminary investigation, to be supplemented by rigorous proof using the method of exhaustion. Again, the Greek emphasis was on form rather than variation (except Stoic speculation), and thus the concept of the infinitesimal was never firmly established.

Another example of Archimedes' use of the two methods, exhaustion and mechanical argument, can be found in his *Quadrature of a Parabola*. Using the method of exhaustion, he inscribes a triangle in a parabolic segment; then within each of the two smaller segments he also inscribes triangles. Continuing the process, he acquires a series of polygons with an ever greater number of sides and demonstrates that the area of the nth polygon is given by the series $A(1 + \frac{1}{4} + \frac{1}{16} + \cdots \frac{1}{4}^{n-1})$, where A is the area of the inscribed triangle having the same base and vertex. In this case Archimedes finds the sum of n terms and adds the remainder by the sequence $A(1 + \frac{1}{4} + \frac{1}{16} + \cdots \frac{1}{4}^{n-1} + \frac{1}{3}$ times $\frac{1}{4}^{n-1}) = \frac{4}{3}A$. Thus, as the number of terms becomes greater, $\frac{1}{3}(\frac{1}{4}^{n-1})$ approaches zero or "exhausts" at $\frac{4}{3}A$.

There was always a gap in Greek mathematics between the real finite exhaustion and a rigorous passage to an ideal limit which is the heart of the infinitesimal. Archimedes also circumscribed polygons on a curve, yet he still maintained the view of Euclidean proportions, greater than and less than, and based his proofs upon the reduction to the impossible. While anticipating the infinitesimal calculus, he cannot be said to have thought in its fundamentally abstract terms.

The calculus has been called one of the most powerful tools in mathematics for the investigation of nature, and it was conceived by Newton and Leibniz to solve physical problems. Archimedes did use mechanical methods as supplements to geometrical proofs, yet again these were simply methods of discovery. His work on the spiral assumes the uniform velocity of a point along a line, and some have characterized his determination of the tangent to the spiral as differentiation. However, in Greek geometry there was no concept of a curve as corresponding to a function, nor the notion of instantaneous velocity, which the differential calculus was meant to solve. Archimedes comes tantalizingly close, yet it would have taken a break with Greek mathematical tradition to achieve it, for any attempt to conceive an actual infinite or instantaneous speed would lead directly to the paradoxes of Zeno.

The most famous story (probably apocryphal) about his mathematical genius was his discovery of a forgery. The king of Syracuse had ordered a crown made of gold, but when it was delivered, he suspected that it contained a baser metal, silver, and sent it to Archimedes asking the mathematician to determine the forgery without destroying the crown. Archimedes was pondering the problem while taking a bath and suddenly discovered the solution. He was so excited that he jumped up from his bath and ran naked through the streets of Syracuse shouting *Eureka!* (I have found it!).

Archimedes had noticed in his tub that the amount of water which overflowed was about equal to the amount his body displaced. Thus he discovered the principle of hydrostatics which says that a body immersed in water displaces water equal in volume to the space the body occupies. The corollary is that bodies of equal weight are not necessarily of equal volume, and therefore he found that a pound of silver occupies more volume than a pound of gold. Archimedes, so the story goes, formed two masses of silver and gold, both the same weight as the crown, and measured their displacement in water. Then, immersing the crown, he discovered that it displaced less water than the silver and more than the gold. Using his measurements, he was able to calculate the ratio of silver to gold in the crown, and legend has it that the villain was found out.

One of Archimedes' propositions in the treatise *On Floating Bodies* says that a solid heavier than fluid will, if placed in the fluid, sink to the bottom, and the solid when weighed *in* the fluid will be lighter than its true weight by the weight of the fluid displaced (Proposition 7). Archimedes here implies the concept of specific weight, which is defined mathematically: weight is regarded quantitatively in relationship to a given volume. Thus his approach may be termed "static" or geometrical as opposed to Aristotle's dynamical consideration of weight in movement.

Geometrical reasoning was also used by Erastosthenes of Syene (275–194 B.C.), a librarian of the Alexandrian Museum, to determine the size of the earth. Tradition has it that Erastosthenes was a poet, historian, mathematician, astronomer, geographer, and possibly a friend of Archimedes. He knew that Alexandria was due north of the city of Syene by about 500 miles (measured by the daily trek of camel trains), and that at the summer solstice the sun was directly overhead at Syene. At the same time in Alexandria, near the same meridian as Syene but due north, the angle of the sun was roughly seven and a half degrees. Since the sun is so far away, the lines from the sun to both Alexandria and Syene may be considered parallel. Therefore, extending the line from Syene to the earth's equator, the angle formed between this line and a perpendicular extended from Alexandria is about 7½°, or about ¼₈ of 360°F. It follows that the arc between the cities, which we already know to be 500 miles (Erastosthenes' units were stadia), is ¼₈ of the entire circumference of the earth's 360°. Multiplying 500 by 48, we get the circumference of the earth as 24,000 miles; Erastosthenes' estimate is usually given as 252,000 to 250,000 stadia, and while the length of the stade seems to have varied over time, the length of his stade not being totally clear, the method works and would indeed give a fairly accurate estimate (see Figure 5.1).

Later geographers, using poorer methods, or simply erring, reduced the size of his estimate, which found its way into popular works. Columbus probably acquired much of his geographical knowledge from these popular and varying accounts; though not an academic, he could read Latin. Thus, if he did indeed use the smaller figure, the Orient on his maps would have been much closer to Europe than it actually was; and this was combined with reports by "good" authorities also cited in such accounts—Aristotle, Seneca, Pliny, etc.—that the western ocean connected the two and, perhaps, did not cover three-fourths of the earth's surface, as others had supposed. Whatever the case, Columbus still sailed over the seas of a central earth, accurately measured or not, enclosed by the physical spheres. But long before the time of Columbus an ancient Greek astronomer, an "ancient Copernicus", had suggested an alternative to the geocentric universe.

The key phrase which describes astronomy after Aristotle is "to save the phenomena." There was some question whether or not Eudoxus actually conceived of his spheres as truly physical. Aristotle certainly did, for his additions to the theory were designed to account for its mechanical operation. Yet, the theory itself had some serious drawbacks; indeed, it never fully accounted for the irregularities of the planets and retrograde motion. The system had one problem which seemed to stand

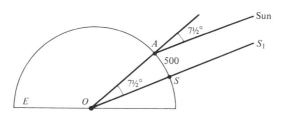

Figure 5.1

out from the others: Eudoxus and his followers had placed the planets on spheres concentric to the earth, which meant that the distance between the earth and the planets cannot vary. However, the planets appeared brighter and therefore closer to the earth when they retrogressed. Although one might accept the premise of a spherical and geocentric universe, the details of the system could not account for such discrepancies.

Nonetheless, Aristotle's physics was allied to his cosmology, and to reject the basic premises of this cosmology—circular motion in the heavens and a stable earth—would mean, in effect, giving up his physics which agreed so well with common sense. Any change in astronomy would necessitate a change in physics, and this astronomers were loath to do; after all, they were concerned with charting the heavens and not motion upon the earth. "Saving the phenomena" meant that it was quite immaterial whether a certain hypothesis was actually physically true—all that mattered was that it gave a better account of the observed planetary motions.

Two rather radical alternatives were put forward to correct the system, contemporary with and a generation after Aristotle. Both would have forced a major change in the basic framework of Greek cosmology and were therefore rejected. The first was put forward by Heraclides of Pontus, who was born around 388 B.C. and probably received instruction from both Plato and Aristotle in Athens. Heraclides departed from his contemporaries in two important ways. First, he apparently held that the earth rotated upon its axis from west to east, completing a single revolution in approximately one day. Second, he said that Mercury and Venus revolve not around the earth but around the sun, although the evidence for this assertion exists only in later commentators.

Heraclides evidently suggested the rotation of the earth as another means of accounting for the apparent movements of the planets, yet he gave no geometrical descriptions, and it appears that this idea was simply a suggestion. Some ancient commentators said that he suggested another motion, perhaps some sort of inequality in the rotation, to account for the irregular velocity of the sun. Since the length of the seasons is unequal—the sun taking longer to go from the vernal (spring) equinox to the autumnal equinox than from the autumnal to the vernal—Callippus had added an extra sphere to those given by Eudoxus for the sun. Heraclides may have simply "suggested" that it might be *possible* to account for this in some other way. The supposedly heliocentric theory of Mercury and Venus accounted for the fact that these two planets are morning and evening stars, always moving in some way in relation to the sun. It would have also explained their change in distance from the earth.

The second alternative was suggested by Aristarchus of Samos, who lived about 310–230 B.C. It is generally agreed that Aristarchus actually put forth a heliocentric hypothesis. His only extant work, *On the Sizes and Distances of the Sun and Moon*, contains no hint of this theory. The treatise is thoroughly geometrical in the style of Euclid, and the ratios of sizes and distances are calculated as falling between upper and lower limits. Aristarchus knew that the moon's light is reflective, and when exactly half the moon is illuminated, the angle formed by a line from the sun to the moon and from the moon to the earth is a right angle. The observer on the earth can measure the second angle there, which Aristarchus estimated to be 87° (its true value is 89° 52′), and complete the triangle. Since trigonometry had not yet been invented, using Euclidean geometry he could only estimate the distances from these measurements. Hence he said that the sun is more than 18 and less than 20 times more distant than the moon (it is 346).

What we know of his heliocentric hypothesis is related by other authors. For example, Archimedes names Aristarchus in *The Sand-Reckoner* as having put forward the theory that the earth travels around the sun, which is the center of the universe. It is possible that Aristarchus was familiar with the system of movable eccentrics (to be mentioned shortly) applied by Apollonius and was struck by the fact that the phenomena could be represented in precisely the same way if he simply made the earth travel around the sun and left everything else unaltered. Since we do not have the hypothesis in his own writing, we can surmise that it was merely a suggestion. By the time of Aristarchus the tendency of Hellenistic astronomy was toward more detailed mathematical representations rather than the mechanics of the universe. The major objections to the heliocentric theory seem to indicate that most astronomers were no longer interested in altering the fundamental Aristotelian physical structure of the cosmos. Even in Ptolemy we find criticisms of Aristarchus which rest upon Aristotelian objections: the earth is the natural center of heavy elements and cannot move; objects moving through the air would be affected by the earth's rotation; there is no apparent change in the stellar parallax. A moving earth simply went against all the dictates of established science and, we might add, religion—the Stoic Cleanthes thought that Aristarchus ought to be charged with impiety for giving the Hearth of the Universe (the earth) motion. Galileo was not the first to be so charged!

Even before the time of Aristarchus the emphasis in Alexandrian astronomy had already shifted to quantitative models, which used complex geometrical constructions to represent heavenly phenomena. The culmination of this trend was Ptolemy's *Almagest*, written around A.D.150. Ptolemy acknowledged his indebtedness to his Alexandrian forerunners, and it is probably best to view the *Almagest* as the synthesis of this long tradition. The mathematical basis of this work was certainly one of the outstanding achievements of Alexandrian astronomy and is known today as trigonometry. Its methods can be traced back to Hipparchus of Rhodes who was active about 150 B.C., roughly three hundred years before Ptolemy.

The astronomers of Alexandria had the good fortune to be in possession of Babylonian star records which greatly antedated Greek observations. Using these, Hipparchus was able to show the precession of the equinoxes, which is due to the wobble of the earth spinning upon its axis (unbalanced by its equatorial bulge) and tracing a slow circle against the stellar background. Since the revolution of this circle is very slow, about 1 degree in 72 years, Hipparchus could have discovered it only by comparing his own observations with those of Babylonia. In short, Babylonian records were combined with Euclidean geometrical methods, resulting in a rigorous mathematical reasoning which replaced the vague theories of the earlier astronomers. Let us briefly consider the basic idea of trigonometry, which, although in somewhat different form, was the method used by Hipparchus and Ptolemy for computing angular distances in their systems.

In a right triangle, since one of the angles is 90°, the other two combined must equal 90°. To know one of these angles is to know the other. In two right triangles, if one of the acute angles is equal to the other, we have similar triangles, no matter what their respective sizes. Euclidean geometry says that if two triangles are similar, the ratio of any pair of sides in one equals the ratio of the corresponding sides in the other. Hence knowing the angle other than the 90° angle gives these ratios as well. In short, the ratios can be computed for any angle in the triangle and given in tables, in Ptolemy's case called "tables of chords."

Besides this facile method, the principal new device adopted by Hipparchus and later Ptolemy (which may have been developed by Apollonius) to replace the concentric spheres was the *epicycle*, a small circle which rotates uniformly about a point on the circumference of a larger sphere encircling the earth, called the deferent. Both are located on the ecliptic, so that the diurnal motion of the planet (its daily rotation) is produced. The rotation of the deferent carries the planet through its annual journey, which, of course, varies for each planet. The epicycle accounts for both retrograde motion and the changing position of the planet, moving close to or away from the earth (see Figure 5.2).

The combined motion of epicycle/deferent produces a looping motion. If the epicycle is constructed larger relative to the deferent, the size of the loops will be increased. A faster-turning epicycle will give a greater number of loops in one journey around the ecliptic. A small epicycle with a period twice that of the deferent will produce an elliptical

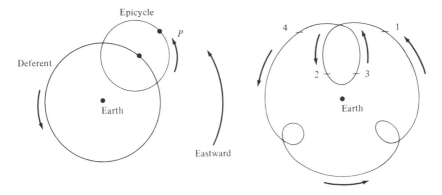

Figure 5.2 The epicycle-deferent and its looping motion.

orbit, and we may place a smaller epicycle upon an epicycle to account for further irregularities of the planets. In short, by appropriate variations in the mechanism, the system can be made to fit an immense variety—but not all—of planetary motions.

We know very little of Claudius Ptolemy's life: his own recorded observations taken at Alexandria span the years A.D.127–151. His fame as an astronomer rests chiefly upon the *Almagest*, which was originally entitled *He Mathematike Syntaxis*, or *The Mathematical Composition*. Arab translators referred to it as "The Greatest," prefixing the article *al* to the Greek *megiste*, and hence it was known in Europe as the *Almagest*. Ptolemy was also a geographer; beyond this, he wrote on harmonics and optics as well.

As denoted by its original title, the *Almagest* is a collection of mathematical procedures and tables with rigorous proofs of the geometrical theorems involved. The trigonometric functions, or Table of Chords, are found in Book I and are presented in the sexagesimal system, exemplifying the Babylonian influence. The preface to the work begins with an exposition of Aristotle's classification of the sciences, adhering to the distinction between sublunary physics and perfect celestial movement. Therefore, astronomy falls into the genre of the mathematical sciences, which, Ptolemy says, serves to prepare the way to theology, since the heavens bespeak of divinity. Next, he discusses general cosmological assumptions derived from Aristotle—the closed and spherical universe, the stationary earth, and the primacy of circular motion. In Book IX, where Ptolemy begins his treatment of the planets, he clearly states that his goal is to explain how apparent irregularities are produced by regular and circular motions, which are proper to the nature of divine things. Such demonstration, then, will *save* the appearances by showing how they can be accounted for by *a priori* mathematical principles.

The *Almagest* is divided into thirteen books, which, in a way, gather the entire tradition of Greek astronomy into one great, complex system. In this it resembles Euclid, although Ptolemy's modifications of the epicycle/deferent theory are basically his own inventions. It is impossible to present a thorough overview of the treatise which does it justice.

Ptolemy found that he could account for the variable velocity of the sun (from equinox to equinox) in two ways: either by placing the sun on a smaller epicycle which rotates once to the west for a single eastward rotation of the deferent, or by a single deferent which is *eccentric* to the earth—that is, a deferent circle whose geometric center is not at the earth but another point displaced from the earth. Ptolemy chose the eccentric solution for the sun on the grounds of its greater simplicity, having one motion instead of two. In this way the center point is placed toward the summer solstice at a set distance from the earth, thus giving an extra six days from the vernal to autumnal equinox when viewed from the earth (see Figure 5.3).

Here, then, is the third addition to the Ptolemaic system, the use of eccentric circles, which for the planets are used in combination with the epicycles. The system becomes even more complicated, for Ptolemy found that in some cases it was necessary to place the center of the eccentric on a small deferent, or even on a second smaller eccentric! This gave rise to the mechanism of *movable eccentrics*. Ptolemy discovered that for the moon the epicycle added by Hipparchus agreed with lunar observations in the syzygies (conjunctions and oppositions with the sun) but not for observed longitudes near the quadratures. In these cases the apparent diameter of the epicycle seemed enlarged, bringing the moon closer to the earth. Ptolemy therefore placed the lunar eccentric on a small deferent around the earth (see Figure 5.4).

Yet for all this, Ptolemy still found problems. Although he was able to give good accounts of the sun, moon, and planets using the epicycles, eccentrics, and movable eccentrics, he discovered that the celestial bod-

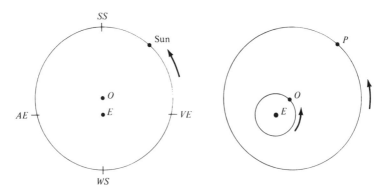

Figure 5.3   Figure 5.4

ies were moving at an observed nonuniform angular velocity. This problem gave rise to another mathematical addition to the system (already complex), called the *equant*. In simple terms, the equant was designed to maintain uniform planetary rotations by measuring these angular velocities not from the earth, but from a point, the equant, displaced from the earth as well as the center of the eccentric. The rate of rotation is uniform, then, not from its geometric point but with respect to the equant. It was this device which drew one of Copernicus' major objections to the Ptolemaic system.

The entire geometrical structure of epicycles, deferents, eccentrics, movable eccentrics, and equants was applied to each of the heavenly bodies in various combinations to give a reasonably good accounting for the appearances. However, two major criticisms of the system stand out, although some might argue that they are really over technical minutiae. First, the whole notion of the equant violates the principle of *regular* circular motion of the planets. Second, the magnitudes Ptolemy assigned to the circles which determine the course of the moon meant that the moon's apparent distance from the earth varies as much as 34 to 65 or about 1 to 2—which also meant that its apparent diameter is at times twice what we actually see!

If simplicity was one of Ptolemy's criteria, he may, perhaps, be also accused of inconsistency. Ptolemy himself argued that the very notion of simplicity, derived from sublunary experience, may be inadequate to the heavens. For him it was axiomatic that celestial motions are simple, even though his constructions designed to "save" them were complex. In the end Plato's "likely story" had been made more rigorous and mathematical and yet, in essence, it was still a likely story.

SUGGESTIONS FOR FURTHER READING

BAILEY, CYRIL. *The Greek Atomists and Epicurus.* Oxford: Clarendon Press, 1928.

BALL, W. W. ROUSE. *A Short Account of the History of Mathematics.* New York: Dover, 1960.

DREYER, J. L. E. *A History of Astronomy from Thales to Kepler.* New York: Dover, 1953.

HEATH, SIR THOMAS. *Aristarchus of Samos: The Ancient Copernicus.* New York: Dover, 1981.

———— . *A History of Greek Mathematics,* 2 vols. New York: Dover, 1981.

LLOYD, G. E. R. *Greek Science After Aristotle.* London: Chatts and Windus, 1973.

SAMBURSKY, S. *The Physics of the Stoics.* London: Routledge and Kegan Paul, 1959.

6

~~~~~~

Greek to Latin

Science in the Age of Rome

For all of its accomplishments, Greek science was a fragile thing—and no wonder: it was the accomplishment of only a handful and barely reached the common folk. In Alexandria the rulers gave state support to the Museum and its community of scholars for reasons of glory rather than their own scientific curiosity. Even Athens could be hostile to philosophy or simply tolerant. When interest was aroused, as among the influential upper classes of Roman society, science was studied in abbreviated and handbook form, which most often resulted in gross misrepresentation.

Roman science, as the German historian Heinrich von Treitschke remarked, was basically Greek science written in Latin. The period which saw the highest attainments of Hellenistic science was also the time when the transmission of Greek philosophy to Rome took place. It is generally accepted that, after reaching its apogee with Ptolemy and Galen (to be treated shortly), the rational science of the Greeks went into a slow and then precipitous falling off, nearly fading away from about A.D. 500–1000 with the breakup of the Western Roman Empire. The seeds of its decline may be found in its transmission to Rome.

The Romans, as has been often observed, were a practical people, and the knowledge they valued was mainly of the applied type. While they surpassed the Greeks in feats of engineering, in law, and in political and military organization, the Romans themselves created little original science. But then they were too busy conquering the Greeks—and the rest of the Mediterranean world. The Romans found much to admire in the Greeks, not only their science, but their art and literature. Greek became the language of the intellectuals in the Empire much as Latin would be the language of scholarship in the Middle Ages. Wealthy Romans developed a passion for Greek learning, yet this predilection was more of a fad than a serious search for knowledge. Thus arose the demand for handbooks, compilations of Greek knowledge designed to titillate the curious more then educate the serious. Greek science became a pastime. Since knowledge was valued for its practical applications, the emphasis was placed upon the moral and ethical philosophy of Greece. In such an atmosphere the popular science embodied little concern for consistency, rigor, and a thorough comprehension of ideas. At the same time the Roman world was infused with oriental mystery religions whose various cosmologies became intermingled with Greek rationalism.

By way of transition to this state, it is necessary to look at Hellenistic biology and the imposing figure of Galen, who stands with a foot in the Alexandrian world and another in the Roman. What we know of Alexandrian biology and medicine is found mainly in the pages of Galen and the Roman encyclopedist Celsus, who lived about A.D. 30. There is roughly a three-hundred-year gap in the documentation of Greek medicine from the Hippocratic *corpus* to Celsus, and only through quotations in Celsus and other writers can we piece together the changes which occurred during this time and up to Galen.

The two most important biologists of the early Alexandrian period were Herophilus and Erasistratus, who probably worked during the first half of the third century B.C. Celsus called the followers of Hippocrates Dogmatists, yet this description should not imply sterile imitation. It appears that the Dogmatists believed that no true physician who is ignorant of internal organs ought to attempt cures. Therefore, dissection was emphasized and for the first time legalized in Alexandria. What seems to us more shocking is the apparent fact that dissection was also performed upon *living* men, criminals supplied to the physicians by the state. Thus the Christian theologian Tertullian of the second century called Herophilus a "butcher" who hated mankind for the sake of knowledge.

Herophilus made important contributions to anatomy, describing the eyes, brain, major vessels of the body, the heart, and genital organs. He gave accounts of the duodenum (the first section of the small intestine, apparently named by him) and the liver. He also recognized the brain as the center of the nervous system, yet he claimed that the optic nerve was hollow. His comparison of the membrane of the eye with a net gave the Greek work *reti*—"net like"—to what is still known today as the retina. Probably through dissection (or vivisection) he was able to distinguish between the sensory and motor nerves, the tendons and the ligaments, the main chambers of the heart, and the blood vessels, especially what he called the arterial vein (which we know as the pulmonary artery).

Herophilus recognized the importance of the pulse rate for diagnostic purposes, identifying abnormal rates with the colorful terms "antlike" or "gazellelike." Here we can only wonder at the possible influence of ancient Egyptian medicine, which knew of the connection thousands of years earlier. He also discovered the ovaries, whose structure and function he compared with the male testes—certainly an important achievement.

His contemporary, Erasistratus, took an amazing leap by using mechanical ideas to explain organic processes, in a way similar to the means used by Harvey to discover the circulation of the blood. Erasistratus rejected the concoction theory of Aristotle, saying instead that food is propelled by muscular action into the stomach, where it is pounded into *chyle*, and then is squeezed through the walls of the intestine into the blood vessels. He also used the concept of a vacuum (again contradicting Aristotle) to explain how tissues were evacuated and filled in the process of digestion.

Erasistratus clearly appreciated the difference between arteries and veins, yet he held that only the veins contained blood while the arteries contained air. Here the question arises: If he dissected, how did he explain the gush of blood he must have witnessed when he cut into an artery (assuming ancient physiology was the same as modern!)? Let us call upon the vacuum! Erasistratus said that when an artery is cut, air escapes, and blood rushes in to fill the "abhorrent" vacuum. Nonetheless, he was the first to discover the operations of the four main heart valves, each acting as a one-way valve, and he knew that the heart operated like a muscle. Yet, he confused its operation with respiration by saying that air is drawn into the left ventricle through the pulmonary veins at each diastole (the relaxation of the heart) and is pumped out into the arteries at each systole (contraction).

The last great Greek representative of biology and medicine, and possibly of creative Greek science as a whole, was Galen, who was born in A.D. 129 at Pergamum in Asia Minor. It is said that Galen wrote more than any other Greek. Although many of his writings have been lost, there is still a considerable *corpus* of works attributed to him, numbering some twenty volumes of about one thousand pages each. His father, Nicon, was an architect who educated his son in mathematics, logic, grammar, and philosophy. According to tradition, Nicon was persuaded by a dream to teach his son medicine when Galen was about 16.

Galen traveled about the Roman world, visiting Smyrna, Corinth, and Alexandria. He returned to Pergamum, where he was appointed a surgeon to gladiators. He also visited Rome for three years but became disgusted with the envy and jealousy of the Roman physicians. He bitterly complained that there were no longer real seekers of truth, but only those who were motivated by money, political power, and fame. It seems evident that his protest represents the conflict between the pure man of science and those who are more inclined to the world of affairs and public success. Yet, he did become a court physician to Marcus Aurelius and also to Lucius Verus, probably in Germany.

Galen was a keen observer and experimentalist who took little on trust and described only that which he himself investigated. He did believe that the physician ought to know philosophy, and he held to the four elements and the four qualities. Although he relied heavily upon his own experience, he held that in medicine, as in geometry, there were certain maxims upon which reason and logic might erect a firm scientific structure. Hence he was not a simple empiricist, for he recognized the necessity of scientific reasoning mixed with observation and experiment. His method of experiment was mainly dissection, and while he did not dissect humans, he did work upon a great number of animals including monkeys (the Barbary Ape), dogs, horses, pigs, and even elephants, lions, bears, reptiles, and fishes. His work was so extensive that in the Middle Ages the word *experimentum* was almost synonymous with medicine.

According to Galen there were three central organs of the body. The liver, which was the source of the veins, produces the *natural spirits* found in all living organisms and the agent of growth. The heart is the seat of the *vital spirits,* responsible for movement and muscular activity. Finally, the brain is the center of the *animal spirits* and the organ of the nervous system, the nerves originating both in the spinal cord and in the brain. Remarkably, Galen knew that one side of the brain controlled the opposite side of the body. He was uncertain whether the animal spirits were a fluid or what we would call a stimulus, and thus he distinguished motor and sensory nerves by differences in texture, not by the nature of the stimuli transmitted by them.

Against Erasistratus, whom he criticized often, he held that nutrition involved more than a mere mechanical process. Food is turned into chyle, but then it undergoes concoction or digestion (*pepsis*). In the liver, then, food is finally converted into blood and receives the natural spirits. The blood flows into the body, where it is absorbed as nutriment. In the heart, blood mixes with air brought in from the lungs and becomes vital spirits, a new fluid different from blood. In the brain it is changed a third time into the animal spirits.

Galen's most important discovery, another critique of Erasistratus, was that in the living animal the left ventricle of the heart and the arteries contained blood, not air. Yet, in holding that the liver was the source of the veins, this marvelous discovery did not result in any theory of circulation.

Now Galen knew that the heart acted like a pump and he also knew that it was filled with blood. How, then, does the blood pass from one side of the heart to the other? Being aware of the valves, Galen knew that the tricuspid valve allows blood to enter the right ventricle but prevents it from backing up into the right atrium and vena cava. He also recognized the coronary blood vessels, so he knew that the blood in the right ventricle was not used up by the heart itself. Pneuma, the spirit of life found in the air, entered the body by the act of inspiration. Passing from the lungs into the pulmonary vein, air entered the left ventricle and was mixed with venus blood, giving rise to the vital spirits. The systole drove vital spirits into the aorta and blood back to the lungs. The innate heat of the heart was responsible for this "elaboration," and noxious vapors from that "cooking" returned to the lungs through the pulmonary vein to be breathed out. Blood ebbed and glowed, the arteries contracted with the heart, and the diastole was an active expansion: all required blood in the left heart. But how?

Enter the septum! The septum is a solid membraneous partition between the two sides of the heart. Galen's careful observation revealed that the septum was pitted, and with this tiny piece of information he proceeded to solve the puzzle. Why should the septum be pitted and these tiny holes be for naught? Philosophically, Galen held that nature did nothing by chance, and he applied his own precepts to his physiology. The pitted septum must really be a *perforated* septum, and blood passed through these minute holes from the right side of the heart to the left. It was reasonable to *assume* that the blood flowed through the septum, which would account for the volume of blood in the left heart. Galen might have easily inferred these pores in the septum from the analogy of capillaries. However, his inference became a dogma—which was to be the fate of much Greek science in Rome and medieval Europe.

Galen offered many fruitful observations which, had they been improved and expanded, would certainly have furthered the understanding of biology and medicine. While Galen believed that the purity of the air was important for health, he also held that plagues were due to impure air, a suggestion which might have served as a beginning for further research into communicable diseases. On the other hand, his theory of the humors led him to prescribe bleeding and cold drinks as the chief remedies for fever. While he was, on the whole, skeptical of magical cures and sometimes referred to other physicians as magicians and liars, he yet pondered the virtue of amulets, wondering if perhaps they acted by some unknown and marvelous antipathy against disease. Thus we find him suggesting that the wearing of an eagle's tongue will cure a cough! He also believed that the heavens affected health, and he held to the doctrine of critical days.

Indeed, speculation is often a valuable tool in scientific discovery, and we cannot fault Galen for speculating. Even the most fantastic speculations may lead to wonderful results. Errors, as we shall see, are just as important to science as *what we label success* (and tomorrow our "success" may be error). Yet dogma in *both* error and success is stifling. The decline of Greek science occurred when both the red meat and the fat of such speculations became apotheosized and made into closed doctrines.

Synthesized accounts of scientific ideas need not be necessarily inferior or misleading, and when they are done rigorously and with intellectual integrity they serve the important role of disseminating knowledge to the general public. In a certain sense Euclid's *Elements* is an example of what may be called a creative textbook, and writers such as Hero, Pappus, Proclus, and Simplicius are valuable references for the study of Greek science. Even the most gifted scientists may provide worthwhile popular accounts of their work which do much to enhance the level of public comprehension. However, there is a fine line between such serious presentations of science and the desire to merely appear erudite— that is, to simply glean as much as possible from every field in order to appear learned.

Posidonius, a native of Syria, became almost as important a source as Aristotle for later textbook writers and was himself one of the Greek masters of the handbook tradition. He was born in Syria about 135 B.C. and traveled to Egypt, Athens, and Spain, finally settling at Rhodes. He went to Rome on a diplomatic mission (representing Rhodes) and became quite well known among the Romans, his lectures being attended by Cicero and even Pompey.

Posidonius was a Stoic who regarded Rome's universal empire as the fulfillment of God's providence and a divine Stoic commonwealth (fortunately he did not live to see the imperial Rome described by Tacitus). He wrote about Aristotle and composed a book on mathematics, but his most influential work was a commentary on the *Timaeus*. During his stay in Spain he kept careful records of the ocean tides, noting that they were fullest at the new moon when the moon and sun are in conjunction, and at the full moon when they are at opposition. His predecessor, Seleucus, had also expounded the lunar theory of the tides, but Seleucus was one of the few ancients who accepted Aristarchus' heliocentric theory. Posidonius adopted the lunar explanation of tidal phenomena, yet he could not follow Seleucus in supporting the views of Aristarchus. Nevertheless, the lunar theory demonstrated to Posidonius the fundamental sympathy of celestial and terrestial phenomena, fitting well into Stoic physics.

The Roman author usually looked for moral and ethical lessons, holding these to be the goals of even the exact sciences. Also, it was a common practice among the Latin compilers to cite the same sources used by the Greek textbook authors, acknowledging them as their own without ever having read them! One can easily imagine the ultimate results of such habits.

Few if any of the Roman writers were able to follow the subtleties of theoretical Greek science, and some were even hostile. Cato the Elder, who lived from 234–149 B.C., had nothing but contempt for the Greek tradition. His work on agriculture was purely practical, little more than a manual, filled with recipes, advice, and typical Roman rhetoric. Vitruvius, living in the first century, wrote a book entitled *On Architecture* which dealt mainly with Hellenistic architecture, a subject the Greeks would probably have considered *techne*. However, Vitruvius maintained that the architect ought to be a man of letters and a student of philosophy, music, and medicine, besides being a draftsman and geometer. Yet at times he uses 3 and 3⅛ for the value of π, which makes one wonder how much of a geometer he himself was! Cornelius Celsus wrote in six of the liberal arts, covering agriculture, the military arts, oratory, philosophy, and jurisprudence; however, his only surviving book is *On Medicine*. His book is thorough and knowledgable, earning him a place beside Hippocrates and Galen among subsequent medical students. In fact, the great von Hohenheim, whom we shall meet in a later chapter, called himself "Paracelsus," meaning more eminent or greater than Celsus.

One of the most notable Roman achievements was the calendar reform initiated under Julius Caesar, which utilized the expertise of Hellenistic astronomy. The so-called Julian calendar adopted a solar year of 365 days, adding a leap year of 366 days every fourth year. This year was still slightly longer than the seasonal year, which runs from one vernal equinox to the next at the rate of 365 days, 5 hours, 48 minutes, and 48 seconds. Therefore, in 400 years the Julian calendar was too long by some three days, and the vernal equinox was a day earlier about every century and a third. Nonetheless, the calendar lasted until 1582, when Pope Gregory XIII ordered a modification, dropping about ten days from that year to bring the vernal equinox back to March 21. The reform made the century years regular years, not leap years, except when those century years were divisible by 400—the year 2000 will be a leap year. The Gregorian reform did not go uncontested. When it was adopted in England in 1752, riots broke out among the peasantry demanding back the lost days!

One of the first great Latin encyclopedists was Marcus Terentus Varro, who was well read in many fields and had a great reputation for learning. Varro lived from about 116–27 B.C. and supposedly wrote 620 books, enough to insure anybody's reputation! Varro is probably most important for his attitude toward learning, for it seems that he considered an expert to be one who could master the greatest number of authorities. Eventually even Varro's encyclopedia became too theoretical for most Roman intellectuals, and it was made into digests and adaptations which superseded the original.

Varro's great contemporary Cicero was famous for his cultured ways and intellectual accomplishments but was hardly a scientist. He did, however, translate the *Timaeus*, and he imitated *The Republic*, substituting his own "Dream of Scipio" for Plato's vision of Er. Cicero had no doubts that any individual would be able to encompass the entire range of human knowledge, which seems to have been the ideal of the Latin handbook writers. For example, he confidently states that although mathematics may be difficult, so many have gained perfection in it that we must conclude anyone who applies himself will be successful in the science.

One of the most elegant Latin authors was the poet Lucretius, who brought Epicurean atomism to Rome. Living in the first century B.C., Lucretius claimed that he was the first to expound these doctrines in Latin. He faithfully reproduced the atomic theory outlined in our previous chapter, and it would be redundant to follow his book, *On Nature*, in detail. In Book V Lucretius deals with astronomy and follows the Epicurean emphasis upon the senses. Any theory, he says, may suffice as long as it does not contradict perception. If there is more than one theory, why Lucretius is quite content to give us the alternatives as if they were equally true.

Trusting his senses, Lucretius says that the sun and moon may well be the exact size that they appear to us. The moon *may be* the source of its own light; the moon *may be* carried in a weaker current of air than the sun so that the constellations pass it quicker, accounting for its apparently faster speed. He even suggests that the sun may be a new body at every rising! There may be many worlds like ours, inhabited. The phases of the moon may be due to another invisible body borne along with her . . . and so it goes!

Lucretius did suggest that the universe will some day pass away to be replaced by another, either similar to or different from this one. Surprisingly, he then presents a theory of evolution which superficially sounds quite Darwinian. Many kinds of living creatures, he said, have perished and have been unable to produce offspring, and those which have survived have done so by some virtue—ferocity, speed, craftiness, etc. In other words, Lucretius gives us a theory of survival which resembles a crude kind of natural selection. Yet, his *modus operandi* is the chance combination of atoms in which ill-suited organisms are doomed to extinction. There is no descent of species, no selective variation, nor is there anything close to the detailed investigation of minute adaptation over immense geological periods which is the heart of the Darwinian theory.

Probably the greatest encyclopedist of the Romans was Pliny the

Elder, who lived from A.D. 23 to 79. By his own account he treated some 20,000 topics, gleaned from 2000 sources of about 100 authors, with his total number of authorities somewhere around 473. The encyclopedia, *Natural History*, survived intact. It is a tremendous compilation in 37 books containing both solid knowledge and the fantastic, the latter probably accounting for its popularity in the ancient world. Pliny held that nature existed for the purpose of moral edification, and he believed that no book was so bad that it did not contain something of value. One can well imagine the kind of work which arose from such a conviction.

Pliny was not a scientist himself, yet he had an honest respect for his sources. He did use the word *experimentum*, sometimes to mean ordinary experience and at other times a test. Nature to him was a vast collection of separate facts to be gathered and listed; in other words, nature was like a dictionary. He died at the age of 56 while observing an eruption of Vesuvius, probably getting too close to the dangerous fumes.

Pliny subscribed to Stoicism, calling the sun the soul and mind of the universe while the universe itself was God—vast, eternal, and divine. We find him giving no mathematical determinations of planetary movements. Rather he provides astrological discussions combined with physical theories. Stations and retrogradations are caused by beams of light striking the planet; Venus emits a genital dew which stimulates the conception of plants and even animals! His geography is a tedious cataloging of place names, hardly matching the work of the greatest ancient geographer, Strabo, of the first century B.C. Strabo's geographical survey of the ancient Roman world included the physical, archaeological, and human resources. But Pliny did discount the common notion that a branch of the Danube emptied into the Adriatic.

Pliny wrote a great deal about medicine and zoology, often combining skepticism with credulity. While he objects to the excessive claims of magical cures and powers, he does hold to the special virtues of animals and plants in healing. For example, we hear that goat's flesh roasted on a funeral pyre cures epilepsy, divinational powers are gained by eating the still-palpitating heart of a mole, and a person can cure a cough by spitting into the mouth of a frog! Sympathy and antipathy seem to be the basis of medicine for Pliny. He also tends to humanize animals, even to the point of crediting them with the discovery of certain cures. The hippopotamus, for example, discovered the use of bleeding as a cure by cuttings its leg on a sharp reed. *Natural History* speculates on just about everything, including the powers and limitations of God—God cannot do certain things such as commit suicide.

After Pliny the handbook tradition degenerated further into digests and adaptations of earlier Greek and Roman textbooks which, more often than not, were themselves outright plagiarisms of still earlier works. One of these handbook writers, Martianus Capella, who lived in the fifth century, wrote a popular work allegorically entitled *Marriage of Philology and Mercury*. This book established the famous seven liberal arts which would become the basis of medieval education. Capella's *quadrivium* (the word itself was first used by Boethius), or four mathematical disciplines, included geometry, arithmetic, astronomy, and music. The *trivium* consisted of grammar, dialectic, and rhetoric. When it came to the content of these subjects, Capella showed a great deal of confusion. In describing Euclidean geometry, for example, he stated that "a point is that whose part is nothing."

It is difficult to assess just how much Greek rationalism penetrated the various levels of Hellenistic and Roman society. To be sure, the old pagan religions persisted among the common people. These religions, however, had lost much of their vitality; they had been reduced to mechanical rites and emotionless observances which, in the end, failed to fulfill religious needs. There are a variety of reasons—political, social, and even economic—for the growth of mystery religions in the Roman Empire. The Hellenistic period saw a gradual merging of Greek and Near Eastern cultures, and the establishment of Roman hegemony over the entire region provided further impetus to such mingling. Far from being completely Hellenized, the Near East exerted its own influence upon the Greco-Roman world. From Augustus to Diocletian the government of Rome evolved into a centralized state very much like the oriental despotisms of the Near East. Many persons found political careers closed to them and sought solace in the study of science or religion. Strange religions, like Greek science, became a fad.

One of the first eastern religions to enter Rome, about the second century B.C., was the cult of the *Magna Mater*, the Great Mother goddess, from Asia Minor. From Egypt came the cult of Isis, evolving into a composite religion of Greek and Egyptian influences, stressing moral and physical purification. From Syria came the cult of the "Invincible Sun," which combined with Chaldean astrology, teaching that the soul returned to the heavens after death to live among the divine stars. From Persia, by way of Alexandria, came the cult of Mithra, the ancient spirit of light who became the god of truth and justice in the religion of the Persian prophet Zoroaster. According to this religion the world is the scene of a perpetual struggle between good and evil, light and dark. All degrading instincts must be resisted, since the entire world is drawn into one or the other of these two rival camps.

Taken as a whole, the mystery religions offered two things the old pagan religions lacked: mysterious methods of purification meant to bring the believer into a mystical state of rapture, and the idea of immortality and union with the deity as the reward for piety. The combination of such beliefs with Greek rationalism produced various hybrid intellectual systems like astrology and alchemy, and also served to enhance, however indirectly, the mystical trends of Pythagoreanism and Platonism.

Astrology, as we have seen, reaches far back into Mesopotamian times and was certainly well known to the Greeks. Astrology's relationship to astronomy had always been close, for astrologers had to be able to trace the movements and positions of the planets whose influences supposedly enabled one to predict terrestrial events and individual characters. Aristotle generally held to the influence of the celestial regions upon the sublunary world, although he did accept chance occurrences. The Stoic principle of univeral sympathy saw the human being as a microcosm deterministically affected by the universe.

Using the Babylonian sources, even serious astronomers like Ptolemy could not help but take an interest in Chaldean astrology. Some, like Posidonius, became convinced of astrology, but it was Ptolemy who gave it scientific support in his famous astrological work, the *Tetrabiblos*. While he remained critical of excessive claims, Ptolemy believed that the powers of the stars could be confirmed by the mathematical results of serious astronomical science. With Ptolemy, astrology became scientific.

The seven planets were often associated with determinate qualities. Jupiter made people apathetic, Mercury was good for business, Venus—of course—nurtured love. Seasons, months, days, and years were all personified as forces governing every change in the cosmos. The idea of cosmic cycles, the Great Year of 432,000 years, was infused by astrological determinism, giving rise to the concept of the exact reproduction of each cycle through time—like Nietzsche's eternal recurrence. How well such doctrines must have fit the sense of fatalism which flourished in the empire!

The influx of these mystery cults also stimulated renewed interest in Pythagoreanism and Platonism. Neo-Pythagoreans tended to see Pythagoras himself as an inspired prophet, and Pythagorean mathematics in turn became a type of number theology, numerology. Sacred books and hymns were composed, and even Homer and Plato were annexed to the movement. Nicomachus, who wrote an *Introduction to Arithmetic* in the first century, was himself a Pythagorean, and besides this manual he supposedly composed a work on the mystical properties of numbers.

The religious element had always been present in the thought of Plato. While a treatise like the *Timaeus* might be taken as a comprehensive manual on cosmology and astronomy, as time went on the otherworldly aspects of Platonic philosophy became the primary objects of study. One of the most important Neo-Platonists, not to mention a true philosopher in his own right, was Plotinus, who lived from A.D. 205 to 270. Plotinus made the most complete and logical presentation of the metaphysical doctrines of Platonic thought, based upon a spiritual hierarchy of being. Taking the One or Good of Plato, he made it entirely transcendent, the one reality, complete and self-contained. Everything, spiritual and physical, is the by-product of the One's self-contemplation. Next on the scale comes the Divine Mind (*nous*), a product of the One, whose object of contemplation is the totality of the Forms. The Forms as the archetypes of all things are themselves living intelligences, united in thought with the Divine Mind. Spirit is true reality, and thus the Soul (*psyche*), which is the radiation of Mind, is the life of the universe, reaching down to the lowest levels of Being. Even brute matter is ensouled, but its soul is asleep, on the very edge of nonexistence.

Plotinus himself was critical of magic, accepting some of its effects while maintaining that the rational Soul is free from it. He leaned toward natural causes as explanations for phenomena, holding, for example, that diseases are not caused by demons but by natural causes, and that the movements of the heavens do not cause events but simply indicate the future course of happenings. Further, Plotinus distinguished different kinds of control exercised by the stars over inanimate, animate, and rational beings, leaving a large field open to the human will.

A brief word, too, must be said about the growth of alchemy as a product of the spiritual forces in late antiquity. Alchemy in simple terms is the transmutation of baser metals into gold and silver. Early chemistry, especially in Egypt, was a well-developed technical art by the time of the Greeks, with a long tradition of metallurgy, jewelry, glassmaking, dyeing, and other such practical activities. It is impossible to go into the details here, but the early chemical recipes and directions were generally instructions for making alloys, imitations of gold or silver, presented empirically with no evidences of occult theories or the obscurity of language which would muddle later alchemy. Yet, such techniques would provide the alchemist with his experimental and technical activities.

The Greek word *chemeia* first appears in the fourth century, probably designating the arts of metalworking. Under various influences, however, such as Plato's conception of matter in the *Timaeus* and especially Aristotle's view of the elements as undergoing constant transformation, it soon came to be believed that such transformations could be brought under human control. Another influence came from a sect called the Gnostics, later a heretical branch of Christianity, who supposedly possessed secret and mysterious knowledge, accessible only to initiates and given by divine revelation.

As usual, the tendency in early alchemy was to attribute the discovery of the art of transmutation to ancient authorities, including half-mythical figures and gods. A corpus of books dealing not only with alchemy but a host of other scientific subjects was attributed to the god

Hermes Trismegistus (the thrice-great Hermes) or his Egyptian counter-part Thoth, the patron god of the arts and sciences. The so-called Hermetic literature held that the magical arts, given by divine revelation, were a source of truth even within the exact sciences. Hence we begin to see a mixing of mystical ideas with the solid technical expertise in the early chemists. Zosimus of Panopolis, living in Alexandria during the third century, united the practical recipes of the metallurgists with the mythical symbolism of transmutation. While Zosimus retains a familiarity with the techniques of the chemical laboratory, many of the later alchemists gave the impression in their writings that they lacked even the most basic skills, becoming lost in their mystical philosphy. The word *alchemeia* comes from the Arabs, and we shall see much more of it later. Its origins, however, may be found in the increasingly supernatural intellectual world of late antiquity.

And so the world of antiquity was to end with a partial eclipse of Greek science and a loss of many of its sources. Yet, the sun was about to rise on a new civilization, a cultural milieu in which human aspirations were drawn to the religious and transcendental realms of intellectual activity. Again the questions had changed, and with that change the Greek heritage was not really lost; it was transformed. To this creative transformation, the Christian medieval world, we must now turn.

SUGGESTIONS FOR FURTHER READING

STAHL, WILLIAM H. *Roman Science: Origins, Developments and Influence to the Later Middle Ages.* Madison: University of Wisconsin Press, 1962.

STILLMAN, JOHN M. *The Story of Alchemy and Early Chemistry.* New York: Dover, 1960.

THORNDIKE, LYNN. *A History of Magic and Experimental Science During the First Thirteen Centuries of Our Era,* Vol. I. New York: Macmillan, 1929.

7

~~~~~~

Athens and Jerusalem

Pagan Science and Christianity

"What indeed has Athens to do with Jerusalem?" These were the words of Tertullian of Carthage, a father of the early Christian Church. They seem indignant, angry, and we sense that Tertullian is revealing his innermost feelings. Imagine him sitting at his desk as he wrote them. On the one hand he has Scripture, the absolute truth revealed by God. All it requires is faith, and for faith the rewards are tremendous. On the other hand let us hypothesize that Tertullian's library also contains books on Greek science. But the two hardly agree. Which is he to follow? Should he accept Greek science, which may cast doubts upon his faith? Should he undertake the arduous task of reconciliation, attempting to bring Scripture and science into harmony? Or should he, in one breath, simply deny the hard-won results of Greek science, saying that it is better to be ignorant of them than to lose one's faith and hence salvation? After all, what does reason have to do with revelation, the Greek tradition with the Hebrew, or simply, Athens with Jerusalem?

Medieval Europe was heir to both traditions. To be Christian is to accept Scripture, and Scripture is fundamentally the word of an omnipotent creator, revealed to human beings, which is at heart superior to the truths of human reason. Although Greek rationalism came down to the West initially in fragments, it was difficult to ignore. But the two traditions often conflicted. Reason was sometimes taxed by Scripture. Science often made revelation doubtful. Most people would probably de-

sire a *modus vivendi*, some means of easing the tension. Then there were
the Tertullians who threw up their hands. Why bother? God has given
us all we really need to know. Science is no longer necessary!

Fortunately the Tertullians did not dominate the intellectual life of
the Middle Ages. Yet, the problem of reconciling revelation and reason
was there, sometimes obtrusive, at other times simmering in the background. It is easy to forget that, until the Renaissance, the Catholic
Church in Western Europe was much more than an intellectual and religious institution. It was the art, the history, even the entertainment of
people's lives. Everyone (except heretics and Jews) was born in the
Church, educated, married, and buried in its bosom. The ultimate goal
of life was salvation; all else was more or less auxiliary, including science. In short, Christianity's most important influence was goal related.
Beyond all this, there was the pure physical wealth and political power
of the Church. All these factors influenced the way in which science was
pursued. Indeed, there were individuals who studied science for itself,
but many of the predetermined values they inherited came from theology.

While there was a lively, critical, and sometimes creative scientific
discussion throughout the Middle Ages, it occurred within a classroom
built upon Christian principles. The greatest of these presuppositions
was the reality of the transcendental world. No Christian could doubt
the veracity of the supernatural; rather, it was the natural world which
was contingent and the science of that world which was uncertain. It is
Plato all over again—or even the mythopoetical tradition. Indeed, Christianity itself was in some ways a little of both—Hebrew myth and Greek
reason.

The first Christians were Jews, and many of the Jews living within
the Roman empire had become Hellenized, which means that they
spoke Greek. Although Christianity was initially a Jewish reform movement in Palestine, resembling the messianic Essenes, it was among Hellenized Jews that the movement spread beyond its homeland.

Christianity, however, was made into a universal religion by Paul,
and it can be counted among the mystery cults which spread through
the Hellenized Roman world. In many ways the success of Christianity
was due not only to its historical character—the Hebrew prophecy and
Jesus himself—but also to its use of the Greek tongue as a vehicle. Like
all languages, Greek carries an entire world of concepts, categories, and
subtle meanings. According to tradition Paul visited Athens and debated with Stoic and Epicurean philosophers within the Greek philosophical medium. He called upon the Unseen God of the philosophers,
quoting Greek poets, in a way calculated to impress the philosophical
mind. From its very birth as a world religion Christianity was infused
with Greek reason.

Yet, there was much in Greek philosophy, especially in the Greek exact sciences, which was far removed from the Judeo-Christian cosmology. Before Paul's day the problem of revealed truth and rationalism had already been addressed by Hellenized Jews, of whom the most important was Philo of Alexandria. In the early years of the first century Philo discussed Mosaic cosmology and pagan philosophy. We should take note here that the cosmology of Genesis has a strong overall similarity to the cosmology of Mesopotamia and even that of Homer and Hesiod. Like theirs, its account of the creation and structure of the world is mythopoetic, meant to appeal not to a person's reason alone but also to the imagination and emotions. Philo was perhaps the first to recognize that rather than being a destructive rival, Greek philosophy might well be used as a constructive helper to interpret the older tradition.

Since mythopoetical science presented the structure of the universe in symbols which admit many meanings, we need not always take the symbol itself as literal truth. Reason alone might aid in the discovery of the primary *meaning* of the symbol. While at certain times we must accept the accounts of Moses as literal truth, we must also use our reason to interpret the symbols of the myth when it becomes allegorical. Here, then, is the first welcome path out of our dilemma: when the literal account of Scripture conflicts with reason, we are to assume that the true meaning is allegorical. Philo naturally found a ready ally in Plato. This should not surprise us. Philo's goal was not to discover the reality of nature for itself; his task was to use science as a means of arriving at spiritual truths, much in the way Plato viewed the practice of mathematics as teaching the mind to think abstractly. In fact, Philo went as far as to say that Plato was Moses speaking Attic Greek!

Although Philo had shown the way, the allegorical road could be treacherous. Heresy was waiting to beckon the unwary. The Gnostics claimed that the God of Israel was not the God of the universe but a lesser divine being equated with Plato's demiurge. The only true Christians, they claimed, were those who received *gnosis* or the spiritual knowledge to recognize the true source of Being. Some Gnostics, again following Plato, held that the essential part of every human being was spirit, thereby dismissing the bodily resurrection of Jesus. Others even saw his material existence as an illusion! Gnosticism could also lead to magic. The Gnostic teacher Simon Magus (literally "magician") appears in stories as St. Peter's archenemy, claiming that he, Simon, was Christ and Jesus a wizard. Thus pagan critics like Celsus were ready to see Moses as a magician and the Hebrews (hence Christians) as being addicted to magic and superstition.

The early Christian fathers had to tread carefully indeed! Clement of Alexandria, who died in 215, held that Greek science did not make revealed truth more powerful but instead guarded the word of God against the assaults of sophistry and heresy. Clement did not totally reject the allegorization of Scripture but cautioned against reasoning without faith. Faith is beyond demonstration; to reason without faith, however, is to fall into error. Thus he accepts the Greek planetary spheres and equates the structure of the world with the Tabernacle. The Ark, he says, represents the eighth sphere of the fixed stars, which is the dwelling place of peace and righteousness, the world of pure mind or God.

Origen of Alexandria had a more comprehensive knowledge of science. Many of his arguments, directed at pagan critics of Christianity such as Celsus, were formulated within the purely rational tradition of philosophic discourse. An outstanding difference between Christian cosmology and Greek science, especially the science of Aristotle, was the doctrine of creation and Aristotle's positing of an eternal world and primary matter. Origen drew the distinction between the eternal world of spirit and the corruptible world of matter. God created a primal substance, which is spirit and eternal. God created the material world of elements, which is imperfect and changeable. All physical nature strives for the most perfect and complete existence. This striving is symbolized in Scripture by the final passing away of the world. So, the world will end—yet we may still call it eternal!

Another physical problem in Genesis was to bedevil many Christian thinkers. On the second day of creation God forms a separation in the midst of the waters which is called the firmament (recall the *Enuma Elish*). This separation is into waters above and waters below. God calls the firmament above "heaven." On the third day He separates the land from the waters *below* the firmament. Are we to understand that there are waters *above* heaven? The ancient Egyptians believed that the Milky Way was the celestial Nile, yet such mythopoetical images contradict the rational theory of natural place. Origen, therefore, interpreted the passage as pure allegory. The celestial waters really mean the power of angels, as opposed to the inferior waters which represent the abyss of evil and demons. The creation story actually enjoins us to separate our spirits (celestial waters) from corruption. For some, this interpretation went too far. Confronted with the censure of reason, Tertullian snorted that faith is certain *because* it is absurd! To know nothing that is in opposition to faith (an opposition to which reason often leads) is to know all things.

After the Emperor Constantine's conversion and the Council of Nicea in 325, which helped to establish official Church dogma, the Christian writers seemed to deal with pagan science more confidently. Commentaries of Genesis became popular. Indeed, the level of these writers' "science" was low, but for many the inability of the various philosophical schools to agree upon physical principles illustrated the inferiority of reason when compared to the unity of Scriptural revelation.

If God is incorporeal spirit, then how did He create corporeal matter? Gregory, the bishop of Nyssa, and his brother Basil of Caesarea identified primal matter with the material of a physical body. All matter, said Gregory, is given to us through certain qualities like color and size. These qualities are not ponderable in themselves, yet taken together they present us with matter. Are not such qualities nothing more than intellectual concepts? Hence God first created concepts—not too difficult to imagine of an incorporeal being—and through the union of concepts there arises corporeality.

An attempt was made also to give the waters of the firmament a more realistic explanation. St. Basil held that the crystalline firmament of the last sphere was analogous to rock and hence capable of supporting the waters in their "natural place." St. Gregory added that the outer convex surface of the final sphere is full of valleys and mountains which serve to hold the celestial waters. St. Ambrose of Milan even found a way to make the celestial waters play an indispensable role in the cosmic scheme. The heavenly waters, said St. Ambrose, serve to *cool* the axis of the universe, which is heated by the perpetual motion of its rotation!

Such *ad hoc* solutions underlined the pressing need for a more rigorous systemization of the role reason played in Christian faith. This was accomplished by St. Augustine, the bishop of Hippo in North Africa. Like no Christian before him, St. Augustine used Greek philosophy, especially Plato and Plotinus, to supplement and enrich his faith. His influence in the Middle Ages was second only to that of the Bible. He lived during the time of the Western Empire's decay, and his greatest work, *The City of God*, was composed to answer pagan criticisms of Christianity. In 410 Rome was sacked by a German tribe called the Visigoths, and pagans generally blamed this disaster upon the abandonment of the old gods for Christianity. Augustine saw in the crumbling Empire the basic fact of the material world's contingency. Like Plato, he emphasized the absolute reality of spiritual being as opposed to the changing uncertainties of material nature. With this doctrine he not only answered the pagans—Rome was bound to decay, as do all worldly things—but constructed the first truly Christian theory of knowledge.

The essence of knowledge for St. Augustine was eternal truth, which is ultimately God. The human soul as the image of God knows itself, and therefore its own self-knowledge is a reflection of divine light. All truth, both scientific and religious, comes from God. Thus the goal of all knowledge is to return to its source. Error creeps in when we reason without a clear vision of this goal—when we reason without faith. Certain knowledge, in short, is the reward of faith. Unless you believe, you shall not understand!

St. Augustine, then, is interested in science insofar as it provides formulas to reach the ultimate goal of all knowledge, God. There is a natural hierarchy of knowledge, just as there is a natural hierarchy in the world. Those who simply weigh the elements, count the stars, and measure the skies are inferior to those who have knowledge of the Creator—of Him who ordered all things in measure, weight, and number. Yet, the rational theories of the scientists are to be preferred to the "ravings" of the ignorant. As long as we keep in mind the ultimate goal of knowledge, science may actually prove useful in the battle against false beliefs and heresies.

Thus St. Augustine was prepared to use rational arguments against Stoic determinism and astrology. The position of the stars, he said in the *City of God*, is a kind of statement which signifies future events but does not cause them. We are to reject causal astrology, not only because it contradicts human free will and the judgments of God, but also because it is absurd to reason and experience. How can astrologers account for the apparent diversity in the life of twins who are born under the same constellations at the same time? The influences of the heavens may produce changes in the material world—for example, the seasons and the tides—yet such influences do not cause a person to sin, to make poor decisions, or to be successful. Only God has perfect foreknowledge of these things, and, although He knows we will sin, His knowledge is not the cause. Faith, then, avoids the pitfalls of unaided reason. Remember, astrology was actually applied astronomy.

St. Augustine endeavored to use reason in the defense of faith whenever he could. He mustered a number of arguments for the celestial waters. If men are able to make boats which float out of heavy materials (heavier than water), certainly God can operate in the same way. Physicians admit that phlegm's natural place in the body is the head, thus by the microcosm/macrocosm analogy the waters of the heavens exist in the highest place of the universe. The celestial waters may really be vapor, producing rain. Saturn is the coldest planet, although because of its greater orbit we should expect it to be hotter than the sun. How is it cooled? Need we guess? There is also evidence that Augustine knew Pliny, and he expressed his doubts over Pliny's descriptions of strange races. Yet, if such things do exist, they still must be part of God's divine plan and must exist for some unknown divine reason. Augustine could find no rational explanation for the existence of Antipodes (people who live on the other side of the earth), and he even wondered how animals came to be found upon islands after the Flood—perhaps they swam, or were transported aboard ships, or were taken by angels on God's command!

Curiosity about the natural world was certainly not lacking among Christian philosophers like St. Augustine. On the other hand, they had no desire to further the cause of science for itself. The best minds of the age were concerned with theological issues and how to reconcile the science they possessed. Besides the Latin handbook tradition, this situation also accelerated the decline of Greek science in the West.

In the eastern half of the Empire there was a greater desire to preserve scientific texts. The disruptions of the Germanic invasions in the West caused a cleavage between the Greek and Latin parts of the Empire. The Greek-speaking East preserved a great deal of its ancient scientific heritage along with the structure of the Roman Empire itself— the Eastern or Byzantine Empire which included Greece. In 529 the Eastern Emperor Justinian closed the Platonic Academy in Athens because of its paganism, and many of its members fled to Persia. One Neoplatonist, John Philoponus, returned to Constantinople from Persia and converted to Christianity. John's conversion, however, did not seem to limit his freedom of thought.

Philoponus may have influenced the West only indirectly through the Arabs, yet his ideas reflect a higher level of scientific knowledge and actually prefigure scholastic criticisms of Aristotle in fourteenth-century Europe.

In his work on cosmology, Philoponus noted that it was not the goal of Moses in Genesis to give rigorous physical explanations for natural phenomena. Rather, Philoponus held that the purpose of Scripture was to conduct people to knowledge of God, which is not the same as the goals of science. However, John did speculate on the scientific meaning of the creation story. The heavens, he said, which were created on the first day are not the same as the firmament. The firmament, formed on the second day and packed with stars on the fourth, refers to the visible heavens of the astronomers. The primitive sphere of the first day is destitute of stars and exterior to the firmament. But God creates nothing superfluously, so the primitive heaven is nothing less than a ninth sphere, which accounts for the precession of the equinoxes discovered by Hipparchus. This idea of the starless sphere would reappear again in the West as the *Empyrean Sphere,* an intellectual abode full of light where angels dwelled. Whether it moved or not was a question of debate.

John also had a rational explanation for the celestial waters. If the upper heavens are crystalline and invisible, how do we explain this transparency? Now, if we assume that the upper heavens are made of air and water, then we may also assume that these elements are transformed from a fluid state into a solid. And Moses, in order to explain this transparent solid, gives it the name "firmament," which really describes its passing from a liquid to a solid state. In fact, without the celestial waters there could be no crystalline spheres.

Nonetheless, Philoponus was not merely an apologist; he also questioned some of the principles of Aristotelian science. To imagine that motion could not take place in a vacuum, or that a vacuum could not exist, places some rather severe limitations on God. Interestingly, John's criticisms are purely physical and rational. Aristotle himself asserted that weight is the efficient cause of downward motion. Let us imagine two bodies of different weights falling through the same medium. The heavier body cuts through the medium better because of its greater downward tendency, and this tendency is something possessed by each body in and of itself. Now, let us imagine the same two bodies in a void. What changes? Certainly not the downward tendency, for it is *intrinsic* to the body. Even without resistance, the bodies in question will still possess this tendency. Therefore, during free fall in a void a body will fall in a certain definite proportion to this tendency and hence in a given time—call it the original time. Thus free fall in a vacuum *will not* be instantaneous! After all, a body cannot be at two extremes in the same instant, even in a void.

John seems to conclude that in a vacuum bodies fall in original times inversely proportional to their weights. Yet, in a proposition which is rather difficult to reconcile with the above, John says that if you allow two weights to fall from the same height, you *will see* that the ratio of their times is not dependent upon the ratio of their weights! This sounds almost identical to Galileo's legendary experiment at Pisa, and we can only guess what John was thinking. Perhaps he believed that resistance canceled differences in weight, or that he was talking about specific weights—different weights of an identical substance—like Archimedes. Nevertheless, it is clear that he imagined motion in a vacuum and abstracted from experience.

Philoponus also had some strong doubts about Aristotle's theory of projectile movement. To think that air rushes in behind a projectile and then pushes it is incredible, for we are actually asking the overworked air to accomplish three separate things: it is pushed forward, it moves back, and it pushes forward once more! How can air avoid being scattered into space like vapor? Equally preposterous is Aristotle's second explanation—that movement communicates a motive force to air, which carries on the motion. Philoponus snorted at this, too. If a bowstring imparts force to air, then what need is there for its contact with the arrow?

John decided that projectile movement was really due to a kind of incorporeal kinetic force impressed upon the object (like heating a piece of iron) which accounts for continued motion until the force is drained away (like the cooling of iron) by resistance present in the body's weight and the medium. His fellow Neoplatonist, Simplicius, said that in projectile motion the impressed force overcomes the body's natural tendency. As this force dissipates, the body slows. Finally, when its own impulse becomes greater than the impressed force, it begins to fall. But something of the impressed force lingers, so downward motion begins slowly, gathering speed as the force drains completely away.

Philoponus and Simplicius show an ability to continue and improve scientific discussion even against the background of theology. Religion, rather than adversely affecting the internal development of science, exerted an influence in the realm of goals and values. And at times this could lead to fruitful ideas.

But in the West, knowledge of the Greek language and much of Greek science was lost as the barbarian tribes settled across Western Europe. Boethius, who was put to death under the Ostrogothic king Theodoric, was one of the last of the Latins to have a thorough knowledge of Greek. His translations of the Aristotelian logical treatises would remain the only sources of logic left to the West until the eleventh and twelfth centuries—and even many of these were lost! Plato was known mainly through the Roman commentators and fragments of translations. There were some popular Christian cosmologies, but these systems portrayed the world in terms similar to the ancient cosmologies of the Near East.

The encyclopedia tradition also persisted among Christian writers. One of the most popular encyclopedias, written in the early seventh century, was the *Etymologies* of Bishop Isidore of Seville. Isidore's chief source was probably Pliny, yet he was careful to substitute the authority of Scripture whenever he could and rarely took a stand on controversial issues. He simply repeated what the philosophers had said, neither agreeing or disagreeing. Like many of his predecessors, Isidore believed that the function of science was to elevate the mind from the corporeal world to God. Pursuing the derivation of words, discovering their origin, seems to have been his idea of knowledge. It is as if words themselves have a transcendental and more real existence than the things they signify. And for Isidor, as for all Christians, the transcendental world was prior to the material; corporeal existence was therefore full of allegorical significance.

Isidore's interest in mathematics was centered upon the significance of numbers in Scripture. Six, for example, is a perfect number, since it equals the sum of its factors (1 + 2 + 3), but it is most important because it refers to the six days of creation. In astronomy he accepted the geocentric spherical universe and the spherical earth, yet when he came to divide the world into zones he completely forgot the necessity of projection, and his zones appeared side by side in circles drawn upon a plane surface. The inferno, like the heart of an animal, is in the center of the earth.

Isidore's zoology (if we may call it that) was taken mainly from a popular work entitled *Physiologus,* probably written in Alexandria during the first century. The *Physiologus* tended to allegorize animals and become a book of "natural wonders." Isidore did the same, although at times he was critical. Still, there are fabulous accounts. Beavers castrate themselves when they see a hunter, since they know their testicles are useful in medicine: hence they escape with their lives! Isidore said that lion cubs are born asleep and remain so for three days until they are awakened by the father's roaring. The *Physiologus* had maintained that they were born dead and awakened after three days, signifying, of course, the resurrection of Jesus. So Isidore has made an improvement here! He also wrote on physiology and anatomy—like Pliny, he wrote on everything. There are many fabulous races: satyrs, cyclops, dog-headed men, and one-legged sciopodes who live in Ethiopia. The ever-present Antipodes, Isidore said, lived in Libya!

Another encyclopedist who had a great reputation for learning was Bede. Bede was born in northern England in 673 and died in 735. He is famous for his *Ecclesiastical History*, yet he also wrote scientific treatises which exhibited a wide knowledge of Pliny and other Latin authors. Bede obviously had a thorough grasp of mathematical astronomy, for he was able to calculate the date of Easter from 532 to 1063. His famous chronology, which was used in the *History*, may be responsible for the Western dating of events from the Incarnation.

During the late eighth and early ninth centuries there was an attempt to improve the cultural and intellectual level of Western Europe in France. Charlemagne of the Franks, who was crowned emperor by the pope in 800, endeavored to collect manuscripts and preserve the writings of men like Boethius, Isidore, and Bede. Alcuin of Northumbria took charge of this educational scheme, yet the continued adherence to tradition, mainly that of the encyclopedist, severely limited any fresh development of thought or science. Intellectual stagnation had reached its darkest period.

The only truly original thinker of the age was John Scotus, whose writings dealt mainly with theology. Scotus was an isolated figure, a metaphysician who believed that reason's categories could not be applied to God. God is more than Being, more than qualities, for He includes all opposites. God is superexistent and nonexistent, Being and nothing, and only "negative theology" could properly speak of Him. John's writings were condemned in 1225. While it is probable that he died in France about 870, legend has it that he went to England, where he was stabbed to death by angry students wielding their pens. Reason did have its limitations!

SUGGESTIONS FOR FURTHER READING

BARRET, HELEN M. *Boethius: Some Aspects of His Times and Work.* Cambridge: Cambridge University Press, 1940.

BREHAUT, ERNEST. *An Encyclopedist of the Middle Ages: Isidore of Seville.* New York: Columbia University Press, 1912.

GILSON, ETIENNE. *Reason and Revelation in the Middle Ages.* New York: Charles Scribner's Sons, 1938.

GRANT, EDWARD, ed. *A Source Book in Medieval Science.* Cambridge, Mass.: Harvard University Press, 1974.

JAEGER, WERNER. *Early Christianity and Greek Phaideia.* New York: Oxford University Press, 1969.

LAISTNER, M. L. W. *Thought and Letters in Western Europe A.D. 500 to 900.* London: Methuen, 1957.

PELIKAN, JAROSLAV. *The Emergence of the Catholic Tradition 100–600, Vol. I, The Christian Tradition: A History of the Development of Doctrine.* Chicago: University of Chicago Press, 1971.

8

The Cosmic Garden

*Islamic Science
and the
Twelfth Century Renaissance*

One of the Epistles of the Brethren of Purity, a Muslim sect of the tenth century, relates a charming little story entitled "Dispute Between Man and the Animals." What gives man the right to dominate the earth? ask the animals. The animals challenge each human claim to superiority. Man in most cases is a weak specimen when compared to the physical beauty, power, and even cunning of the animals. Not even science nor the physical command of nature gives humans a right to dominate. But the animals are finally forced to concede one factor; among human beings there have been sages and prophets who have achieved knowledge of God, who have realized the deepest purpose in all existence. Man in this way is God's regent upon the earth and is also the representative of all earthly creatures before the Divine Throne. This may be a source of human pride, but it carries an awful responsibility.

It is significant that the animals accede to human domination of the earth only when they recognize that human beings alone gain knowledge of spiritual realities. Thus we have a profound insight into the character of Islamic science. There is a natural hierarchy of knowledge from the physics of matter to the metaphysics of cosmological speculation, yet all knowledge terminates in the Divine. All phenomena are creations of Allah, His theophanies, and nature is a vast unity to be studied by believers as the visible sign of the Godhead. Nature is like an oasis in the bleak solitude of the desert; the tiny blades of grass as well as the most magnificent flowers bespeak of the gardener's loving hand. All nature is such a garden, the cosmic garden of God. Its study is a sacred act.

The fundamental concept of nature's unity and the absolute oneness of God is reflected by the Koran itself. Koran means *Recital;* it is neither history nor myth but a series of over six thousand verses in which the basic theme is the omnipotence of Allah (God) and the command to submit (Islam) to the Divine Will. The Koran sets down a stern code of conduct, simply and direct. Like the Christian, the Muslim accepts the reality of the transcendental world, yet unlike Scripture, the Koran makes no pretense at a cosmological system. When stories are related, even those from the Judeo-Christian tradition, they are meant to emphasize the power of Allah and the fearful price of refusing to submit. Therefore, the science of the visible world is, on the whole, unencumbered by revelation. In fact, the faithful are commanded by Allah to study nature, for nature is His metaphor. Reason could, of course, run afoul of religion, and Islam did have its Tertullians. But the command to study nature and the underlying concept of its unicity reflecting the oneness of God stimulated a keen interest among Muslims in science. The first Muslims were Arabs. By the eighth century, however, a vast Islamic empire had been established, reaching from Spain to the borders of India, and including many non-Arabic peoples. The Omayyad dynasty of Damascus ruled this immense territory until 750, but the fall of the Omayyads to the Abbasids brought about a political fragmentation of the empire. Muslim territories were united linguistically, since Arabic was the language of the Koran. Straddling such diverse cultures, however, the science they eventually transmitted to the West was a massive synthesis of Greek, Persian, Hindu, and even Chinese thought. Thus not only did they preserve and give to the West a great deal of its forgotten Greek heritage, they also assimilated and enriched that tradition with non-Greek sources.

In Spain the Omayyad dynasty still held sway and produced an elegant scientific center of learning at Cordova. In the east the Abbasids built a new capital at Baghdad and gathered there an array of scientists in order to enhance the glory of the dynasty, much as the Ptolemies did in Alexandria. From nearby Jundishapur, a famous Persian medical school and Hellenistic preserve, the Abbasids brought physicians and scholars to Baghdad. Nestorian Christians still practiced the Greek tongue and were employed by the Abbasids to translate Greek scientific works into Arabic (the first translations of Aristotle in the east had been into Syriac about 450). How different all of this was from the West, where, when coming to a Greek phrase, the medieval scholar simply replaced it by the word *grecum*—"It's all Greek to me!"

The Arab reception of Greek science was not uncritical, as it often had been among the Latins. The very existence of other traditions provided alternative and sometimes contrasting views. While the primary element of Arabic science was Greek, Arab commentators felt free to criticize their Greek masters. Allah exhorts the faithful to study nature, not necessarily Aristotle's accounts of nature. If Aristotle provides us with a method which seems to work well, so much the better. Indeed, the Arabs never broke completely away from the Greek tradition; their science remained basically Hellenistic. Yet, they were not bound hand and foot to it.

From the late eighth to the early tenth centuries the translation and assimilation of Greek science proceeded at a rapid pace. The doctrines of Pythagoras, Plato, and the Neoplatonists were especially interesting to the Brethren of Purity and Jabir ibn Hayyan (called Geber in the West). Jabir lived from about 720 to 815, making him one of the earliest Islamic philosopher-scientists. The Jabir *corpus* is a vast collection of alchemy, cosmology, numerology, and astrology, as are the epistles of the Brethren.

Jabir saw the Aristotelian qualities of matter in terms of numerical proportions. Although this concept was fundamentally mystical (numerology), Jabir did stress the concept of balance and quantitative mesurement in his alchemy. He also introduced the concept of the *elixir*, probably Chinese in origin, which was foreign to Greek alchemy. The elixir can be vegetable, animal, or mineral (in mineral form it is known as the Philosopher's Stone), and it is the agent of transmutation. In a way the elixir is a mystical rendering of the modern chemical catalyst! For the Chinese the elixir was used to "cure" the metals, converting imperfect metals to gold by bringing about the proper proportions in their natures. To Jabir the elixir also symbolized an inner transformation of the human being, bringing the inner spirit out of the frozen crust of the outer form. Thus the elixir creates in the microcosm the perfection that it achieves in the macrocosm. Likewise, the four qualities give rise to the two basic substances, sulphur and mercury, which correspond to the principles of male and female. Again we feel the influence of China, where the two basic principles of existence are called the yin and yang—the active and passive, earth and heaven, man and woman. The Chinese *I Ching* (*Book of Changes*) incorporates the principle of change into a grand mathematical system based upon combinations of the yin and yang.

The science of numbers according to the Brethren of Purity was the very heart of Universal Unity and substance in the Pythagorean sense. It was the "first elixir" and the highest alchemy. Thus they stressed the study of mathematics as a means to uncover the ultimate reality and fabric of nature. Like Jabir, the Brethren saw beneath the changing Aristotelian qualities a fundamental mathematical harmony which reflected the unity and oneness of Allah and His creation. That their mathematics was basically Pythagorean mystical numerology does not reduce the importance they assigned to it in the study of nature; on the other hand, the magical and mystical features of their mathematical meditations on nature made it more than simply a useful tool.

The Arabic predilection for mathematics led to some important advances within the field itself. Hindu symbolism, the practice of using a separate symbol for each number from 1 to 9, found its way into Arabic texts. Positional notation in base ten had become standard among the Hindus, and zero was treated as a number (the Greeks had simply used it to represent the absence of number). The Hindus had also introduced negative numbers and even developed ways of operating with irrationals. Hindu arithmetic was basically independent of their geometry, and so they simply ignored the logical problems of using irrational numbers in their calculations. Without compunction and unencumbered by the philosophical distinctions which had obsessed the Greeks, the Hindus blithely applied to irrational numbers the same procedures used for rationals. The Arabs adopted the Hindu numbers and method of treating irrationals, although they rejected negative numbers.

The Arabs are probably best known, however, for their algebra. In the ninth century the mathematician al-Khwarizmi wrote a number of important mathematical and astronomical works, including a treatise *Al-jabr w' al muquâbala*. In this context the word "al-jabr" meant "restoration"—in effect, to restore the balance of an equation by elimination. For example, $2x^2 + 5 = x^2 + 30$ can be simplified by the restoration $2x^2 = x^2 + 25$, and further simplified by $x^2 = 25$, which can then be solved, $x = \sqrt{25} = 5$. The example also demonstrates the meaning of *w' al muquâbala* or "simplification." The Arab algebraists did not u se such symbolism; rather they stated their problem in words. Al-Khwarizmi refers to the unknown as a "root," like the root of a plant, from which comes our term.

Interestingly, al-Khwarizmi justified his solution of quadratic equations by geometrical arguments, which probably indicates Greek influence (Euclid, Archimedes, and Apollonius were all translated into Arabic and well known). Omar Khayyam, who is probably more famous for his poetry, took an important step by solving cubic equations geometrically—using hyperbolas, ellipses, and parabolas and their various intersections. In the long run the combination of the two techniques and the Greek emphasis upon deductive geometrical proofs would severely limit Arabic mathematics, but their algebra found its way into the West in the twelfth century. The importance of Arabic numerals (from the Hindus) and algebra in western mathematics is obvious.

The Arabs were quick to master Ptolemaic astronomy, improving observations and calculations. Rather than using the table of chords, they computed trigonometric tables as arithmetical identities. They constructed tables of sines and cosines, adding the ratios which are today known as tangents and cotangents. In the late ninth century the astronomer Abul-Wefa introduced two new identities, the secant and cosecant, thus completing the set of four basic trigonometric functions. In the eleventh century al-Biruni applied spherical trigonometry to the determination of geographical longitude, thereby founding the science of geodesy (recall Isidore). The Arabs constructed observatories and invented the instrument known as the astrolabe. The astrolabe is a kind of celestial computer, like a slide rule, and with it the Arabs were able to determine with far greater accuracy celestial altitudes, times, and even the heights of mountains and the depths of wells.

In studying the Ptolemaic system, the Arabs asked: "Are these mathematical spheres physically real?" "Yes, of course," answered Thabit ibn Qurra in the ninth century. "But how? How do we make them agree with Aristotle's celestial machine?" The astronomer al-Haythan (Latin Alhazen), who lived from 965 to 1039, had the answer. The planet may be found to lie between a giant convex circumference and a smaller concave circumference, both concentric to the earth. Between these two, we can distinguish solid orbs, eccentric to the earth, which can be made to carry within their concavity a spherical epicycle. Alhazen also added a ninth "starless" sphere, assigning to it the diurnal rotation carrying all the other spheres. The sphere of the fixed stars, he said, was responsible for precession, moving one degree in a hundred years (the modern figure is 72).

The Koran says that all things perish save the face of Allah. Al-Biruni studied such change in the material face of the earth itself, the records preserved in rock strata. He determined that the Arabian Steppe was once a sea and that the Ganges Plain in India was a sedimentary deposit. Great changes had occurred upon the earth *before* the creation of humans, and he recognized that fossils might very well indicate the extinction of species. But the concept of the Great Chain of Being was a sacred theory embedded in the very idea of nature as a divine hierarchy, so such observations did not lead to any theory of evolution. However, the Muslims did consider the Chain to be one of temporal succession in which some forms of animals precede others in different cosmic times while the Chain itself remains above time. Therefore the existence of fossils proved to be no problem for al-Biruni nor did it threaten his religion.

Ibn Sina (called Avicenna in the West) was certainly one of the greatest Arabic scientists and philosophers. Born in 980, he supposedly mastered *all* the science of his day by the age of sixteen! Indeed, by the time of his death in 1037, the range of his writings was so vast that we may well believe his youthful propensity for learning. He often compared the universe to rays of the sun; particular beings were points of these rays issuing from the source of Being, God, just as the sun illuminates the world. Thus in certain areas of his science he was ready to follow Aristotle or Galen as he did in his medicine, yet his overall cosmology drew upon Neoplatonic ideas. He actually integrated the two in order to achieve conformity with Islam.

Ibn Sina was no slavish follower of Aristotle. In discussing projectile motion he takes up the cause of Philoponus. The projectile receives from the mover an "inclination" or *mail* which resists gravity and change in the motion. Violent *mail* or "impressed force" differs according to the weight of the body, and in the absence of resistance *mail* and its generated movement would continue indefinitely as in the void. Gravity is simply "natural" *mail*. Now, his statement positing the indefinite persistence of motion in a vacuum is certainly a departure from Philoponus and almost sounds inertial. Yet, the concept of *mail* is still within the Aristotelian qualitative tradition of dynamics, not the quantitative and abstract Newtonian inertia. What is more, the word *mail* actually means "desire," and ibn Sina also considered the existence of a psychic *mail* which linked motion to the love of God. Psychic *mail*, in fact, is almost identical to Aristotle's Final Cause—love of God, which moves the spheres, is the Prime Mover in the universe.

The views of Philoponus were also considered by another Muslim philosopher, ibn Bajja (called Avempace in Latin), during the twelfth century. He believed, like Philoponus, that the medium resisting violent motion is a factor to be *subtracted* from the motion. Such doctrines would eventually reach the West through the Muslim commentator Averroes (we shall hear more of Averroes in the next chapter) and would become topics of debate down to Galileo. Indeed, Galileo's mechanics has a long history, which in some ways reaches back to Philoponus and the Arabs.

Through the agency of ibn Sina, the medical tradition of Galen and Hippocrates was also transmitted to the West, but again with the particular Muslim flavoring. While ibn Sina systematized the basic medical principles of both, his starting point is the premise that the human being is a spiritual, psychical, and physical entity whose health is tied to the harmony of all. Now, ibn Sina held to the doctrine of the four humors and the four qualities which underlie the humors. Health, however, is not simply a question of body function—the balance of the physical humors—it is also a question of the individual's temperament, uniquely determined for each person by the mixture of qualities and affected by a variety of circumstances: climate, food, rest, exercise, even emotional or spiritual states of mind. While the physician must be aware of anatomy—which for ibn Sina was Galenic since Islam disapproved of human dissection—it was more important to know the individual's specific temperament. It is interesting to note that the Muslim physicians, with their religious emphasis upon the unicity of nature, would take such a highly individualistic view of treatment in medicine. The general tendency of modern medicine is the opposite: to apply generalized cures on the assumption that all bodies are basically alike. And surgery for the Muslims was a radical cure to be used sparingly; the individual should be coaxed and aided into a restoration of health. The mental and physical states were of equal importance, and in this sense we return to the unity of nature reflected by the unity of the microcosm.

It might well be argued that the flowering of Arabic science had much to do with the Arabs' geographical position and the availability of the ancient sources. It is true that they were interested in the science of the physical world, yet their motivation to study nature was different from that of the Greeks. Their obsession was not to understand the cosmic garden for itself, but rather to find within its workings evidence of the divine gardener. Thus, while they improved and expanded their inherited Greek tradition, the Arabs lacked any real motivation to drastically alter it. Their purpose was not to construct a better physical science, but to demonstrate how the science they possessed could lead to the ultimate goal of all knowledge. As it did for Christians, the transcendental world actually determined for them the goals of this-worldly science. Perhaps this was the key to the dilemma of reason and revelation: to use reason in order to exemplify the truths of faith based upon the assumption that the garden and the gardener were one. In this way the Arabs might be called the first scholastics.

Yet, there may be something deeper here. Pierre Duhem noted that the Christian assertion of a single divine law governing the universe accomplished, by the way of theology, a unification of the celestial and terrestial realms whose separation had been the basic premise of Aristotle. The unity of natural law was implicit in the creative act of God. Without pronouncing judgment, is it not possible to see this idea more clearly in Islam? Nature, both earthly and celestial, is a metaphor of Allah's oneness. A differentiated universe operating by seemingly diverse natural laws is only a first approximation, for beneath this flow exists the supreme reality, which is no less than the will of Allah. Earth is not heavy by its own nature—natural phenomena follow laws which, at heart, are not found within the phenomena themselves. Modern science need not ask *if* nature is one, wrote Henri Poincaré, but *how* it is one.[1] By the Will of Allah, answered the Arabs.

Of more immediate importance to us is the transmission of Arabic science to the West. In comparison to that of the Arabs, the natural science of the West was indeed meager. Yet in the twelfth century the pic-

---

[1]Henri Poincaré, *Science and Hypothesis* (New York: Dover, 1952), p. 145.

ture changed drastically. Historians have labeled this change the Twelfth Century Renaissance.

Consider the level of learning in Europe before the twelfth century. A great deal of the "old logic" of Boethius has been lost; what formal education one can find is inside the monasteries. Spain is in the hands of the Muslims, as is Sicily, where the Arabs ruled from 902 to 1091. Indeed, Christendom is in close proximity to Arab learning, yet until the twelfth century there is little contact. We see the lonely figure of Gerbert of Aurillac, the future Pope Silvester II, traveling to Barcelona to study Arabic mathematics and astronomy. It is the end of the tenth century when he returns to Rheims, where he "dazzles" his contemporaries with his newly acquired knowledge. But few follow his footsteps to the source. The century turns, and we find a great debate being waged over the question (inherited from Boethius) of the reality of universals. We hear of Anselm studying logic in order to show the noncontradictory nature of revealed truth. He even uses his logic to prove the existence of God: where does the idea of an infinite Being (in a finite world) come from if not from that Being Himself? Among many there is a growing interest in philosophy.

At the end of the eleventh century the Crusades begin, and while the crusaders are not scholars but men of action, they do bring back tales of the rich civilizations in the east. Toledo has fallen to the Christians in 1085; twenty years earlier Sicily had fallen to the Normans. Slowly Latin contact with Arab culture is increasing; it is not an intellectual contact—it is hostile and warlike—but it is contact nonetheless.

The twelfth century opens, and the use of philosophy increases. One of the most important textbooks of the age is Peter the Lombard's *Sentences*. Reason aids us, says Lombard, in faith, for the image of God is found in the mind even before we become believers. Also by the twelfth century the School of Chartres has become a center of Platonism. God worked in the world through Platonic "images," and the world is a reflection of these images (ideas). In short, the Western scholars were literally wringing dry the precious sources of Greek philosophy they possessed.

The twelfth century also witnessed the highest point in the development of the medical school at Salerno in southern Italy. As early as 1077, Constantine the African came to Salerno and translated the medical texts of the Arabs. Besides Galen and Hippocrates, Constantine may have introduced some of Aristotle's biology, for he believed that the practice of medicine was related to philosophy. Thus Salerno may have been one of the first purely liberal arts schools, and by the twelfth century the medical course also included three years of logic. It was the harbinger of the twelfth-century universities.

We find the first use of anatomical texts at Salerno as demonstration aids for dissection (mostly of pigs). In one such text we are told that the term "anatomy" means "correct division." Pathological and physiological discussions are intermingled, along with asides to philosophical and even cosmological significance of certain organs. The authors of the texts are not hesitant to deny or correct other anatomists, although their respect for the authority of the ancients is still very great.

The teachers at Salerno were aware that on some points Galen and Aristotle did not agree. Galen says, for example, that the brain is the center of sensation and motion, but Aristotle says the center is the heart. The Salernian author points out that the varied states of the heart are accompanied by variations of moral virtues, which are internal characteristics of the animal, and that motor and sensory functions are external characteristics dependent upon the internal. Thus the heart is the source of them. Yet, the author does agree with Galen that the liver is the seat of digestion. Another text states that the heart is the center of the body as the sun is the center of the world!

A very useful piece of information deals with heart size. Animals with hearts that are large in proportion to their bodies are usually timid and fearful, since their heart is not as hot as a smaller one confined to a narrow space and boiling fiercely. Such hot-hearted creatures are bold. But lest we be misled, the boldest are those animals which have both large hearts and great heat. As we all know, the greatest courage is the province of great hearts!

Salerno may be used to illustrate one of the reasons for the outburst of interest in Arabic learning during the twelfth century. Accurate anatomical texts greatly enhanced the practical work of the physicians, and such texts were to be found among the Arabs. The same is true for other areas of knowledge. In an age which considered astrology to be applied astronomy, nearly every ruler had a court astrologer and was interested in accurate observations. The Church was also interested in astronomy and the construction of accurate tables for the dating of religious holidays. By the twelfth century, for many reasons, Western Europe was well prepared to receive the science of the Arabs—and receive it did.

Toledo became one of the chief centers of translation. Its Archbishop Raymond actually established a school of translation, and at Toledo one of the greatest translators of them all, Gerard of Cremona, made translations of more than 72 Arab works. Spain quickly became the center of astronomy, and tables computed with Toledo as meridian became standard throughout Europe in the thirteenth century. Arab instruments and methods played an important role in the preparation of these tables. Gerbert had revived the abacus, the old Roman counting table, yet the use of Roman numerals made it too clumsy for astronomy. Then, in 1126, Adelard of Bath brought out Arab trigonometric tables, and in 1145 Robert of Chester translated the *Algebra*. Later, in 1202, one of the greatest mathematicians of the age, Leonard of Pisa, began to write his own treatises on algebra. In the course of the century Arabic numerals, methods of computation, and even instruments, such as the astrolabe, had all come to the West.

In Sicily and southern Italy the newly established Norman kingdom encouraged translations. Here some knowledge of Greek persisted, and a few manuscripts were brought from Constantinople. Norman rulers themselves took an interest in the new science, and in the thirteenth century the remarkable Frederick II actively pursued knowledge. Frederick's book on falconry was highly critical of Aristotle, castigating the Greek philosopher for relying too much on hearsay and not enough on personal observation. Frederick's menagerie, his collection of exotic animals, was the wonder of the age. He was indeed an exceptional man, impatient with textbook learning, wanting to see things for himself. Legend records him shutting up a man in a winecask in order to prove that the soul died with the body. He had two men disemboweled alive in order to examine the effects of rest and exercise upon digestion.

Some of the translators found in the materials a stimulus to personal inquiry. Adelard of Bath, who was instrumental in giving Arabic mathematics and astronomy to the West, sought to explain this new science in a work entitled *Natural Questions*. The book is a collection of 76 problems ranging from natural history to physics, a combination of the old medieval lore and the new science. In some ways it recalls the Roman encyclopedists, yet Adelard is not afraid to mix authority with his own speculations.

Natural Questions includes references to the *Timaeus*, as one might expect, but Adelard also cites Aristotle's *Physics*. He is perhaps one of the first Western writers to do so. Like Aristotle, he denies the vacuum, saying that the sensible world is a plenum of the elements. Then he describes an experiment with a vessel full of water in which a hole is made at the lower end and the upper opening is sealed. After a certain interval, the water comes forth because liquid is porous and admits air into the upper portion, says Adelard, and the air replaces the evacuated water as the elements readjust themselves to their natural places. To Adelard this is experimental proof that nature abhors a vacuum. Also, a body falling through a shaft dug through the earth will stop at the center. The reason is the principle of sympathy—earth shuns fire, and the upper space around the earth is fiery in all directions, so earth (the falling body) seeks the point which on all sides is equally distant from the surrounding fire. The earth is, therefore, the center and the bottom of the universe. Centuries later Newton would show that a homogenous sphere attracts a body as though its entire mass were concentrated at its central point, but his principles of gravity (not causes) were certainly not occult ideas of sympathy.

The twelfth century also saw the rise of universities, which replaced the monasteries and cathedral schools as the centers of learning. The intellectual revolution was thus accompanied by an institutional one. The word "university" originally signified a corporation or guild, a society of masters like a craft guild. The universities of Paris and Oxford probably had their origins in the twelfth century. Bologna became famous for its school of law, Montpellier for its law and medicine. In the early thirteenth century we find universities at Padua and Naples in Italy and at Salamanca in Spain; in England we find the origins of Cambridge. No one was allowed to teach without having followed the school of a master, and slowly the various faculties—theology, law, medicine, and the liberal arts—were incorporated into distinct bodies. The modern university is a direct descendant of the medieval.

Such was Western Europe in the twelfth century. The "new logic" of Aristotle had replaced that of Boethius. Scholars now had in their hands his scientific treatises, together with the Arab commentaries on them, which served to round out the great Aristotelian system. Ptolemy, Euclid, Archimedes, Galen, and the Arab scientists slowly found their way into the curriculum of the Faculty of Arts in the new universities. The new materials could be confusing, owing not merely to their intrinsic nature (consider the difficulties the Ptolemaic system would present) but to the very fact that they had been translated from what were essentially translations of the original Greek sources. Passing through so many hands and over such a great period of time, the very technical content of the science itself was in some cases seriously altered. And there were very few in Europe who initially possessed the expertise to master the rigors of a Ptolemy or an Archimedes. Still, the new science was a vast improvement over the old.

Yet, consider the deeper problem such an infusion created. Certainly medieval scientists were a diverse lot—arguing, criticizing, holding contrary opinions in many areas and defending them with great vigor. But at least technically they were all Christians, members of the Universal Church. While the rational common sense of the Aristotelian system appealed to them, they now had first-hand evidence that it contained many propositions clearly contrary to their faith. The twelfth-century Jewish philosopher Moses Maimonides, attempting to synthesize the Hebrew tradition and the science of Aristotle, entitled his work *The Guide for the Perplexed*. Indeed it was a perplexing situation.

SUGGESTIONS FOR FURTHER READING

CORNER, GEORGE W. *Anatomical Texts of the Earlier Middle Ages: A Study in the Transmission of Culture.* Washington, D.C.: Carnegie Institution, 1927.

HASKINS, CHARLES HOMER. *Studies in the History of Medieval Science.* New York: Frederick Ungar, 1924.

————. *The Renaissance of the 12th Century.* Cleveland: World Pu blishing Company, 1957.

NASR, SEYYED HASSEIN. *An Introduction to Islamic Cosmological Doctrines.* Boulder: Shambhala, 1978.

————. *Islamic Science.* Kent, England: World of Islam Festival Publishing Co., 1976.

PETERS, F. E. *Aristotle and the Arabs: The Aristotelian Tradition in Islam.* New York: New York University Press, 1968.

9

$\diamond\!\!\!\diamond\!\!\!\diamond\!\!\!\diamond\!\!\!\diamond\!\!\!\diamond\!\!\!\diamond\!\!\!\diamond\!\!\!\diamond\!\!\!\diamond\!\!\!\diamond\!\!\!\diamond\!\!\!\diamond\!\!\!\diamond\!\!\!\diamond\!\!\!\diamond\!\!\!\diamond\!\!\!\diamond$

The Errors
of the Philosophers

"But these Greeks hardly agree, with each other or even with themselves!" It is easy to imagine something like this statement, a faint echo from the thirteenth century, drifting to us across the ages. Here is our phantom scholar contemplating Ptolemy and Aristotle. Aristotle's spheres are concentric, physical, his heavens mechanical—a ponderable (aether *is* an element) machine. Ptolemy talks about epicycles, eccentrics, equants (whatever they are), which are nowhere found in Aristotle. Yet, Ptolemy uses Aristotelian physical arguments to prove that the earth lies in the center of the universe at rest. But then he gives the impression that his complicated mathematical constructions are only meant "to save the phenomena!" And here are the Arabs—oh!

Our ghostly scholar is also a Christian, maybe even a theologian. What is this? Aristotle rigorously demonstrates the eternity of the world; his immortality of the soul is practically nonexistent, his Prime Mover a mere shadow! How reassuring it is to know that the laws of nature are rational and completely regular—but doesn't this eliminate miracles? No accident or property may exist apart from substance. So how can the bread and wine of the eucharist *actually* become the body and blood of Christ while still maintaining the superficial appearance of its natural substance? Then there is Averroes—oh!

Wealth can sometimes be a burden, even intellectual wealth. Coming from so many sources, the new science was certainly inconsistent on many points; even Aristotle could be vague and seemingly contradictory in his details. And the great commentator, Averroes, did not help matters.

Averroes (ibn Rushd) was born in Cordova of a distinguished legal family in 1126. He was initially interested in Greek medicine, but soon Aristotelian philosophy became his dominant passion. And what a passion it was! When Aristotle spoke, it was as if reason itself had pronounced judgment, for Averroes saw in Aristotle a man who had achieved the highest reaches possible to the human mind. The main task, as Averroes saw it, was to clarify and explain the system in its purity, even the points upon which Aristotle himself was vague. All criticisms of the beautiful system had to be answered.

Thus Averroes accepted without question the eternal universe in which matter, motion, and time have no temporal beginning or end. Ptolemy's system may be a convenient way of "saving the phenomena," yet true astronomy must be based upon physical principles—Aristotle's spheres. Personal immortality is an impossibility; motion in the void is absurd—Averroes argued each point as he believed Aristotle would have.

Although Averroes himself was a Muslim, he made no effort to reconcile Aristotle with his orthodoxy, nor did his reason bow down before his faith. Indeed, he seemed to hold the two apart, coming perilously close to what in the Middle Ages was called the doctrine of the double truth: two incompatible assertions are held to be true at the same time.

It is possible that Averroes did not believe in the doctrine of the double truth, for in a treatise which did not reach the West (until the nineteenth century) he posited three types of people: the masses who accept the authority and literal word of Scripture, the theologians who are satisfied with probable arguments, and the scientists who require the absolute demonstrations of reason. The latter he warned to keep their demonstrations to themselves, lest they destroy the faith of the former. The purpose of revelation is to teach right practice and knowledge of God. On the other hand, the duty of the elite is to remain silent, refraining from allegorical interpretations which serve only to confuse the undisciplined and the uneducated.

Yet, silence in Western Europe was hard to come by. Averroes himself had sharpened the contrasts between Aristotle and faith, and his own enthusiasm for such a beautifully rational and complete system had done much to hasten its integration into the universities, especially into the liberal arts faculties. Aristotle became "The Philosopher" and Averroes "The Commentator." Some, like Siger of Brabant, were completely swept away and held that the obnoxious principles which were in opposition to faith could not be disproved by reason and were, in fact, necessary truths. Revealed truth could not, therefore, be the province of reason, only of faith. All else, the science of the physical world, belonged to reason. For many this was really the doctrine of the double truth, and Siger and his like were labeled Averroists who simply paid lip service to faith.

All this did not sit well with the theologians. After all, the theologians themselves employed Aristotelian logical concepts to elucidate articles of faith, following the older tradition if not the Greek heritage of Christianity itself. Deductive logic demonstrates truth by *necessity*. Hence if the truths deduced from faith were necessary, the contradictory truths of natural science must be merely probable. But this is to lose all scientific significance, for science becomes nothing more than a plaything of the human mind.

The tension between the theologians and the Masters of Arts increased during the thirteenth century. The theologian Bonaventure, who himself developed a metaphysical system which held that all forms of knowledge were "handmaidens" of theology, attacked such doctrines as the world's eternity and the impossibility of personal immortality. Giles of Rome published a book entitled *Errors of the Philosophers* which criticized Aristotle, Averroes, Avicenna, and even Moses Maimonides. The end result was the great Condemnation of 219 propositions issued at Paris in 1277 by bishop Etienne Tempier (we shall treat the specific articles of the Condemnation in Chapter 11).[1]

Yet, we should not lose sight of the fact that besides being a highly integrated and satisfying system, Aristotelianism *could be made compatible* with Christianity. This was the task of scholasticism: to effect a separation of Aristotelianism from its unchristian elements, granting its autonomy in questions of physical science while integrating it into a larger metaphysical whole. Theologians like Albert the Great and Thomas Aquinas felt themselves free to criticize specific points—unlike Averroes, who accepted everything the Greek philosopher had uttered—yet their comprehensive reconciliation created a definite reluctance in themselves and in others to reject the entire edifice. The Condemnation of 1277 directly influenced the scholastic discussion of Aristotelian science in the fourteenth century, seriously undermining some of its most basic principles. Yet this occurred only after the attempt had been made to Christianize Aristotle, and this program did have an influence upon the course of medieval science.

Albert of Cologne, better known as Albert the Great (Albertus Magnus), was a man of broad interests and keen powers of observation. He

---

[1]Earlier in the century, in 1210, the provincial synod of Sens decreed that Aristotelian books on natural philosophy were not be be read in Paris. In 1231 Pope Gregory IX modified the ban and ordered the treatises purged of error, and in 1245 Pope Innocent IV extended the ban to the University of Toulouse. Yet by the time of Giles and his *Errors*, between 1270 and 1274, it appears that all of Aristotle's available works were included in courses at the University of Paris.

may have received a medical education at Padua, which probably influenced his writings, yet he also joined the Dominican Order and took a degree in theology at Paris. Albert hoped to establish the validity of Aristotelian science within its own limits, cutting away the things which clashed with his faith, and grafting on his own observations where Aristotle had left gaps or was wrong. His *Book of Minerals*, for example, was the first real systematic attempt to establish a science of minerology. It was written, as Albert himself says, because of the general lack in both Aristotle and Avicenna. The scientific premise is Aristotelian, faithfully adhering to the doctrines of natural place, natural movement, the four elements, and the four causes. Yet, the subject matter is unique, and there is a great deal of what can only be called Albert's own personal observations. He often concluded his discussions by stating that "This is a matter of experience" or that a certain point could be proven by anyone "who wished to experiment."

Albert clearly held that alchemical transmutation was not a matter of natural science, for it did not depend upon scientific demonstration. Rather, it is a matter of experience in the occult and the supernatural. In a separate treatise dealing with alchemy, Albert described how he had studied the books of the alchemists and found them devoid of knowledge. Yet he did seem to accept the possibility of transmutation, and he possessed a working knowledge of some basic compounds as well as an appreciation of practical experience. He may have understood the principle of incubation, for he knew that a dung pit could be used as a natural incubator (the reason, which he did not know, is that the thermophilic bacteria produce a temperature of 50 to 70°F). Albert even gave some practical advice to would-be alchemists: the alchemist should be patient, not associate with politicians, and begin operations with plenty of funds.

His *Book of Minerals* shows that Albert apparently accepted Avicenna's account of quicksilver and sulphur as the basic principles producing all species of metal—a common belief in Arab alchemy from the time of Geber. Therefore transmutation is like medicine; it "heals" the metals of their corruption, and the alchemist is like the physician who assists nature in the restoration of health. It is not art but nature itself which performs the work—Albert even compared the elixir to yeast in bread. Thus, while Albert has said that alchemy is an occult art, he has yet given a *natural account* of its process; in essence, his "magic" is based upon natural causes. Like Aristotle, Albert seems to be telling us that nature operates according to necessary laws. As a Christian he must accept miracles; on the other hand, it appears that God (and the alchemist) works magical events through natural causes. While we may not comprehend the Divine Will, we are still free to investigate the workings of that Will upon physical phenomena. Hence it seems to be the *essence* of the occult which lies beyond the realm of science and not its phenomenal instrumentation in nature.

Albert's scientific works covered a vast range of subjects: geography, biology, botany, astronomy, optics, and mathematics. The "otherworldly" label which is often applied to the Middle Ages, implying that medieval people had little interest in nature, seems to be inapplicable to Albert. In fact, the statement may not apply to medieval people in general. Look at the great cathedrals, says historian Lynn Thorndike: the artists who chiseled the figures of animals, plants, and birds knew nature better than the scholar (of course, they carved mythological beasts too). With reverence and faithfulness they carved into stone their observations of the natural world. They were "Darwins with a chisel."[2] Yet we must not forget that they were still building cathedrals and not theories. Albert was a theologian, and his metaphysics was as important to him as his science. Like the great cathedrals, his thought encompassed both realms.

Because Albert did not construct a system, his fame has been eclipsed by his brilliant student, Thomas Aquinas. Aquinas has come down to us as the scholastic *par excellence*, the first to truly purify the Aristotelian system and create a vast, specifically Christian metaphysic in which natural science and revelation were each assigned their autonomous spheres while ultimately springing from the same principle. For Aquinas there simply cannot be two separate or contradictory truths; all knowledge and hence all truth come from a single source.

In Aquinas's view, every finite being is made up of act and potency, and the essence of what a thing is exists only in the particular actuality of the existent being. Being takes its name from the verb "to be," an *act* of being. Physical being as an act of becoming (potential) must be contingent upon a cause, for an uncaused contingent thing is a contradiction. But in every finite being essence comes into being with existence and depends upon it. The entire physical universe, therefore, hangs upon an act of being, and this act presupposes a totally subsistent Being in which essence and existence are one—God. All potential qualities exist in God actually, thus He is fully transcendental. Yet, the act of Being which infuses these qualities into the world makes for an immanent creator. All must reflect the Creator, since all creatures participate in being, though in varying degrees. Human knowledge of being derives from the senses and the rationalistic science of Aristotle. But God as pure Being is the source of both movement and existence; in short, God is the radiating center of reality. Hence if nature is rational, God must be rational, and in the end all knowledge must derive from Him.

[2]Lynn Thorndike, *A History of Magic and Experimental Science*, Vol. II (New York: Macmillan, 1929), pp. 536–37.

Underlying all science, in whatever form it assumes, is the ultimate principle of being—existence as such. The sciences study specific beings—parts of Being, so to speak—and therefore should not be confused in either methods or content with that science, metaphysics, which studies existence *per se*. Thus metaphysics does not depend completely upon the content of Aristotelian science, for no single scientific methodology could monopolize the sum total of Being in all its complexity. Aquinas wrote many commentaries upon Aristotle's physical works, amending the philosopher when he conflicted with Christianity, but also defending his positions. Perhaps it is true that Aquinas established metaphysics and science as autonomous realms, answering fundamentally different questions, yet totally compatible under his analysis of Being. However, we must not forget that this synthesis was accomplished by his basic acceptance of *Aristotle's* scientific principles. Historically, at least, the two realms were interdependent.

It is no wonder, then, that Aquinas in adopting Aristotelian logic to a profound metaphysic would see these same principles operating in the physical world. Thus the external form of his universe is Aristotelian, and its inner significance on the activity of pure Being is his own reconciliation of reason and revelation. Unlike the French mathematician Laplace five centuries later, Aquinas in his science still had need of "that hypothesis"—God's action in the universe.

In the *Summa Theologica* Aquinas adopted the opinion of Maimonides that Ptolemy's constructions are hypotheses to save the appearances but that it is possible they may be saved some other way. Some modern Thomists regard this as a distinction between probable empirical theories and necessary philosophical demonstrations. Yet, the assertion may also be meant to rescue the material Aristotelian spheres. It is significant that Aquinas distinguished three heavens: the Empyrean, the crystalline (the firmament of Genesis), and the sidereal sphere, which consisted of the seven planets and the fixed stars.

The most famous textbook on astronomy in the thirteenth century, the *Tractatus de sphera* of John Sacrobosco, uses only a single *Primum Mobile* as the outermost sphere which moves east to west. Sacrobosco also uses eccentrics, epicycles, and equants for every planet except the sun. Bernard of Verdun accepted these constructions but held that they account for the variation of the *material orbs* which carry the planets and not the planets themselves. Like the Arabs, he assumed that the epicycle is located within the thickness of the eccentric orb. Aquinas, too, may have sought to reconcile Ptolemy and Aristotle by adapting the reality of the Aristotelian spheres to the hypothetical geometrical rendering of the phenomena by Ptolemy. Aquinas also considered the theories of Heraclides of Pontus and Aristarchus of Samos, only to reject them by Aristotelian arguments.

Aquinas, of course, could not follow Aristotle and Averroes in positing an eternal world without beginning or end. Nonetheless, he did see in their arguments a certain persuasiveness and thus decided that the question did not fall into the domain of science. Aristotle's arguments, according to Aquinas, do not conclude with strict necessity, yet the arguments raised against him are not necessary either. Infinity and eternity are simply beyond the realm of physical science, since we experience neither in the physical world.

Then there was the problem of motion in the void. One of the most influential treatises of the Middle Ages was Averroes' Commentary on the *Physics*, specifically Comment 71 on Book IV. Here Averroes stated the Aristotelian problem of motion and tied it to the rarity and density of the medium in terms of a ratio. There was some confusion as to the nature of this proportion. If we take the density of the medium, the proportion will be inverse ($V/V' = R'/R$), or if the rarity of the mediums is compared the proportion will be direct ($V/V' = R/R'$). Nonetheless, without a medium we would have the unthinkable—instantaneous motion.

The most important part of the Commentary, however, is Averroes' account of Avempace's theory that the medium is simply to be subtracted from natural movement. One argument raised by Avempace, as reported by Averroes, is that since the circular movement of the celestial bodies occurs in the heavens where there is no resistance, then, if we accept Aristotle's ratio, the stars should move instantaneously—which they obviously do not.

Averroes objected to this by calling upon the relationship between mover and moved in natural motion. In the celestial bodies and in animals the efficient cause of motion is the form (incorporeal intelligences in the spheres), which is wholly distinct from the moved material. Resistance in the heavens arises from this relationship. On the other hand, the form of an inorganic (simple) body is not distinct *actually* from that body, and the form *does not* act upon its matter as in the heavens. To imagine motion in the void would be to treat the form of an inorganic body as a distinct reality separate from that body. In fact, Avempace is a Platonist, admitting the existence of an indwelling separate force which moves a body. The condition of being moved is being moved by something external to the body through contact, and this is accomplished by overcoming external resistance. Because the gravity of a body is intrinsic to that body, the body itself does not resist its own weight but requires a resisting medium, as common-sense experience shows. To explain physical motion by ideal or imaginary power and in impossible situations (the void) is, for Averroes, against all dictates of science.

Surprisingly, when Aquinas came to this commentary he supported Avempace's theory without naming him! One of his arguments for Avempace was entirely physical. Like a plenum, a void is extended and dimensional, hence movement from one distinct point to another requires that parts of the void be traversed in succession, and this takes time. Now Aquinas still followed Aristotle in holding that the medium is necessary for violent (dynamic) motion—in order to continue it. Nonetheless, he took an important step by defining the state of "being in motion" as a condition of the mobile body in relationship to a spatial frame of reference.

Aquinas remained a metaphysician and did not follow through his own implications. To ask why a heavy thing moves downward, he declared, is to only ask why it is heavy. The inclination (Aquinas says *impetus*) to natural motion exists in the body *per se*, and it is a passive principle or *potency* to move which is imparted by the "generator" (Creator?) which makes bodies heavy or light. In fact, it appears that Aquinas understood Averroes as having placed separate intelligences in the spheres as the sole efficient causes of their movements! While Aquinas may accept the infusion of intelligences or angels in the spheres, he cannot assign them the basically creative power of God. Forms are the *means* by which natural bodies are moved, and forms are integrated into matter through the bestowal of existence by God. Again, to say that it is impossible for the void to exist places some rather severe limitations upon the power of God. Can we not say, therefore, that the motivation behind Aquinas' scientific demonstrationss—theories which may seem to us to be leading in fruitful directions—springs from his metaphysics? Thus the very content of his science is on a certain level intertwined with his metaphysics. The two are not as autonomous as even he himself may have believed.

Near the end of life (he died in 1274) Thomas Aquinas had what can only be called a mystical experience. All of his written works, he said, were as "straw" in comparison to what was revealed. There is an interesting alchemical treatise entitled the *Aurora Consurgens (Rising Dawn)* which some historians have attributed to him. Its mystical symbolism suggests that, if indeed Aquinas wrote the treatise, it may have been somehow connected to this experience. The elements and the alchemical processes are all presented in a vague and mystical language combining religious forms with chemical operations. In fact, the author of the *Aurora Consurgens* seems to be using alchemical terms as vehicles to express spiritual truths which have little to do with the ultimate goal of physical transmutation. Gold seems to represent the uncorrupted spirit, the divine idea or form imprisoned in corrupting matter, and the process of transmutation signifies the raising of the alchemist's soul, his intellectual form, to a mystical *gnosis*. The actual transmutation of gold seems to be but the physical counterpart of the real spiritual transformation. It is possible only because such a liberated soul can work magical effects in the physical world.

The psychologist C. G. Jung believed that the real goal of alchemy was to bring the unconscious mind of the alchemist into a union with the conscious mind. This desire was expressed through the alchemical language of opposites. The tension between the two aspects of the human mind, the unconscious mind which is mythical and irrational, and the conscious mind which is rational and scientific, is resolved by gaining a psychic equilibrium Jung labeled the *self*. The alchemist strove for this realization of the self, which was identified with a rebirth of the soul from the darkness of the material world. It was "a healing self-knowledge and deliverance of the pneumatic body from the corruption of the flesh."[3]

Might this not be said of Thomas Aquinas, even if he did not write the *Aurora Consurgens?* Perhaps this tension was true of the whole Middle Ages—the tension between faith and reason. By granting them autonomous domains, Aquinas in the end sought to unify the two through his concept of Being. Thus, the very content of his science, his Aristotle, however different from that of the Greek philosopher himself, made up the stone with which he built his metaphysical cathedral. Change the stone and the cathedral itself is altered. Or we could put it the other way around: to change the cathedral would require new stone.

SUGGESTIONS FOR FURTHER READING

COPLESTON, F. C. *Aquinas.* New York: Penguin Books, 1955.

GOHEEN, JOHN. *The Problem of Matter and Form in the De Ente et Essentia of Thomas Aquinas.* Cambridge, Mass.: Harvard University Press, 1940.

GRANT, EDWARD. *Studies in Medieval Science and Natural Philosophy.* London: Variorum Reprints, 1981.

MARLING, JOHN M. *The Order of Nature in the Philosophy of St. Thomas Aquinas.* Washington, D.C.: Catholic University, 1934.

THORNDIKE, LYNN. *A History of Magic and Experimental Science*, Vols. 2–4. New York: Macmillan and Columbia University Press, 1929–1934.

————— . *The Sphere of Sacrobosco and Its Commentators.* Chicago: University of Chicago Press, 1949.

---

[3]C. G. Jung, *Mysterium Conjunctions, Collected Works,* Vol. 14 (New York: Bollingen, 1963), p. 90.

10

~~~~~~

The Gate and the Key

Empiricism, Mathematics,
and Experiment

Those who desire to gain an education in one of the sciences today inevitably find themselves enrolled in what is commonly called a "lab science." This phenomenon is familiar to most students in modern higher education (and painfully so when it comes to scheduling the lab). You attend the lecture in which the basic theory and vocabulary of the science are presented; then you enter the laboratory to "do" the science. You learn to measure, experiment, and observe. At first the two aspects of the course seem rather incongruous. Under the microscope the strange forms can be disconcertingly at odds with the clearly drawn diagrams in the textbook or on the blackboard. Hopefully, if you persist, a pattern will emerge. And who can forget the sudden elation when the experiment succeeds just as the lecture said it would? Just as often, though, it fails—and that we never forget!

In a way, Aristotelian science was studied out of the textbook, and the basic activity centered around the exegesis of the text: disputes and discussions. The ultimate goal was to discover *why* things happened, their causes, and to find out what deductions could be made from accepted or inducted principles. Yet, there was another tradition in medieval science, vaguely similar to the lab part of our modern course. There was an emphasis upon observation, measurement, and even experiment. To be sure, it took a secondary seat to the study of the text, and its practitioners hardly got their hands dirty. It did, however, stress the procedures we would equate with the lab.

This method was called by Albert the Great the "Pythagorean" approach. It certainly had many ancestors besides Pythagoras, and we may simply call it the practical quantitative tradition of measurement. Observatories, like those in Spain and across the Arab world, were astronomical laboratories in which it was more important to insure the accuracy of tables than to speculate upon the reality of the spheres. Astrology, too, was an applied science. The Hermetic tradition, although fragmentary in the Middle Ages, gave rise to the alchemist's laboratory where experiments were conducted based upon the texts. There was the Archimedean geometrical tradition in statics, and of course the importance of weights and measures would increase with commercial interests during the later Middle Ages. In optics the mathematical treatment of light rays and the theories of vision were combined with Aristotle and the medical tradition of Galen.

In a basically agricultural society such as Medieval Europe, the most important use of Euclid's geometry naturally would be in surveying. Medieval mathematicians did learn to appreciate Euclid's proofs, and many versions of the *Elements* were circulated throughout Europe. But the purposes for which Euclidean geometry was studied often were defined by its possibilities of application. Besides surveying, Euclid offered valuable aid to the understanding of astronomy and optics. Even the geometrical references in Aristotle and Plato could be better understood via Euclid. Both the *Elements* and *Posterior Analytics* were held to be the best illustrations of true scientific methodology. The actual application of geometry to natural phenomena represented a kind of intermediate science, falling between the rigorous demonstrations of pure mathematics and the logical deductive causality of natural philosophy.

This "intermediate" science did capture the attention of some who studied pure mathematics. Leonardo Fibonacci of Pisa composed a work in 1220 entitled *The Application of Geometry*. Taken mainly from an Arabic text, which was derived from a commentary by Proclus on the *Elements*, the treatise supposedly represents Euclid's method of dividing figures into like or unlike figures. The popularity of Leonardo's book no doubt arose from the problems of surveying, in which the division of different plots of land was a major concern. Whether Euclid himself was concerned with such applications is highly doubtful.

Leonardo's father was a merchant and employed his son in commercial duties, which required Leonardo to travel. It was probably in the course of his commercial occupations that Leonardo encountered Arabic numeration and algebra. It is also probable that most Italian merchants were familiar with Arabic mathematics, since by and large their more important commercial contacts were with the East. Leonardo's *Liber abbaci* popularized algorism (arithmetic founded upon Arabic notation), which had attracted little attention inside the schools. We hear Roger

Bacon admonishing theologians to study the system in order to better acquaint themselves with the art of numbering.

Leonardo was not alone in the field. Both Jordanus of Nemore and John Sacrobosco composed works on the Arabic system, though these were probably less influential than Leonardo's. Jordanus treated linear and quadratic equations. He also employed symbolism derived from a simple shorthand method of using letters to represent line segments or to designate various ratios. Yet, when such developments were studied outside their practical applications, they were usually connected to problems in Aristotelian natural science.

What is called today "statics" was known in medieval times as the science of weights. Again, its practical applications would seem to lie in areas of commerce. There was, however, a theoretical tradition, derived from a pseudo-Aristotelian work (possibly done by Strato) called *Mechanical Problems*, as well as from the methods of Archimedes received through the Arabs. The most significant contribution of the medievalists was to combine these two traditions.

The Aristotelian dynamic tradition represented rest and motion as fundamentally opposed conditions, unlike modern statics, which views equilibrium as a special case of motion. Yet, in *Mechanical Problems* we find the principle of virtual displacements: What lifts a weight W through a vertical distance H will lift a weight KW through a vertical distance H/K. Jordanus applied this principle to the Archimedean law of the lever, adopting geometrical illustrations. He showed that the effective *force* of a weight on any lever arm, bent or straight, depends upon both the weight and its *horizontal* distance from a vertical line passing through the fulcrum (point of support).

Jordanus also treated the problem of positional gravity in the early thirteenth century. According to Aristotle a body has a natural weight and descends, if unobstructed, toward the center of the earth. This was generally known as "natural gravity." Jordanus demonstrated that positional gravity is a component of natural gravity acting along an inclined plane. When a weight is resting upon a plane whose inclination is constant, the ratio of its positional gravity to its free natural gravity, said Jordanus, is equal to the ratio of the vertical component of any given potential trajectory along the plane. In modern notation, Jordanus holds that $F = W \sin a$, where F is the force along the plane, W is the free weight, and a is the angle of inclination. In other words, a weight can be heavier *positionally* when the inclination of the plane is less oblique, and the force of gravity along the plane is inversely proportional to its obliqueness.

Indeed there was, as we can see, some application of mathematics to natural phenomena. Conditioned as we are by modern science, we may well wonder why a full mathematization of nature did not arise in the

Middle Ages. But we should not allow historical hindsight to color our appreciation of the philosophical problems medieval scientists faced.

Recall the Euclidean method of deduction from axioms and common notions. One of the latter states that things which coincide are equal. Thus Euclid proved that two triangles were congruent by placing one upon the other and showing that the given angles must coincide. This may seem perfectly self-evident on paper using *perfect* triangles. But may we say that in the physical world the universal triangle remains uncorrupted through the act of movement? In fact, we might easily contend that such mathematical definitions tell us nothing about their actual existence in the physical world. Where, then, do the principles of mathematics come from? And if universals are immanent in things, as Aristotle said, how can we actually maintain that universals are in any way different from their singulars? The unity of a universal, a triangle, is in all triangles; yet, even if we can show this universal for every triangle, we still have the problem of unity in diversity. There still exists a logical hiatus between the universal or formal definition and the thing which is actually observed or measured—between the lecture and the lab.

This was the methodological problem! Aristotle had never fully clarified the intuitive leap from induction to universal definition. In the thirteenth century the problem was most clearly stated by an Englishman, Robert Grosseteste. Grosseteste was born in 1168 and was educated at Lincoln and Oxford. He went to Paris to study theology and ended his life as Bishop of Lincoln, dying in 1253. His scientific career began later in life, and it seems that his early concept of truth was tied to his theology, as we might expect—the truth of an existent thing was illuminated only in the presence of the First Truth, God.

However, Grosseteste's later studies of Aristotle demonstrated to him that scientific knowledge—which was the definition or discovery of the universal—could be grasped apart from this divine illumination. Only before the Fall was the human mind able to grasp in one act both the essence and singular. In the corrupted material world, knowledge had to begin with the senses, and scientific knowledge could be achieved only through demonstration, whose instrument was the syllogism.

Grosseteste asserted that all demonstration must be through the middle term of the syllogism, which is, in fact, the definition of the thing in question. Consequently, the method of demonstration would not be complete without considering the method of definition. Let us take an example from Euclid. A right-angled triangle is a three-sided straight-lined figure which contains a right angle. How do we arrive at this? Well, we search for these qualities among the various geometrical figures, a process Grosseteste called resolution. Then we simply reconstitute these qualities theoretically in our minds and arrive at the definition. Yet, where in this have we asserted that a *real* right-angled triangle exists with such qualities in the material world? Grosseteste held that the universal premise *must* exist in real things as their form and cause. But the universal is neither one triangle nor many; rather it seems to be a logical entity *outside* the corporeal world!

Confusing? Actually we had no right to use a mathematical example, for according to Aristotle it can give no account of the four causes. Nonetheless, the mathematical example does illustrate the gap between pure theory and the facts that the theory is to explain. Grosseteste is simply saying that the universal premise that is inferred through the process of definition already contains within itself the conclusion, through it contains it only vaguely. Premises and inferred conclusion are only *suggested* by the facts. In short, the premises gained through induction—the qualities of our triangle—are only *hypotheses* when they are generalized into a universal statement about all right triangles. As hypotheses, they can be doubted. A hypothesis is for Grosseteste simply a formal assumption of our expectation. In order to convert our hypothesis into a scientific demonstration which reveals the actual state of its existence we must go out and *experiment*. We must leave the lecture hall for the lab.

Experiment seemed to mean for Grosseteste what it means for us: a controlled procedure designed to verify or falsify a hypothesis. In effect, it is a test of our definition, coming after induction. In fact we may deduce actual consequences not included in the original generalization. Through experiment and observation we may distinguish the true causes from the possible, grounding our demonstration in the factual world.

In Grosseteste's thought, mathematics enters the physical world through metaphysics. He believed that the actual structure of the universe was generated by a self-diffusion of light (*lux*). Not only was this light the origin of spatial extension, it was the primordial cause of all natural effects. And Grosseteste explicitly held that such causes must be expressed by lines, angles, and figures. The intensity of light varies with the distance of its source, and in passing from a rare to dense medium a light ray is refracted toward the perpendicular. Light propagates rectilinearly, and the strongest ray is that which maintains this rectilinear progress. Therefore, light which is bent by refraction is weaker in a dense medium than in a rarer one. Color, heat, and other natural effects are dependent upon the purity of the medium, the strength of light, and the quantity of the rays. Our mathematical example applies after all!

Mathematical optics could well explain and enhance the doctrine of metaphysical *lux*, but mathematical astronomy was another problem. The concept of *lux* resulted in a spherical universe which was dense and opaque toward the center and rare and transparent toward the periphery. Hence celestial light, the strongest and most perfect, became an invisible and spiritual entity in the upper heavens, diffusing into colors and other qualities in the lower spheres. Since optics studied this metaphysical principle, it demonstrated that the most certain knowledge was to be obtained through mathematics and physics, analogous to Grosseteste's earlier concept of Divine Illumination. Yet astronomy was different. Its mathematics tended to corrupt the simple material spheres suggested by the diffusion of light. On physical principles Grosseteste was forced to accept Aristotelian spheres; as a pure astronomer he would have to agree with Ptolemy!

By implication, then, the mathematical hypotheses of Ptolemaic astronomy could be doubted as physically real *even if* verified by observation. So the entire program of experimental science and mathematical investigation bows down before metaphysics. Here we see how culture influences the pursuit of science. While Grosseteste may sound modern at times, his goals were not. The final goal of all natural knowledge was a glimpse of the Creator. Grosseteste's metaphysics directly affected the very *content* of his science, determining, in fact, the structure of nature. And his metaphysics spoke in the same terminology as his science. This is probably the most important influence of all. Conceptual tools acquired in the education of any individual cannot be given up all at once; the language of any science is like a filter through which we see the world. Grosseteste's language was Aristotelian.

Nonetheless, Grosseteste's theories exerted a profound influence upon Oxford natural philosophers who followed. The most important of these in the thirteenth century was Roger Bacon.

Bacon is seen by many historians as a lonely figure, standing outside the academic fold. He even suffered condemnation and imprisonment. Yet, he drew upon Grosseteste's light metaphysic, the obsession with optics, and the idea of experimental science. However, he went far beyond the Bishop of Lincoln in stressing what he considered a basic fault of scholasticism: the separation of pure thought from living experience.

In opening his *Opus Majus* he harshly attacked the pedantry and false conceit of those in academic authority. These scholars, he said, actually conceal their ignorance through an ostentatious display of knowledge. Not only was this destructive to science, but Bacon went to the extreme by claiming that *all* the evils of the human race sprang from it. And to drive home the point, Bacon stated that he had learned more useful things from the unlettered than from all his "illustrious" teachers. No wonder he incurred the wrath of the schoolmen!

There has been a tendency to view Bacon as something of an inspired prophet, an anomaly of the Middle Ages who foresaw the tremendous technological possibilities inherent in the full application of mathematics and experiment to science. He called mathematics "the gate and key" of the sciences. He discussed the principle of the burning mirror, showing by the aid of diagrams the double refraction which multiplies the power of light rays. His discussions of magnification seem to indicate a knowledge of the optical principles necessary to construct a telescope; he even claimed that Caesar scouted Britain from the shores of Gaul using such a device! He contemplated all sorts of machines and even hinted at the possibility of firearms. Speculating upon the future, he did not hesitate to predict that future ages would surpass his own in knowledge and in fact would be amazed at the ignorance of his day.

Hindsight tends to draw such statements out of context. But as we read, we discover the figure of the antichrist stalking his writings, and he often seems to identify the Mongols as the legions of the antichrist. Frequently he states that the antichrist will use inventions such as burning mirrors. For Bacon the most important application of mathematics, besides optics, seems to be astrology. He admits that astrologers are not able to predict particular events with certainty; the heavens incline, not compel, the rational soul to action. Yet, all inanimate things are caused by the heavens "without contradiction." Mathematics not only enables us to establish chronology, the calendar, and other practical things, but it gives us astrology, which is necessary for the physician, the businessman, and in fact for everyone for all actions in life. Evil mathematicians are able to summon demons (as Faust did?), and even the Mongols proceed by astrology, which is why such small and weak men (according to Bacon) are able to conquer so much of the world!

His obsession with light and the multiplication of rays led him to assert that the ebb and flow of the tides were in fact caused by the multiplication of moon rays, which comes from the fact that the moon is the source of its own light! It was also Bacon who, following Alhazen, gave a complete description of the solid celestial orbs, thus reconciling Aristotle and Ptolemy. Again, by his assuming separate orbs for the planets and giving them a thickness so as to produce convex and concave surfaces, the outermost convex and innermost concave surfaces are made concentric to the earth while the others are eccentric. Between the surfaces of this middle or eccentric orb is a concavity containing a spherical epicycle, either like a solid globe or like a wheel of the two surfaces. It was commonly held that God moved the spheres as efficient and final cause, yet their erratic motions seemed to indicate that they had wills of their own. Hence, said Bacon, like the soul in the body, angels are the agents of movement in the heavens.

Bacon, of course, stressed the importance of experimental science, and his arguments remind us of Grosseteste. Experiment and observation can be used to correct false authority. Thus Bacon denies the unfortunate self-castrating beaver and identifies the musk-producing glands. Yet again, experiment has other uses which seem far more important.

Alchemists' lack of success is due not to false principles but to the difficulty of such "experiments" and the complicated labor involved in them. Further, Bacon seems to blame the apparent failures of alchemy on the fact that few people take an interest in experimental science! The same holds for magic—experiment is able to separate illusions from the real thing. And we can be sure that the omnipresent antichrist will use experimental science!

Bacon is neither pure Faust nor enlightened forerunner of modern science. He can condemn magic for its excessive claims, yet he often accepts its principles, such as "like produces like" and "as above so below." He relates these principles to mathematics and experiment, feeling that he has given them a firmer scientific basis. In a sense his laboratory science is that of Hermes Trismegistus. Indeed the word *laboratorium*, the Hermetic origin of laboratory, is composed of *labor* and *oratorium*, labor and prayer—like Bacon's science.

Bacon's work upon optics originated from a long tradition which included the nature and propagation of light and the nature of visual perception. While Plato had posited a visual ray projected from the eye to a luminous object, Aristotle believed that light was a state of the transparent medium which resulted from the presence of a luminous body. Since the eyes are composed mainly of water (which is also transparent), they *receive* light and color from the medium, actually becoming part of a homogeneous chain. Galen also adopted the medium as the instrument of vision, yet he maintained that a Stoic visual pneuma issues from the brain, through the optic nerves, and to the eyes from which it issues out into the surrounding medium and is *joined* with the illumination.

Euclid and Ptolemy, as might be expected, adopted a mathematical approach to optics. In his *Optica* Euclid postulated (like Plato) that visual rays proceed rectilinearly from the eye and, further, that they collect to form a visual cone. The nature of vision can therefore be reduced to geometrical discussions of perspective, angular separation of the rays, and the position of the object within the visual cone. On the other hand, where Euclid posited discrete rays, Ptolemy believed that visual rays form a continuous cone. Vision is strongest along the axis of this cone and becomes weaker as the rays diverge from the central axis.

Among the Arabs, al-Kindi adopted the visual cone and held that many such cones project from every *point* on the surface of the eye. Avicenna defended what can be loosely called the "intromission" theory of Aristotle by asking the mathematicians some rather disturbing physical questions. How can a ray issuing from the eye reach the stars? How can something as small as the eye produce cones large enough to fill the entire visual field? But it was Alhazen who took the geometrical analysis of light rays and applied it to the intromission theory. Light cones issue from the visible object.

Still there is a problem. If we reverse al-Kindi's light cones or pyramids, we have an object emitting these pyramids in all directions and a massive number of points (apex of the pyramid) falling upon the eye. How do we explain a coherent image in the presence of such a multitude? Alhazen postulated that only perpendicular rays may pass through the cornea without refraction, and the collection of such rays constitute a pyramid having the object as base and the center of the eye as its apex. So what about all those nonperpendicular rays? Alhazen suggested that these are refracted as they pass through the humors of the eye and are perceived *as if* they were perpendicular. Hence they "reinforce" the perpendicular image.

There are numerous difficulties here, not the least of which is *how* the incident ray as a refracted ray in the eye can be perceived to issue from the same point as the perpendicular. Further, Alhazen holds that the ray is refracted in the eye by the glacial humor (lens). But the glacial humor is, according to Galen, the seat of the visual power. So does vision take place on its surface? If this is true, then incident rays are perceived away from the perpendicular, and the theory falls. Not until Kepler devised the theory of the retinal image and the focusing power of the lens were these difficulties met. And it is interesting to note that Kepler, as we shall see, was a kind of mathematical Pythagorean who early on used light rays to drive the planets around the sun!

Bacon took Alhazen's theory and added to it Grosseteste's metaphysics of visual species and their multiplication, as well as Aristotle's transformation of the medium. Bacon, in fact, virtually agreed with everyone! Alhazen's mathematical intromission theory accounts for vision, yet the eye also acts along these "hypothetical" rays, enobling the medium and rendering objects capable of stimulating sight. Yet, once an object is "excited" by the incorporeal species of sight, vision results through an impression on the eye in the manner of Alhazen. Thus the phenomenon of vision is "saved," just like the physical spheres of Aristotle and the mathematical mechanisms of Ptolemy.

Those who followed Bacon, notably John Pecham and Witelo, more or less continued the work of synthesizing the various optical traditions. Pecham diverged from Bacon in saying that the visual power moderates rather than ennobles the medium, finding Bacon's concept too animalistic. Witelo denied the actual physical existence of visual rays and held like Alhazen that lines from the object to the eye are imaginary (hypothetical). On the other hand, Witelo was quite prepared to discuss light in metaphysical and even Neoplatonic terms, labeling it "the first of all sensible forms." All of these men, from Grosseteste to Witelo, discussed other optical phenomena: the rainbow, reflection, the problem of image formation in mirrors, and so on. In this sense, optics was one of the physical sciences in which mathematics played a major role.

One of the most significant uses of the experimental method and mathematical explanation was in Theodoric of Freiberg's account of the rainbow. Grosseteste had believed that the rainbow's formation was due to refraction of the sun's rays through successively thicker layers of mist. Albert the Great is usually credited with the recognition that individual drops of water play a significant role as well. Theodoric adopted both theories; however, he demonstrated mathematically that the rainbow and its colors could be explained by assuming both the refraction and reflection of light inside each individual drop. By means of optical geometry he was also able to account for the phenomenon of color reversal in the secondary rainbow.

Most important, Theodoric explicitly stated that he used models— translucent crystal spheres (stones)—to represent each drop, and he measured angles with an astrolabe. It has been noted that such a procedure was one of the most important scientific triumphs of the Middle Ages. In essence, Theodoric's testing of model raindrops represents the methodology of the experimental laboratory. Further, it may be said that Theodoric actually imagined that the models gave a true picture of the rainbow as it was in reality. His hypothesis, mathematically conceived and experimentally verified, was meant to be a true representation of things as they actually were.

Yet the rainbow is not the cosmos; the drops are not the planets. Optics was still a specialty, indeed a mathematical field within the larger Aristotelian division of the sciences. Innovations within any such specialty do not necessarily affect the larger climate of thought. Few felt any pressing need to radically alter that climate; the questions they asked and the answers they gave were still phrased in its language. The motivation to ask new questions, to reconsider the terminology, and to seriously entertain ideas science had held absurd came from a source outside science itself. Outside the lecture hall, trouble was brewing.

SUGGESTIONS FOR FURTHER READING

CROMBIE, ALISTAIR. *Robert Grosseteste and the Origins of Experimental Science 1100–1700*. Oxford: Clarendon Press, 1953.

————. *Medieval and Early Modern Science*, 2 vols. New York: Doubleday, 1959.

LINDBERG, DAVID C. *Theories of Vision from Al-Kindi to Kepler*. Chicago: University of Chicago Press, 1976.

————, ed. *Science in the Middle Ages*. Chicago: University of Chicago Press, 1978.

MARRONE, STEVEN P. *William of Auvergne and Robert Grosseteste: New Ideas of Truth in the Early Thirteenth Century*. Princeton, N.J.: Princeton University Press, 1983.

11

~~~~~~

Medieval Skeptics

In 1277 Pope John XXI instructed the Bishop of Paris, Etienne Tempier, to investigate the University of Paris. The Pope was obviously concerned with some of the doctrines being taught there by the Masters of Arts. In three weeks time Bishop Tempier believed that he had found ample reasons for the Pope's concern. It appeared to him that the philosophers were teaching "abominable" errors, which they merely labeled as doubtful. In effect, did they not hint that these errors were probable, or perhaps truthful? Did they not in reality hold the doctrine of the double truth and through their dangerous discourses lead the innocent into error (shades of Averroes)? As far as the Bishop was concerned, they did! Thus he issued a blanket Condemnation of 219 propositions along with the penalty of excommunication for holding any one of them. Tolerance for the errors of the philosophers had ceased!

Many of the condemned propositions bear directly upon Aristotelian science. In fact, many lie at the very heart of the system, especially in the realms of cosmology and physics. According to the Bishop, some of the abominable errors were: that God could not make several worlds; that God could not produce something new; that God could not make a vacuum by moving the heavens rectilinearly; that the elements are eternal (and the world too); that nothing should be believed unless it could be asserted from things which are self-evident. If such propositions were actually held—and this has been debated—they certainly did impose limits on the absolute power of God. Indeed, some of the propositions attacked the supreme miracle of the Eucharist: that God cannot make an accident exist without a subject or make more dimensions exist simultaneously. And if we are looking for more overt reasons why the theologians reacted with such hostility, consider the following: that theological discussions are based on fables; that the only wise men of the world are philosophers!

But what did it all mean? We are still faced with the strange fact that the basic Aristotelian system of the physical universe remained virtually intact until the sixteenth and seventeenth centuries. The Condemnation itself was only local and was in fact annulled in 1325, owing possibly to the growing acceptance by theologians of the Thomist synthesis. Yet others would use the articles as a starting point to attack the synthesis, throwing it into a maelstrom of doubt.

Then, there is the verdict of the pioneer historian of medieval science, Pierre Duhem. In the Condemnation of 1277 Duhem believed that he had found the "birthday" of modern science. The Condemnation forced the scientists and philosophers to think in new ways, to entertain possibilities Aristotelian science had considered absurd. Yet is it not strange, we might ask, why this reexamination of such fundamental principles still left the system basically intact for another two centuries? Stranger still, the Condemnation was *ex cathedra*, and we should normally expect such intolerance to hinder and complicate serious rational science. Further, the emphasis was upon the power of a supernatural agent, God, at the expense of actually weakening scientific certainty about *natural* causation. Thus, the great historian of the Scientific Revolution, Alexandre Koyré, believed that the influence of the Condemnation was exaggerated—it had pruned the branches but left the roots intact.

The Condemnation was influential in that it did interject a healthy dosage of doubt into the standard science. But doubt is not rejection, nor do scientists immediately throw up their hands and throw out their science even if some of its basic presuppositions are questioned. Yet, almost invisibly, it can be transformed, and it would require a great many other changes, many external to science, before someone would proclaim the transformation a revolution.

In essence the Condemnation had decreed that philosophy was not equipped to deal with faith, and thus it was up to the philosophers and theologians to show how this decree was rationally so. This led to a profound critique of knowledge, which had consequences for medieval science. The name most closely associated with this critique is William of Ockham, a Franciscan, who was born around 1280–90 and died of the plague in 1349. His philosophy, which is generally called nominalism or radical empiricism, has led some to identify him as a skeptic and as one of those who contributed to the destruction of the scholastic edifice. This may well be open to question, yet he was skeptical of human knowledge which transcended sensible cognition. In a sense he did not actually destroy the scholastic enterprise; rather, by means of his realistic epistemology, Ockham restricted it.

Although William of Ockham did not study directly under John Duns Scotus, who died in 1308, the seeds of his thought may be found in the older man. Duns Scotus was a metaphysician, but he held that to speak of God, or prove His existence as Being *per se,* is beyond human reason, which cannot escape the physical order of things. Like Grosseteste, Duns Scotus held that natural science begins with the direct apprehension of singular effects, but while we may contemplate the world of the contingent using reason, we cannot leap the unimaginable gulf to the infinity of God except by faith.

Ockham took the concept of the direct apprehension of individuals and applied to it an extreme economy of thought. Science is a series of statements *about* things. It uses concepts which stand for groups of things (universals) and logically relates them to particular objects. This is called "abstractive cognition," as opposed to the direct experience of the singular, called "intuitive cognition." The abstracted universal is simply a "sign" derived from the experience of singulars. But no abstraction can be conceived without being derived from the experience of particulars. In short, only *things* are said *to be* with certainty; abstractions are contingent and probable statements *about* things which do not carry the same ontological reality.

Ockham was not skeptical of the ability of science to discover causes and connections. Rather, he refused to posit scientific conclusions as *necessary a priori* statements apart from objects. Therefore, such "self-evident" doctrines as the nonexistence of the void, the eternity of the world, or no action at a distance are only hypothetical, not strictly necessary. What transcends experience cannot be verified by intuitive cognition, and hence it is not absurd to imply things about faith which, because of the limitations of human experience, must always remain beyond experience. Faith and metaphysics could no more be refuted than demonstrated.

Ockham was willing to accept conditional evidence and hypothetical necessity in the natural world—I can say without much doubt that fire *will* burn me. Yet, he separated natural knowledge from supernatural, for any universal statements about metaphysics are only signs which cannot be verified. They are, so to speak, simply logical games. He certainly did not deny the possibility of God's intervention and of the supernatural enlightenment of mystical experience. Neither did he go as far as Nicholas of Autrecourt, who, like Hume, would question even the empirical causal propositions of science, seeing these too as fictions. Ockham held that we *do* have the right to posit natural causes for natural effects. We do not have the right to extend them to what cannot be observed. And finally, it is preferable to state our analyses of nature with economy: "What can be done with fewer assumptions is done in vain with more"—Ockham's famous "razor."

Such a philosophy could tend to cast doubts upon the conclusions of science, which are, in essence, abstracted from experience. For example, Nicholas of Autrecourt felt that perhaps motion and change could be described better by the jostling of invisible atoms, yet he did not have to claim certitude for this theory. Others, like Jean Buridan and Albert of Saxony, argued that general principles inducted from experience were legitimate and did carry a degree of certainty. However, the emphasis was upon the logical consistency of ideas in contrast to existential reality. Thus, these thinkers were free to consider alternatives to Aristotelian principles once thought impossible, leaving to one side the question of absolute reality.

It was quite popular among the fourteenth-century scholastics to pose questions *secundum imaginationem*, "according to the imagination." Hypothetical possibilities could be worked out with a great deal of rigor, but whether or not such theories actually reflected or explained nature need not be considered. We should remember that Aristotle had sought the true workings of the world. Nominalism was a philosophical justification for the Condemnation, yet to entertain imaginary possibilities and show their logical consistency *was not* Aristotle's goal. Therefore, Aristotle's details could be criticized and made more rigorous, but the motivation was lacking in nominalism to completely overthrow his system or discover alternative ways of explaining nature as it was in reality.

Ockham's influence was profound in the fourteenth century, especially in discussions of change and the laws which govern motion. Take the problem of motion, for example. According to Ockham, any true scientific statement must reduce to a statement about singular, empirical things. In order to have motion, we must have *something* which is moved, just as in order to speak of any quality, say white, we must have before our eyes *something* white. A body moves by acquiring a certain distance in a certain time, just as a body acquires degrees of the quality "whiteness" or heat. Science does not concern itself with the metaphysical essence of white or movement, but with white things and moved bodies. In fact, what physical science treats is the *intensity* of change in some body—that is, *how* the body moves or *how* it becomes whiter. With the emphasis upon singulars, the description of the acquisition or loss of some quality is in essence mathematical—the gain being additive and the loss subtractive. In short, nominalism led to quantification, a mathematical treatment of qualitative change and motion. The overtones of this for science are obvious.

Duns Scotus was perhaps the first to discuss the problem of how qualities such as heat vary in intensity. The traditional terms signifying such variation were *intension* and *remission* of forms. The problem may have also arisen from considerations of transubstantiation in the Mass. How does the apparent wine and bread of the Eucharist acquire the divine qualities of Christ's blood and body? We already know the cause—a miracle. Walter Burely held that bodies acquire a series of forms, a form being anything which admits of variation—heat, color, density, velocity, etc. The question is: At what instant is a certain intensity or remission of the form present in the body? How did the substance change?

Such problems were discussed in fourteenth-century Oxford by a group of scholastics collectively known as the Mertonians (from Merton College, Oxford). In positing the distinction between being in motion (the quantitative state of a body like the intension of form) and the condition of being moved (a force acting upon a body) Ockham made the first truly clear distinction between kinematics and dynamics. The Mertonians were quick to pick up on this distinction as well as the analogy between the variation of forms and velocity. One of the earliest Mertonians to apply a sort of quantitative analysis to motion was Thomas Bradwardine in 1328. Bradwardine accepted the Aristotelian proportionality of motive power and resistance, yet he gave it a quantitative interpretation and shifted the argument to *how* uniform motion occurs.

Bradwardine applied some mathematical reasoning to what he understood to be Avempace's law. Aristotle had shown in the *Physics* that if a certain mover moves a certain mobile body through a certain distance in a certain time, then half the potency of the mover will move half the body through an equal distance in an equal time. Let us say that the motive power is 4 and the thing moved is 2; then according to Avempace we will have $4 - 2 = 2$. But if we take half the motive power and half the body moved, we get $2 - 1 = 1$, and the implication is that more work is done by a motive power against a resistance (mobile body) than is done by the sum of the parts of the motive power against the parts of the resistance corresponding to them, since in the case of the whole the excess is 2 while for the half it is 1, when they should be equal.

Consider, however, Aristotle's proportion, $V = F/R$, where F is the motive power and R is the resistance. Bradwardine saw that this relation mathematically contradicted the assumption that when the motive power and resistance are equal, or resistance is greater than the motive power, there should be no motion. Say we have a motive power greater than resistance by the ratio 2/1; there will be some motive power which is half the original, 1, and if the resistance remains the same, 1, the proportion is also 1 and we have motion, when, in fact, we should have V

= 0. Even increasing the resistance beyond F will give us motion, which implies that any motive power is capable of "infinite capacity."

Thus Bradwardine decided that Aristotle's function must be *exponential*. That is, in our notation, if we have a given motion determined by $V = F/R$ and we want twice (or n) times this V, the ratio must be multiplied by itself. Say we double the motive power, so that $2F' = F$, while the resistance remains the same, $R = R'$. The new motion should be $2V = V'$, and this is accomplished mathematically by $F'/R' = (F/R)^{2/1}$. The exponent 2/1 actually represents the ratio of velocities, V'/V, so Bradwardine's function may be written as $F'/R' = (F/R)^{V'/V}$.

Bradwardine also considered motion in the void, and again his work had nominalist overtones and employed mathematical reasoning. Elemental bodies, he thought, are nowhere found in nature in their absolute purity; observable bodies are always compounds. We might well wonder, therefore, what the relationship is between these elements (earth, water, air, fire) *within* the compound and how this relationship applies to motion. Instead of simply viewing motion as determined by the predominance of a single element, we could think of it in terms of a ratio of elements in a compound. Hence a ratio of heavy to light elements within a compound, say 4/1, would cause the body to fall faster than a ratio of 3/1, and the light elements can be seen as actually *resisting* fall. In the void, then, this "internal resistance" prevents instantaneous velocity, making motion possible.

Bradwardine saw that the quality of a body, its "internal resistance" determined by the ratio of its parts, need not depend upon its gross weight. The same proportion of internal resistance may exist in two mixed bodies of similar composition but of different size or gross weight. Thus he arrived at the conclusion that two bodies of equal internal ratios of elements—equal specific weights defined as the ratio per unit volume—will move at equal speeds in a vacuum. Later Galileo would extend this law to cover all bodies of whatever composition. But note the similarities.

Bradwardine's successors at Oxford—John Dumbleton, William of Heytesbury, and Richard Swineshead—also treated velocities as magnitudes in the same way as qualities. They called uniform motion the traversal of equal distance in *any equal* time intervals, distinguishing uniform velocity from nonuniform, since we may have two movements of equal distance yet one is uniform and the other varying (what they referred to as *difform*). For example, we may have a warm body at one point in time and a hotter body in another. The form of the heat is immutable—the body is never cold, and two cold bodies do not make a hot one. Nonetheless, we have a change of the quality which is continuous *within* the given boundaries of the body. What about this change? Is it regular? Or, is it by jerks—sometimes fast, other times slow? What is its value at any given instant? The same goes for motion that varies over instants of time yet between given points. Like a variable quality, motion requires us to speak of intrinsic instants. Yet, why ask such questions anyhow?

Any talk about instants in motion raises the perplexing problem of Zeno's arrow—the mere positing of an instant even in thought brings motion to a standstill! Surprisingly, though, we *do* experience instants in motion. If I am running through a forest and for some reason collide with a tree (the reason is unimportant), my velocity at the *instant* of collision *does* make a difference, Zeno's paradox or not!

The Mertonians did not use symbolism in their mathematics; they were logicians. But William of Heytesbury in 1335 spoke of the instant in time and position as a distance, indeterminately small, which *would be* traversed by a body during some time interval with a velocity it possessed at this instant. Thus if we make the distance "indeterminately small," *any* length of time, however minute, is sufficient for the body to traverse that distance. In this way we may speak of the instant as a "limit" which is approached in successively smaller intervals yet is never actually achieved—hence motion is not frozen. Heytesbury dealt with such problems "according to the imagination," and it is doubtful he would have seen them as applying to the physical world (one of his examples was Plato and Socrates increasing to a limit where they cease to exist!). Thus he did not develop techniques for actually calculating these values, but he did deal with the logical problems which would arise *after* Newton and Leibniz established the techniques of the calculus.

By far the most important concept developed by the Mertonians dealt with uniform acceleration, their celebrated Mean Speed Theorem. Uniform motion or intensity was defined as traversing equal spaces in equal times. Difform motion varies and can do so in infinite ways. However, we may consider the *rate* of difform motion. A motion may well vary at a uniform rate, gaining or losing *equal* latitudes of velocity. This is a uniformly difform motion, or uniform acceleration or deceleration. The Mertonians saw that a uniformly difform motion, say acceleration, could be related to uniform motion. A body that is accelerating, from zero or some finite value and acquiring velocity uniformly, during a given period of time will cover exactly the same distance as it would have covered in the same time if the value of its velocity were the mean between its initial and final velocities. That is, the distance covered will be the same in uniform acceleration as if the body moved for the same time at its mean, or mid-time, speed.

One of the conclusions they drew is that a body accelerating uniformly from rest will cover three times the distance during the second half of the time as it did during the first. In effect, the Mean Speed Theorem is a verbal statement of Galileo's law of uniformly accelerated free fall. If we consider the terminal velocity from rest as V, the mean is $\frac{1}{2}V$ in time t, and the distance S is $S = \frac{1}{2}Vt$. Further, the terminal velocity is gained by multiplying the rate of aceleration (a) by time (t), or $V = at$. Therefore, by substitution, we have the formula $S = \frac{1}{2}at^2$. There is no indication, however, that the scholastics actually applied the theorem to free fall. Their motion took place within a bounded system like the qualitative change in a body (Galileo's was unbounded). Free fall was treated separately in terms of dynamics, where speed was related to distance (as it was in Buridan's impetus, as we shall see).

The introduction of modern notation gives the Mertonians's theories a shorthand clarity which in actuality they did not possess. Their descriptions were at times confusing, often verbose, and purely dialectical exercises of the imagination which would appear to Renaissance humanists as sterile logic-chopping. The Mertonians were doing their best to clarify their own Aristotelian textbooks. They were still speaking in the vocabulary of Aristotle. That they expanded this vocabulary into areas Aristotle had not, and considered things he would have thought impossible, does not make them modern physicists in medieval bodies. In the climate of the Condemnation they were simply demonstrating how such impossibilities could be logically conceived without contradiction. Their use of mathematical reasoning did not necessarily mean to them that mathematics was the fabric of nature; rather it was the stuff of imagination. In a sense they were too logical to suppose that this "stuff" might be a real, effective link to the sensate world.

However, we might say that in Paris two new concepts did arise which in a sense altered the language of Aristotelian science. One was the theory of *impetus*, formulated by Jean Buridan; the other was Nicole Oresme's graphing of velocities and intensities.

Around 1320 Franciscus de Marchia posited a theory of impressed force (*virtus impressa*) to explain the motion of a projectile. It is difficult to determine how much the fourteenth-century scholastics knew of Avicenna's *mail*; nonetheless, de Marchia's impressed force was a remedial power left in the projectile by the mover, like heat impressed in a body by fire. This force was self-dissipating in that it existed in the body only for a certain time.

Jean Buridan, who was a contemporary of Bradwardine, said that the mover imparts a certain energy to the body capable of continuing motion, which he called impetus. The stronger the mover, the greater the impetus. Air and gravity serve as resistance, in that they decrease the impetus of the body which in turn slows the motion. But here Buridan made a highly suggestive assertion. Impetus would endure for an infinite time if it were not affected or resisted by some outside force. In short, Buridan's impetus is not self-corrupting; rather it is a permanent state of motion impressed upon the projectile. Hence there is no need to posit intelligences in the celestial spheres, for we may assume that God set the spheres in motion at creation by imparting to them a permanent impetus. Since there is no resistance in the heavens, there is no decrease of this impetus, and the spheres rotate eternally. Impetus applies to rotary motion like that of the spheres or a potter's wheel as well as to rectilinear motion.

Like the Mertonians, Buridan tended to view his impetus in quantifiable terms. A dense and heavy thing like iron is capable of receiving a greater quantity of impetus than a rarer thing like wood (Buridan's examples). An equal volume of iron receives more impetus per unit matter than an equal volume of wood, which explains why we are able to throw a lump of iron further than an equally large piece of wood. On the other hand, a feather, being extremely rare in density, has its impetus quickly overcome by air resistance. And impetus is also measurable by the initial velocity imparted to the projectile. Thus the measure of impetus is determined by the quantity of matter and speed, making it similar to momentum in Newtonian physics. The difference is that impetus is the cause of motion—in a sense, the internalization of Aristotle's external mover.

It appears that most of the scholastics considered gravity as the "heaviness" of a body. If heaviness, then, is the cause of free fall, we must still account for the acceleration perceived in this state. Since weight is constant, Buridan conceived acceleration to be cumulative increments of impetus (accidental impetus) produced by the heaviness of the body as it fell. It seems, however, that Buridan's weight as a force produces increments of impetus before increments of velocity and is not *directly* related to acceleration. Further, he seems to equate the impetus imparted by gravity (weight) with distance rather than with time, as did many other scholastics.

The concept of impetus enduring indefinitely without resistance and with uniform motion is, perhaps, suggestive of inertia. But according to the concept of inertia, rest and uniform rectilinear motion are identical states, and Buridan would probably have found it preposterous to equate his impetus with rest. Impetus, like the medieval intensities of heat or color, is qualitative. It is nowhere near a mathematical definition based on force. Impetus in fact "saves" dynamics—for, like his fellow scholastics, Buridan refers to his explanations as "according to the imagination."

One of the most interesting and imaginative treatments of the intension and remission of qualities in the fourteenth century was Oresme's *On the Configuration of Qualities,* probably written around 1350. Grosseteste and Witelo, says Oresme, used geometry to "imagine" the intensity of light, although they knew that light itself was not really a geometrical magnitude. So why not use geometry for other things as well? Not only velocities and intensities, but such things as beauty, pleasure, spirit, sound, and even celestial influences may be represented by geometrical figures and ratios. For example, the attractive power or quality of a magnet for iron may be due to the accord between the geometrical configurations of each. Why, there is no end! If we are puzzled by the apparent power of words, especially magical words, such powers may be due to their configuration rather than their meaning. This is the reason animals are influenced by words, although they cannot possibly know their meaning.

The intension and remission of qualities naturally imply continuity, since the qualities blend into each other in succession. Yet, as in our problems of instantaneous velocity, intuition suggests that there is a certain discreteness in the continuum. How may we visualize this strange paradox? Take a horizontal line, says Oresme, and call it some total extension, duration, or subject. Now, draw a perpendicular to this line. The perpendicular may be said to represent intensity or velocity—the change of quality the subject undergoes. The entire area is representative of the quantity of the quality, and the perpendicular can be seen to represent the qualitative intensity at any point in the subject.

Let us see how the system works for uniform and difform intensities or velocities. The horizontal represents some subject or time, and the perpendiculars represent intensities (Figure 11.1a). So a uniform velocity or intensity gives us a rectangle: the perpendiculars are all equal and thus the intensity is uniform (Figure 11.1b). The area can be seen as the quantity of the entire change. For a uniformly difform intensity, or uniform acceleration, the figure becomes a triangle, since the lengths of the perpendiculars vary uniformly and the acceleration begins at zero. It is also possible to represent the simultaneous existence of two contrary qualities in the same body by a diagonal, or even to represent difformly difform variations by curving figures.

Oresme used his graphing system to provide the Merton Mean Speed Theorem with a geometrical proof. To do this he had to show that a rectangle whose altitude represents the whole velocity equals in area a triangle which represents a body accelerating uniformly from rest. Thus such a body could be shown to traverse the same distance as a body

Figure 11.1

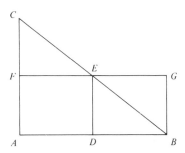

Figure 11.2 Area *ABEF* + triangle *EGB* = rectangle *ABFG*. Area *ABEF* + triangle *EFC* = triangle *ABC*. *EGB* = *EFC*, and when equals are added to equals the results are equal, so triangle *ABC* = rectangle *ABFG* in area.

moving during the same time interval at a uniform speed equal to that of the middle instant of the uniformly accelerated motion (see Figure 11.2). The area of triangle *ABC* equals the area of rectangle *ABFG* since by the *Elements* (Book I) triangles *EFC* and *EGB* are equal. Why? Adding to each triangle area *ABEF* gives us rectangle *ABFG* and triangle *ABC*. Since the area measures the total velocity, which is equivalent to distance, and the perpendicular represents the velocity at a point within the total movement, the distances are equal and $ED = \frac{1}{2}AC$, which is the proof. We shall see nearly the same representation used by Galileo. Not only this, but Oresme also showed that in the case of uniform acceleration from rest, the distances traversed in equal consecutive time periods are as the odd numbers, 1, 3, 5, 7. Again we shall find Galileo saying the same thing.

Yet Oresme had no functional conception of coordinates, no algebraic equations applied to his graphs, no negative values—in fact, no numerical values at all. His was not analytic geometry; it was formal Euclidean. Still, he expanded his method as far as it would take him. He considered the intensity of some quality increasing toward infinity within a quantity which was finite—what modern mathematics calls a convergent series. Indeed, he did speculate on how his configurations might explain natural phenomena, yet unlike Galileo he never claimed that in fact they did. He never left the study for the lab.

The same may be said for scholastic speculations upon cosmology. Some of the articles of the Condemnation naturally led them to entertain possibilities previously thought absurd. By the imagination the infinite power of God could create a plurality of worlds. How is this rationally possible? We could well imagine worlds contained within each other, as did Edgar Rice Burroughs and Jules Verne. Oresme concluded that such worlds were possible, although reason and experience were not equipped to verify them. Of greater significance, however, was the discussion by Buridan, Oresme, and Albert of Saxony of the simultane-

ous existence of separate worlds in space. Such an idea was plainly contrary to Aristotelian natural place and natural motion. Yet, Oresme
claimed that "up" and "down" refer only to *our* experience; they are relative to our earth. There is no contradiction inherent in imagining other
worlds which have their own relative motions. In short, Oresme suggested a concept of separate centers of gravity, each giving rise to a
closed mechanical system.

Buridan also stressed the relativity of motion in regard to the earth's
rotation. For example, given two ships, one moving and the other at
rest, an observer on the moving ship could well imagine the ship at rest.
Hence purely astronomical observation could not decide the issue of
heavenly or earthly rotation! On the other hand, the principle of
economy is better met by a tiny rotating earth rather than a rotating massive celestial sphere. Buridan also marshaled geology to show that the
earth moves in some way. The earth is not a homogeneous body; it suf-
fers erosion and earthquake, altering its density and heaviness. Thus its
center of gravity continually seeks the geometrical center of the universe, which causes a rectilinear shift. In short, the earth's constant
weight redistribution causes its center of gravity to shift! Nonetheless,
true to his impetus theory, Buridan found that the earth's rotation
caused difficulties in explaining the flight of an arrow because of the
motion of the air. Oresme answered this objection by saying that if the
air and all sublunar matter shared in the earth's rotation, we would not
necessarily experience the wind of the rotation or the horizontal movement in general.

The scholastics even considered the possibility of the Stoic infinite
void beyond the last sphere. In this they had an ally in the Hermetic
treatise *Asclepius,* which spoke of such a void as bereft of physical bodies
yet filled with spirit. Buridan saw the possibility of God, the supreme
spirit, being omnipresent in this dimensionless void. God is conceived
as being everywhere in the world, existing in many places simultaneously. But God created this plenum and so cannot be restricted to it, for
God is an infinite power, unlimited and unenclosed. Thus Bradwardine
concluded that God is an infinite sphere, having His center everywhere
and His circumference nowhere. The identification of God with the
infinite will be seen again, even in Descartes.

The fourteenth century was alive with many portentous speculations. Yet, consider the ulterior motives of the scholastics. While striving
to rigorize Aristotle, in demonstrating the possibility that two *reasonable*
conclusions could be fully contradictory, they were in fact demonstrating the ultimate impotence of reason. Having shown that the earth's rotation was rationally possible, Oresme in the same breath rejected it on
the grounds of faith. Probable explanations and reasonable hypotheses
did not carry the same ontological reality as faith. What was certain

about the physical world could be expressed in a paraphrase of Socrates: "I know nothing except that I do not know." And Oresme said this very thing!

Perhaps if we are searching for the true fourteenth-century picture of the cosmos, we must look not to its science but to its poetry. Dante's epic poem, *The Divine Comedy*, takes the reader down through the nine circles of Hell, up again to the earth, to purgatory, and finally through the celestial spheres to the Empyrean throne of God. It is the Aristotelian universe of qualitatively determined regions, but a universe in which the basically symbolic spiritual meaning predominates over the literal physical structure. It mirrors the Christian drama of existence, in which human beings occupy the middle position in the Great Chain. Through *The Divine Comedy* the human imagination does achieve, for medieval people, a true representation of reality.

What was the religious significance of medieval cosmology? Why was it so strong? Let us, "according to the imagination," hypothesize a dialogue between a modern historian and Satan.

HISTORIAN: I was thinking about Dante's cosmology. He was a poet and certainly did not intend to write a scientific treatise. No, the reality he sought to portray was moral; his physical structure was in fact a metaphor for ethical reality. But that is precisely my problem. Pride, as you well know, was a great sin in the Middle Ages.

SATAN: Indeed! For my pride, and the disobedience which came from it, I fell from heaven.

HISTORIAN: Well, yes—but as I was saying; if the structure of the universe was primarily a moral design with the physical particulars subservient to it, why place the earth in the center of the universe? Is that not a very prideful thing—seeing themselves in the center when it should be God?

SATAN: Now you wound my pride! Have you forgotten, my friend, that Dante places *me* in the center?

HISTORIAN: Then I would say that the arrangement becomes even more untenable in moral terms. Perhaps scientific reasons were the primary motivating factors and Dante had to deal with these the best he could.

SATAN: You are forgetting Neoplatonism. Hell was farthest from the One and therefore from Reality, and Dante has me in the very center of Hell. So those who inhabited the earth lived between Reality and Nonbeing. Their earth was thus a place of corruption and death, a terribly humbling abode. It might have been in the center of things physically, but spiritually it was far from the nonspatial moral center, God.

HISTORIAN: Then it was simply convenient to stay with the Aristotelian universe? In that case it would be of little religious significance to take the earth out of the center.

SATAN: It was a relief to me! But I'm afraid that move did take away a great deal from the moral picture.

HISTORIAN: Are you simply trying to confuse things, as is your nature?

SATAN: Oh no! Nothing interests me more than the question of moral significance, and the medieval cosmos was very significant. First, it was exciting, for upon the central earth, between Being and Nonbeing, your ancestors were also in the midst of the great war between myself and the High Throne. And every individual soul was so important that the universal powers of Heaven and Hell fought for it. The earth, corrupt and yet a place of possible redemption, was full of this drama. Everywhere humans looked, from society to their physical picture of the cosmos, they saw signs of this struggle, with themselves in the middle of it.

HISTORIAN: And pride in their cosmic position and importance. We're back to my original problem.

SATAN: Leave the pride to me! Let's just say that the medieval cosmos infused all life with meaning. People gazed up to the heavens and knew that the seven spheres of the seven planets were alive, each singing its own song and each having its own element or quality. And beyond the crystal spheres was the angelic chorus at the foot of God's Throne. So when the soul of man came down into birth on earth, it passed through the seven spheres and picked up at each an element; thus the human being was actually composed of the celestial elements—the microcosm. And so, too, was society a hierarchy in accord with universe, reflecting the cosmic organization. Following the rules of society was a cosmic act, for by doing so the individual was in accord with the universe. Why, even the shape of the earth itself told of the great drama; when I was flung out of heaven I fell through the spheres like a giant comet, smashing into the earth and driving through to the center, creating a great conical trench which is Hell. And what was driven forth on the other side was the mountain of Purgatory, rising from the massive ocean which covered the whole southern hemisphere, the summit of the mountain being Paradise, from which the four rivers flowed in the four directions. All of this was of high moral drama and significance. All of this has dissolved in your world, my friend, and I wonder what you have found to replace it.

HISTORIAN: Yet modern science has given us a truly awe-inspiring universe, far more awesome and dramatic than your puny medieval cosmos.

SATAN: So it has. But have you discovered your mythical center and your limiting sphere? No longer may you look to the physical universe for the patterns and norms which guide your life. You are thrown back upon yourself, to the inward sphere, and there, where all norms are in flux, each individual must come to the lost Garden by a separate path, finding in that terrible loneliness the way to life's significance. I wonder: Will your precious reason bear this redoubtable burden?

Then the fourteenth century did not have to search long in the world for the signs of reason's limitations. Consider the helplessness of physicians in the face of the Black Death (bubonic plague) which struck Europe around the middle of the fourteenth century. They could find causes for the plague in infected air, or in the malignant conjunction of the planets, or in the Jews; one chronicler even blamed the pestilence on a change in women's fashions! Sober reasoning was mingled with wild speculation.

The physicians did not give up. They tried everything: burning scented wood, diets, bleeding, lancing, and a host of other remedies. But in the end people died—in shocking numbers—and the physicians were helpless. How feeble human science must have finally seemed before the piles of corpses and the omnipresence of Death. Indeed, reason applied with imagination could achieve wondrous things, yet as Hamlet said: "There are more things in Heaven and earth, Horatio, than are dreamt of in your philosophy."

SUGGESTIONS FOR FURTHER READING

CLAGETT, MARSHALL. *The Science of Mechanics in the Middle Ages.* Madison: University of Wisconsin Press, 1959.

————— . *Nicole Oresme and the Medieval Geometry of Qualities and Motions.* Madison: University of Wisconsin Press, 1968.

CROSBY, H. LAMAR, JR. *Thomas of Bradwardine.* Madison: University of Wisconsin Press, 1955.

DUHEM, PIERRE. *Le Système du Monde: Histoire des Doctrines Cosmologiques de Platon à Copernic,* 10 vols. Paris: Hermann, 1913–1959.

GRANT, EDWARD. *Physical Science in the Middle Ages.* Cambridge: Cambridge University Press, 1977.

MOODY, ERNEST A. *Studies in Medieval Philosophy, Science, and Logic.* Berkeley: University of California Press, 1975.

WILSON, CURTIS. *William Heytesbury: Medieval Logic and the Rise of Mathematical Physics.* Madison: University of Wisconsin Press, 1960.

12

"Man can do all things..."

The Renaissance

Leon Battista Alberti, a typical "Renaissance man," was born in Genoa about 1404 and died in 1472. An artist by profession, he was also an athlete and a self-taught musician who wrote his own compositions. He also studied mathematics, physics, architecture, even canonical law. Alberti applied his knowledge of the sciences to painting and architecture, subjects which had been simple "crafts" in the Middle Ages. And he was extremely proud of his independence, contemptuous of the schools and their dreary bookishness. To him nature itself had grown old and tired, for few were the great intellects who could broaden knowledge and discover totally new arts and sciences without teachers or models. How great would be our fame if we could do such things! The possibilities are nearly endless, for "men can do all things if they will."

From the perspective of the nineteenth century, Jacob Burckhardt called Alberti and the other great Renaissance Italians "the all-sided men"—*l'uomo universale*. These "universal men" broke away from the categorical constraints of medieval consciousness. They asserted their individuality, their uniqueness apart from the guild or the school. They rediscovered antiquity, free from the filtering veil of scholasticism. According to Burckhardt these all-sided men asserted their subjective independence, which, in turn, allowed them to view nature more objectively. And their antiquity was that of Pythagoras. Plato, Epicurus, the

Stoics, and Cicero, besides that of Aristotle. They were, wrote Burckhardt, "the first-born among the sons of modern Europe."

A great debate has swirled over Burckhardt's modernist conception of the Renaissance. Some historians have even gone as far as to deny the very existence of such an abstraction as "the Renaissance." The problem is even thornier when it comes to the history of science. Pierre Duhem held that modern science was born in the fourteenth century among the scholastics and finally triumphed after a long period of struggle. The so-called Renaissance, in its obsession with antiquity, was a reaction—sterile, without scientific substance, a hindrance. After all, its subject matter was not science but the *studia humanitatis,* the disciplines of rhetoric, history, poetry, and moral philosophy. Even Burckhardt was forced to admit that these subjects and the "humanists" who molded them into an educational program probably did detract from the pursuit of pure science.

But let us reflect for a moment. Besides the imaginative intellectual activity of the scholastics, human interaction with nature occurs on a much wider scale. In any age the practical problems of life inevitably demand activities which call for an understanding of that interaction. The craftsman, the engineer, even the artist deal with nature directly. The merchant uses mathematics to keep accounts, to figure bills, credits, and debts. The idea of negative number, for example, is helpful in computing business transactions. The merchant does not worry about the logical contradiction inherent in the concept of a minus 1. If it works, use it! In short, the goal of understanding nature rationally is secondary to the goal of dealing with nature directly in practical affairs.

The Greeks had maintained a separation between *techne* and philosophy. Any concrete manipulation of nature became for them an art, and finally, philosophy. Among the Greeks, *there was no concept of material progress.* With the possible exception of Roger Bacon, the same thing might be said of the thinkers of the Middle Ages. Given the role assigned to science in medieval culture—the purposes for studying it—the style of the schools was aptly suited to that role. On the whole their science did answer with a reasonable degree of success the questions they asked.

Yet outside the schools the world was changing. National monarchies were being formed, states which challenged the universal Church for people's allegiance and, in a sense, for their hearts and souls. In Italy we find many city-states ruled by self-made men, princes whose legitimacy was based upon their personal popularity rather than the half-religious awe of medieval loyalty. New avenues of wealth were being opened. Commerce was coming to rival feudal ownership of land as a source of affluence. The European world expanded tremendously, following the footsteps of this new commercial class and listening to the tales of Marco

Polo and other travelers. The world outside Europe became interesting, exotic; it beckoned to adventurers who wanted to see its marvels for themselves rather than at second hand through Pliny or Isidore. The Renaissance was the age of seafarers.

There was an increasing contact with Constantinople even before its fall to the Turks in 1453. Byzantine scholars came to Italy and initiated the study of Greek. By the fifteenth century the ancients, including *all* the Platonic dialogues, could be read in the original. The standard texts of Greek science, too, could be read in the original. Thus the Ptolemaic system and Aristotle could be studied as written, and their apparent errors and inconsistencies could no longer be blamed upon faulty translations. Another significant development, the invention in Germany of the printing press, would gradually lead to a greater dissemination of knowledge.

The art of warfare also underwent great changes. The Hundred Years War between the English and the French began primarily as a feudal conflict. When the French finally drove the English out by 1453, the armies had become professional and cannon had begun to replace the catapult. The new ordnance required improved methods of mining, metallurgy, and fortification which could withstand cannonade. New problems arose which challenged engineers, mathematicians, and a variety of artisans. These improvements in turn enhanced the sheer physical power of the new nation-states. The financial burden of war on the national scale, coupled with the improved technology, made these states more and more dependent upon the international bankers and credit market. The interpenetration of the marketplace and the state had begun. The subsistent and local economic patterns of the Middle Ages were gradually swept away by this new rationalization of violence.[1]

The original source of the new wealth was, of course, trade. The Italian city-states had established a monopoly in the Mediterranean, where navigation by traditional methods was fairly easy. But Atlantic states such as Portugal, which desired to break the Italian monopoly of this eastern trade, had to deal with the problems of sailing the vast wastes of the open ocean. This required better instruments and more accurate mathematical techniques. The scale of latitude based upon astronomical observations had to be integrated into a chart drawn from estimated linear distances and magnetic bearings. The determination of longitudes required mathematical projection—geometrically representing the spherical earth on a plane surface.

The use of astronomy in navigation, especially that of the open seas, created the demand for more accurate tables. A book like Sacrobosco's *Sphere* was useless to the navigator. The old translations of the *Almagest*

---

[1]See Chapters 3 and 4 in William H. McNeill, *The Pursuit of Power* (Chicago: University of Chicago Press, 1982).

contained numerous mistakes, errors which might prove fatal on the great oceans. And the Toledo Tables, over two centuries old, contained serious inaccuracies as well. Therefore at the University of Vienna, George Peurbach and his student John Müller (called Regiomontanus) began work to correct these problems. They even went to Rome to examine the *Almagest* in Greek. Regiomontanus completed the work after Peurbach died in 1461, but Ptolemy's own complex system of heterogeneous geometrical constructions were now clearly revealed for all to see.

Everyone knows, of course, that Columbus "discovered" the "New World" in 1492. His fame, however, rests upon a mathematical error, the kind of error the new navigators were seeking to avoid. Columbus adopted the erroneous calculations of the tenth-century Arab astronomer Alfraganus (first recorded by Roger Bacon and discovered by Columbus in Pierre d'Ailly's *Imago mundi* of 1410). According to d'Ailly, the length of a degree at the equator was 56 ⅔ land miles, when it should be 60 land miles. Therefore, Columbus believed that the distance from the Canary Islands to the Orient was a mere twenty-five hundred miles. The idea of sailing west to reach the Indies was as old as Aristotle, but the problem was the immense distance. Nonetheless, by faith and mathematical error Columbus attempted it. What is more, he probably died never fully aware of his great discovery—in his own mind he had reached the Indies, maybe even Eden!

When Columbus reached the "Indies" he probably expected to see the fabled creatures described by Pliny and the others. So did the Portuguese navigators who rounded Africa and reached India. Where were the Antipodes? The Sciopodes? The tropics were habitable after all! And look how fallible Ptolemy and Aristotle were. They didn't even know that a great continent like America (not to mention two such continents) existed. Experience had shown them wrong or simply ignorant. And if they were mistaken in these things . . .

None of these voyages was undertaken to prove Aristotle wrong. The bringing to light of errors of tradition was a byproduct of other pursuits. Consider the use of negative numbers. In commerce they make sense. Yet the sixteenth-century humanist Gerolamo Cardano, a mathematician of considerable talent, believed that negative numbers were fictions—impossible numbers. On the other hand, the military engineer Simon Stevin of Bruges, working with inclined planes and centers of gravity, adopted the Archimedean method of exhaustion. But while the Greeks always had something left over—an infinite number of figures in a finite area was illogical—Stevin said there really was no difference, that the continued subdivision is less than any given quantity, so for all "practical purposes" there is nothing left. Stevin and Cardano both used irrational numbers. Zero was accepted as a number. Stevin probably introduced the decimal for operating with fractions to save labor in figuring weights and measures. About 1594 John Napier introduced logarithms to facilitate trigonometric calculations (logarithms are the power to which a number, called a base, must be raised to obtain a given number). In short, people were dealing with nature in ways which the older scientific tradition would have found irrational.

In a certain sense, Columbus's "discovery" of America was a mistake. His whole venture was based upon false mathematical premises. Paradoxically, such mistakes, by the very fact that they *are* errors, can further understanding. Science does not always proceed reasonably, simply because human beings themselves do not. Seemingly outlandish fantasies do play a role. And here is where the humanists become important in the history of science, although perhaps not for the reasons Burckhardt might have imagined. Their indirect effect upon science derived from a source he mainly ignored: mystical Platonism and magic.

Throughout the Middle Ages we are sometimes able to detect, like a mountain stream of many tributaries, a river of mysticism. The sources of the stream were varied and at times difficult to trace. Raymond Lully, who died in 1315, spoke of a mathematical universal art which by means of symbolic geometrical diagrams could be made to illustrate cosmic creative forces and the truths of Christianity. Astrology, alchemy, and Hermeticism in general were part of this tradition; yet, on the whole, it was fragmented, diverse, and inconsistent. Its fragments were scattered and incomplete. The Renaissance humanists revived this stream from its source in antiquity, and they provided for the first time the extant writings of its ancient originators.

Perhaps there are psychological reasons for a resurgence of mysticism in the Renaissance. The fourteenth century had been a period of unprecedented disasters. The terrible plague, the ravages of the Hundred Years War and the Babylonian captivity of the Church had wreaked profound disillusionment. To the average person the world seemed to be falling apart. Artists began depicting the figure of Death and decaying corpses in horrifyingly vivid scenes. In the fifteenth century there began a near epidemic of demonic possession; in 1484 witchhunting was given official papal support. Flagellants roamed the countryside, and heretical movements gained force. Even among some fifteenth-century philosophers we are able to detect a reemergence of Platonism and mathematical mysticism.

One of these philosophers was Cardinal Nicholas of Cusa, called by Alexandre Koyré "the last great philosopher of the dying Middle Ages."[2] Like Plato, Nicholas saw the material world in terms of constant change and multiplicity. In God, however, all opposites, all contradic-

---

[2]Alexandre Koyré, *From the Closed World to the Infinite Universe* (Baltimore: Johns Hopkins: 1957)., p. 6.

tions and divergent qualities, are gathered in a Unity. How are we to comprehend, asked Cusa, the relationship of the finite and limited to the infinite Unity? For his answer, Cusa, the metaphysician and theologian, turned to mathematics.

Consider the two most opposite forms in geometry, the straight and curved lines. Now try to imagine an infinitely great circle. Contrary to our limited reason, the circumference of such a circle expands and expands until it becomes *straight*—the tangent! In the infinite circle the two opposites coincide. The same holds for motion: a body moving at infinite speed coincides with a body in a state of absolute rest. But our reason completely fails to comprehend such absolutes; they are absurd. Such absurdities teach us what Cusa calls the doctrine of "learned ignorance." This is because mathematical concepts, either rational or irrational, are idealizations, and the world is a mere shadowy Platonic image of the absolute.

From his doctrine of learned ignorance Cusa drew some remarkable cosmological consequences. There can be no *absolute* center of the universe, only a metaphysical ideal center which is God. And this center, in being beyond reason, is also the circumference. The same holds for an *absolutely* finite universe: there is no *absolutely* final enclosing sphere, nor can there be a physical state of *absolute* rest. The inevitable result of such assertions is that the physical earth moves and cannot be the center of the universe in any absolute sense. Any rational account of the universe is strictly relative to the observer, and human cosmology is simply unable to give a completely objective account of the world.

With Nicholas of Cusa the qualitative structure of the Aristotelian cosmos loses all meaing. It is simply impossible for human reason to say with any certainty that the earth is base or the heavens perfect. Hence there is no reason to suppose that the inhabitants of the earth are in any way inferior to the inhabitants of other worlds. If the latter (angels?) are more perfect, it is because they partake in greater measure of the intellectual or spiritual essence of existence. But human beings need not desire another nature, only to perfect their own!

Notice the boldness of the cardinal's thought. The universe is indeterminate (only God can be rightly called infinite), and other worlds may be inhabited. The universe has its center everywhere and its circumference nowhere—the earth moves. It would be rash to see Cusa as a precursor of Copernicus or as holding a doctrine which prefigures the infinite universe of Giordano Bruno. His goals are metaphysical and his thought is mystical, although springing from a basically rational mathematical approach. Like the scholastics, he seems to say that in terms of what we can know, the best knowledge is to admit our uncertainty. Yet nowhere does he pretend that his picture of the universe is simply "according to the imagination." And his mathematical critique is Platonic, not Aristotelian.

There may be something else here, too, if only by implication. Who can say that human beings are necessarily corrupt or that the earth is any less than the stars? Might not our spiritual natures be amenable to improvement, to *progress* if you will? Might we not answer Satan's objections by the exciting idea of progress? A curious thought for the Middle Ages!

Perhaps there *is* something deeper going on here—a basic change in people's attitude toward the cosmos and their place within it. At least this is the opinion of historian Frances Yates. The humanists revived not only Plato and Neoplatonism, but also the magical cosmology of Hermes Trismegistus. Cosimo de Medici in 1462 granted a villa near Florence to the humanist Marsilio Ficino to translate Plato from the Greek. However, in Cosimo's collection of manuscripts was a group of treatises ascribed to Hermes. These were the first translations Ficino made at the so-called Florentine Platonic Academy.

Yates saw in the Hermetic literature a prelude to the rise of modern science. That prelude, surprisingly, was in the nature of its *magic*, which arises from its mystical cosmology. The Renaissance accepted Hermes as a real Egyptian priest, who lived about the time of Moses and whose teachings were the origins for the doctrines of Pythagoras, Plato, and Plotinus. The Hermetic writings are often eclectic and inconsistent, yet there seems to be a central theme: the entire universe is infused by a cosmic spirit, a *spiritus mundi*, passing from God through the celestial spheres, which orders, harmonizes, and indeed maintains the world. But the uniqueness of the Hermetic cosmology comes from how humans themselves fit into this scheme.

Human beings are part of the material world, yet humans are also given divine creative powers. Casting aside corporeal preoccupations, men and women are able to exercise this creative power *over* nature. The key is mathematics. The celestial spirit which penetrates everything does so by measure and symbols. To acquire an intimate knowledge of this mathematical harmony not only ennobles the mind, as Plato held, but enables people to exercise power and control over nature. In fact, the ability to operate and create, control and manipulate natural forces was the supreme goal of the magus. No longer are we merely contemplating the world.

In the *Ascepius* we find the Egyptian magi invoking celestial powers and reproducing them in material images. Inert matter is thus infused by human operations with an animating spirit, a *force*, which *mechanizes* the statue. In the *Picatrix* the magus uses talismans as a kind of instrument to control external events. Ficino's own philosophy mirrors such ideas. For him there is a real connection between symbols and things.

To operate with words and ideas, using the proper formulas, is to operate with real and essential nature. The formulas of mathematics are the great keys to unlock the mysteries of nature's harmony. The keys are not conventional and not imaginary; they are fully *real*. Nature expresses itself through proportions. Humans, by mastering the formulas of such proportions, can *control* things. The magician is a kind of spiritual mechanic.

Pico della Mirandola discovered the same principle in the Cabala, a book of Hebrew mysticism. The Hebrew language, with its correspondence between letters and numbers, provides a method for using magical forces. In the sixteenth century the magician Henry Cornelius Agrippa wrote that the use of mathematical sciences enables the magus to control and direct planetary angels. In the same century Fabio Paolini spoke of "magical machines." And still later, in 1628, Thomas Campanella persuaded Pope Urban VIII to sit in a sealed room containing seven candles in order to neutralize the potential danger of an eclipse. The candles symbolized the celestial bodies, and Campanella set them in favorable conjunctions so that the Pope could absorb their beneficial influences—like a magical light bath!

Yates saw in Renaissance magic a kind of "first step" toward the mechanistic-mathematical philosophy of the Scientific Revolution—the emphasis upon empiricism, the Baconian control of nature, and even the heliocentric theory. Some historians have viewed her conclusions as overstatements, pointing out the inherent problems of theory-choice and the more direct, internal influences of medieval science which led to the revolution. Internally, at least, the influence of Hermeticism is still an open question. What defeats us is the simple fact that we can never truly look into the minds of our scientists. Even when the living hand behind the writings becomes visible and sources are credited, we must realize that is the conscious mind which is revealed. We ourselves are sometimes hardly aware of the unconscious motivations behind our own thoughts and activities.

And perhaps this is the common thread. In Christian theology the idea of progress, of improvement in the human condition, was otherworldly, redemptive, even apocalyptic. In the modern world the idea of secular and material progress is invisibly connected to science. We may doubt it profoundly, still, it is our inheritance from the past. Perhaps it *was* the Renaissance magician who planted the seeds of this idea firmly into the unconsciousness of the West. Drastic change requires a psychological as much as a scientific coming to terms. Progress may not even be a philosophical concept but rather a feeling, vague and ill defined, that change is not to be feared. With the proper knowledge, with a science of reality, change can be controlled!

Certainly the most popular personification of the Renaissance scientist is Leonardo Da Vinci. Yet, the ideas stored in his notebooks had little effect (if any) upon the course of scientific development. What makes him stand out in our eyes is his emphasis upon technical engineering, his use of experiment, and his integration of mathematics into empiricism. But did we not see such things in the Middle Ages? What is the difference? Leonardo's concept of science was not to merely "save the phenomena" but to grasp the design of nature in every detail, to physically seize hold of it. His idea of scientific activity was not to contemplate or criticize the ancients, but to go out and look for himself. Books to him might well be a source of knowledge, yet they were not the *only* source or even the best. He believed that human beings possessed the ability to create things which did not exist in nature but were only possibilities. The imagination, unaided by experiment and experience, was simply a source of empty dreams. Leonardo did not merely welcome change—it was his passion!

Leonardo was an artist. Painting to him was not a simple craft, as it had been in the Middle Ages. It was a science, because it was based upon the noblest of human senses, sight. The artist "sees" nature, grasps it in the mind, and recreates it upon the canvas. The act of recreation requires certain techniques, for painting must be an exact copy of reality. The artist therefore must know the principles of perspective. Leonardo's predecessor in this subject was none other than Alberti. Alberti placed a transparent screen between his eye and the scene and imagined lines of light running from the scene to points on the screen. If the screen is the painter's canvas, then the points make up a section, and lines preceding from this section to the eye create the original scene exactly in sight. Since the section also depends upon the position of the observer's eye and the position of the scene, we may have many sections of the same scene. The painter, then, who cannot look through the canvas must operate by the mathematical laws of perspective in order to paint realistically.

Leonardo adopted the technique of perspective, yet his idea of painting as a science went much further. The artist must look into nature, uncovering it layer by layer until it reveals its secrets. These secrets, Leonardo believed, were to be found by studying its structure. Thus he applied himself to anatomy, mechanics, biology, botany—everything. His penetrating eye saw through skin, tissue, and organs to bone. He dissected and observed, seeking not only form but the very origin of movement in an animated body. Such movement, he said, was based upon mechanics—the centers of gravity and the principle of the lever. Therefore mechanical science was the most useful science, and like perspective it was based upon mathematical laws. Nature, like art, operates by these laws. The scientist, like the artist, approaches nature through experience and mathematics.

Leonardo's belief in the human creation of new things led him to apply the principles he had learned as an artist. He filled his notebooks with all sorts of inventions—flying machines such as a screw helicopter, war machines, architecture, and a host of instruments. Mechanics to him was not a theoretical discipline but a method of application. He studied the mechanics of birds in flight, saw an analogy between flight and swimming, and with this analogy contemplated the possibility of human flight. His anatomical studies sought mechanistic explanations. The heat of the heart was not innate, as Galen believed, but was caused by the friction of rushing blood against the walls of the ventricle and by circulation. He even constructed glass models of the heart. His discussion of impetus sounded at times as if he were groping toward a theory of inertia. He also studied astronomy and believed that the seas on the earth make it shine so that to an observer on another world it would appear to be a star. He even said that the sun did not move!

Still, Leonardo created no new scientific theories, nor was the actual level of his mathematics very high. Like Alberti and many of the other humanists, he was mostly self-taught. No doubt his originality sprang from his independence. While we are tempted to hail him as a forerunner, we must at the same time realize that his influence upon his own contemporary science was minimal. And for all his originality he still maintained some of the old patterns of thought. Motion was a departure from the natural state of rest. Breathing cools the blood. His definition of force as an incorporeal essence even sounds Hermetic. On the other hand, he opposed magic and the idea of bodiless spirits, for a spirit occupying space was in reality a vacuum, and a vacuum was for him impossible in nature.

Nonetheless, like the age he lived in, Leonardo was a symbol of transition. The artists, the engineers, even the magicians were beginning to view nature with new eyes. In this they parallel the scholastics, who were also looking at their Aristotle in a new light. Both were searching for a science that *worked better*. The scholastics, influenced by the nominalist critique of the hypothetical *a priori*, implied that a genuine science of nature must be empirical. On the other hand, the Condemnation tended to uproot their imaginations from the empirical—God can do all things!

Perhaps this is the difference. The artist-engineers, the humanists, and the magicians were directly connected to the world of human power; men like Ficino and Leonardo worked for princes and republics. The science of the future would still find a role for God, yet the place of the Deity in the content of that science was slowly relegated from the main body of the text to a footnote. The process may have started in the Renaissance—"Men can do all things if they will."

Medieval theoretical science had altered the Greek inheritance. Renaissance life had altered medieval values. They are two perspectives of the same scene. Combined, they would foster a new scientific outlook, a change in the human conception of the universe we arbitrarily call the Scientific Revolution.

SUGGESTIONS FOR FURTHER READING

KENSMAN, ROBERT S., ed. *The Darker Vision of the Renaissance: Beyond the Fields of Reason.* Berkeley: University of California Press, 1974.

KRISTELLER, PAUL OSKAR. *The Philosophy of Marsilio Ficino.* Gloucester, Mass.: Peter Smith, 1964.

_____ . *Renaissance Thought and Its Sources.* New York: Columbia University Press, 1979.

KOYRÉ, ALEXANDRE. *From the Closed World to the Infinite Universe.* Baltimore: The Johns Hopkins University Press, 1957.

WALKER, D. P. *Spiritual and Demonic Magic From Ficino to Campanella.* Notre Dame, Ind.: University of Notre Dame Press, 1958.

WIGHTMAN, W. P. D. *Science and the Renaissance,* 2 vols. London: Oliver and Boyd, 1962.

_____ . *Science in Renaissance Society.* London: Hutchinson University Library and Co., 1972.

YATES, FRANCIS. *Giordano Bruno and the Hermetic Tradition.* Chicago: University of Chicago Press, 1964.

~~~~~~

Interlude

A Conversation
with a Mathematician

"You historians!" Dan protests. "All you do is argue. You're a lot like pure mathematicians, you know. You have your facts and methods, I suppose, which are established by your community. So do we mathematicians. But there are so many fields. Sometimes you'll find mathematicians from two widely separated areas arguing over what constitutes a proof—that is, if they're on speaking terms. We'll even argue whether mathematics itself, without human beings, exists!"

"But Dan, how is this like history?"

"Take your so-called historical facts," says Dan with a mischievous wink. "When (and if) you finally agree on them, there is always the question of what they mean and whether their logical consistency is inherent or imposed from without. Is history rational in itself, or is it the historical community which decides? I think you'll find a great deal of argument—just like we argue over the criteria of proof."

"Dan! Are you saying history is relative?"

"Maybe. But there is one fact you cannot escape: history is always written by the survivors. Wasn't it Samuel Butler who said that God cannot alter the past while historians can; that's why He tolerates them?"

"You know, Dan, come to think of it, I've heard some philosophers saying similar things about *science!* I think you'll find some of their ideas interesting.

"The old philosophy of science used to suppose that modern science was basically empirical—testing, observing, experimenting—allowing the facts to dictate the theory. This method is what made the scientific revolution. Nature simply compelled people to reject the old medieval rationalism (a kind of metaphysics in disguise) for hard empirical fact. Since then science has been pretty much an inductivist enterprise. Oh, yes, theories changed and problems arose. But since the sixteenth and seventeenth centuries science has been, on the whole, pretty continuous—accumulating facts and climbing the great mountain of truth with better theories.

"Some philosophers began to doubt this fine picture of science. Surprisingly, they appealed to the history of science. What, they asked, were these great pathfinders actually doing? The great scientists, more often than not, did not proceed like inductivists: they dreamed, fantasized, guessed, thought up theories—they even cheated. Often they simply made up facts to fit a pet theory.

"Perhaps it was Ernst Mach who started this line of thought. Mach, a nineteenth-century scientist, believed that many of his own supposed 'objectivities' were constructed by elements we would call subjective or intuitive. In his history of mechanics Mach tried to show that quite frequently the scientist begins with hypotheses and then operates with experience. Archimedes and Stevin, for example, began with *ideal* levers; Galileo's inclined plane was an *ideal* plane. Only afterward does the physical lever or plane become rational. Mach held that nature is composed of sensations; these are its elements. A fact is a "thought-symbol" of sensation. Nature simply *is;* thought-symbols (facts) are but economical interpretations of relatively fixed or repeated sensations.

"Some philosophers, especially a group called the Vienna Circle, decided that verifiable facts must always reduce ultimately to sense impressions. Their object was to work out an ideal observational language which did not go beyond sense-data and conformed to strict logical rules. This was the true 'scientific method' for verifying theories. So with a language rid of metaphysics, inductivism was alive and well.

"No, said Karl Popper, inductivism in whatever guise could not really verify scientific theories. Then what could? Popper turned the game around. Verification was not the method of 'real' science; what separated science from pseudoscience was falsification! Scientists test theories to see if they can be disproved. Once a scientist has invented a hypothesis, its strength and even its rationality depend upon how well it survives the fires of criticism. Scientists often try to rescue theories by all kinds of *ad hoc* additions. For a thousand years or so they attempted to save Ptolemy with all kinds of ingenious ideas. In the end they failed. Yes, scientific creation may be haphazard and even irrational, and theories may at first be totally removed from brute sense-data (no one *experi-*

ences the earth moving). Still, there is a 'logic of scientific discovery,' Popper claimed: it is falsification. Most important, falsification says that theories *must be* testable. Rather than simply appealing to confirmation, a theory must prohibit certain things from happening, which then may be tested. A theory which is vague, like astrology, explains away any test. A good theory contains within itself the possibility of its own destruction.

"Popper's falsification held to the hegemony of experiment as the engine of scientific progress. Then came 1962 and Thomas Kuhn's *The Structure of Scientific Revolutions*. Any established science, Kuhn wrote, is basically conservative and ahistorical. Textbooks tend to present a body of accepted theory *and* observations or experiments *as if* this was the way nature worked all along. In a sense the acceptance of a theory by a community of researchers has already *predetermined* what they can expect to see in their experiments (like you were saying about mathematics, Dan). What the scientists see depends upon what they look for, and what they look for (and at) depends upon what previous experience (their education) has taught them to see! If you view the world like Aristotle, then the facts will be Aristotelian, and so will your problems, methods, and science as a whole. To change your science is to change reality!

"Science is like an actor who changes masks from role to role. The stage is the same, but the dialogue and costumes are different. Kuhn labeled these roles 'paradigms.' A scientific paradigm is an all-encompassing play which includes not only theories but practices, laws, instruments, methods, and, yes, facts. One paradigm's fact is another's fiction. When problems appear, all sorts of patchwork is done (no member of the community attempts to overthrow the paradigm). If this is unsuccessful, the problem is brushed off as metaphysics (or something unscientific like that). Only another paradigm can overthrow an established one. Such battles between two paradigms are called scientific revolutions.

"Rationality itself may be paradigm determined. Hence there is no real rational judgment to be made between two competitors. Usually it is some crisis too big to ignore that starts the revolution. When the dust clears, however, the older paradigm is forgotten by the newly established community or seen as error. And since the textbooks are written by the survivors—like the history, Dan—the revolution becomes invisible . . . why, the 'facts' were there all along! The survivors no longer speak the old language.

"Kuhn really started something. Apparently he was not a complete relativist; that is, he saw science as a whole as fundamentally progressive. Still, many philosophers believed that they had to defend the rationality of modern science against him. Hardly had they begun when they discovered another heretic in their midst—a real radical, Paul

Feyerabend. Feyerabend also studied the history of science, and for those in search of absolute standards his conclusions were quite shocking. Reason in any given historical period is constantly overruled, methodology is constantly overturned or simply nonexistent. Those scientific 'successes' rationalists prize most (like the rest of this book) could have occurred only when someone rejected the accepted methodology, the facts, and acted in a way that was basically opportunist, irrational, and sometimes without scruples. Proofs and evidence often get in the way!

"Feyerabend even questioned the superiority of science as opposed to other ways of understanding nature. Science's superiority exists only when viewed from within and by its own rules. To be really objective—a word found in all rationalist sermons—we must measure our science beside some *opposing* tradition, taking that tradition on *its* terms. (Technology will not do; after all, mystics visited the stars long before space travel.) To dismiss any tradition out of hand by calling it irrational or unscientific is to practice the kind of dogmatism which would have made our own science impossible. We simply cannot have it both ways.

"Rationalists went to the barricades, as you can well imagine. I'm not so sure they've been successful combating Feyerabend, but I hear that some historians got pretty upset with him, too. On the other hand, a philosopher named Gonzalo Munévar says that it is a good thing science changes so radically (his book was entitled *Radical Knowledge . . .* get the picture?). If science is to be of value, says Munévar, it must promote our 'getting along' in a *changing* world. It must be absolutely flexible and potentially adaptive to sudden or slow change. Flexibility means that there are no general recipes for rationality, discovery, verification, or whatever.

"Well, Dan, I guess philosophers of science are not much different from historians (or mathematicians?). They seem to argue on forever. But then, that's the fun of it all. You know I believe that you're right. History is pretty much of an argument, even at the very basic level of its rationality. I think we need to be a bit more strict with our own presuppositions . . .

"Hey Dan! Maybe the alchemists did turn lead into gold! How wonderful! Perhaps a 'willing suspension of disbelief' *is* the method of history. And of science too! The survivors may write the history, but the survivors also have imaginations. A rigorous imagination? Is that a contradiction? What would a scientist say?"

Dan grins. "Oh, you don't have to tell *me* about rigorous imagination. Or playful imagination, for that matter. Have you forgotten? I'm a mathematician . . . !"

SUGGESTIONS FOR FURTHER READING

DAVIS, PHILIP J., and HERSH, REUBEN. *The Mathematical Experience*. Boston: Birkhäuser, 1981.

FEYERABEND, PAUL. *Against Method*. London: Verso, 1978.

————. *Science in a Free Society*. London: Verso, 1978.

HESSE, MARY B. *Revolutions and Reconstructions in the Philosophy of Science*. Bloomington: Indiana University Press, 1980.

KUHN, THOMAS S. *The Structure of Scientific Revolutions*, 2d ed. Chicago: University of Chicago Press, 1970.

————. *The Essential Tension*. Chicago: University of Chicago Press, 1977.

MACH, ERNST. *The Science of Mechanics*, trans. Thomas J. McCormack, 3d ed. Lasalle, Ill.: Open Court, 1974.

MUNÉVAR, GONZALO. *Radical Knowledge: A Philosophical Inquiry into the Nature and Limits of Science*. Indianapolis: Hackett, 1981.

POPPER, SIR KARL R. *The Logic of Scientific Discovery*. London: Hutchinson, 1959.

13

The Earth Moves!

Let's imagine we could interview a man from the sixteenth century. Suppose our ghost is an educated member of the middle class, residing in a city in Germany. "You mean a burgher living within the Holy Roman Empire," corrects the ghost. To forestall a lecture, we immediately ask our major question: "What, in your opinion, made your century revolutionary?" Most likely the phantom would launch into a lively discussion of Martin Luther's Reformation—angrily if a Catholic, enthusiastically if a Protestant. In either case we would learn that our ghost's concept of revolution is quite different from ours. "A violent break with the past?" the ghost would exclaim, amazed at our ignorance and our lack of the rudiments of learning. "No, no, revolution means a return to an earlier position or norm, like the revolution of the celestial bodies to their original places—a return to original purity which has been violated!"

"What about the other revolution, the astronomer who claims that the earth moves?"

"Oh, you mean that fellow Nicholaus Copernicus, the Canon of Frauenburg. He published a book just before he died in 1543—*De Revolutionibus Orbium Caelestium (On the Revolution of the Heavenly Spheres)* was its title, I believe. I haven't read it; it's too technical for me. But I know that in the preface he explicitly states that mathematics is for mathematicians. So you'll have to ask them!"

"But Copernicus made the earth move! Now that's certainly a radical change!"

We break off the conversation and look at the "revolutionary" (our terms) book. In a dedication to Pope Paul III, Copernicus admits that to ascribe motion to the earth might seem ridiculous to the multitude and "almost" contrary to common sense. It is better, however, to follow the injunctions of the Pythagoreans and keep these mysteries secret. Mathematicians have been unable to agree on the proper mechanism of the spheres, and they are even uncertain of the movements of the sun and moon; thus the poor farmers who depend upon the accuracy of the calendar for the planting and gathering of their crops find that this calendar is unable to forecast such a simple thing as the beginning of spring (by the time of Copernicus, the vernal equinox—the beginning of spring in the northern hemisphere—was about ten days short in the old calendar). Further, they have used all sorts of assumptions—eccentrics, epicycles, equants—and yet have not achieved certainty.

Is it really a problem of certainty? Ptolemaic astronomy did represent the movements of the planets with a reasonable degree of accuracy, and Copernicus admits this "for the most part." As he continues, though, we begin to see another, perhaps more important objection. In constructing a workable system, astronomers have been forced to violate the "first principles of regularity of movement," and they have not been able to fit the *harmony* of the universe together. The picture they have painted of the planetary system is a jumbled melange of constructions which are physically inconsistent with one another. If we compare the system to a portrait, it is as if the artist took the arms and legs from one figure, the body from another, the head from a third, and patched them together. The result is less a living portrait than an artificial monstrosity, hardly "pleasing to the mind." But the world is a creation of God, and as such it must have a certain beauty of form which befits the divine artist.

Consider the equant. To achieve uniform velocity Ptolemy takes a point removed from the center of the deferent, yet for the deferent he takes the center point. Each accurately measures a particular phenomenon, but taken together they are wildly at odds. Then there is the curious behavior of the inferior planets, Venus and Mercury. The centers of their epicycles are always set on a line in the direction of the sun, and the revolutions of these points through the zodiac, or around the ecliptic, all take a year. Their synodic periods, the time between their greatest angular distance from the sun, varies as they move on their epicycles: for Mercury it is 116 days and for Venus 584. In the case of the superior planets, Mars, Jupiter, and Saturn, the centers of their epicycles can be found anywhere in the ecliptic, and the periods of these planets in their epicycles are their synodic periods. Now why should the inferior planets behave as if their sidereal periods were tied to the sun? Further, the outer planets can be arranged by the varying times for each to traverse the ecliptic. Not so the inferior planets; their arrangement is arbitrary.

Surely many astronomers had noted these puzzles. Back in the thirteenth century Alfonso X of Castile and León had remarked that God should have consulted him before creating the universe; Alfonso would have recommended something much simpler! Yet such anomalies were puzzles, hardly an astronomical crisis.

Copernicus, on the other hand, apparently felt a deep philosophical as well as technical discontent with the system as a whole. The purely technical issues were compounded many times by the lack of symmetry in the "world machine."

Born in 1473 on the banks of the Vistula, Copernicus attended the University of Cracow and in 1496 traveled to Italy to study church law. He also studied medicine and astronomy, learning Greek in its scientific context as it had been instituted at the medical school in Padua. Here he probably became acquainted with the humanist revival of classical Greek sources and may have encountered, if only indirectly, Florentine Platonism, Pythagoreanism, and perhaps the Hermetic corpus. Meanwhile his uncle, a bishop, secured his appointment as canon of the cathedral of Frauenburg, a position he held until his death. Although he did not earn his living by astronomy, he undoubtedly continued his study of the subject, and when Pope Leo X invited astronomers to the Latern Council of 1514 to reform the calendar, Copernicus was asked for assistance. The request came too soon for Copernicus; it was probably a decade later, when he was in his fifites, that he wrote *De Revolutionibus*, the book in which he hoped to establish the mathematical harmony and truth of the world machine.

This emphasis upon harmony ought to remind us of the ancient Pythagoreans. It did Copernicus! For he mentions them, and Aristarchus, in an earlier manuscript. Had not some of the Pythagoreans placed the earth in motion around a central fire? Copernicus abhored novelty for its own sake, wrote his pupil Georg Joachim Rheticus; he had really returned to a more ancient cosmological doctrine. We might well view Ptolemy in the same way that Luther at first viewed the Church: Ptolemy's complicated system had seriously disturbed the pristine harmony of the ancient philosophers. So, like the early Luther, Copernicus did not want to overthrow Ptolemy; rather, he desired to purify the system by returning to its ancient wellspring. He was a revolutionary in the sixteenth-century sense of the word.

The *Revolutionibus* does indeed follow closely the technical and organizational style of the *Almagest*—as well as many of its assumptions. The motions of the spheres are circular and uniform; the universe is a closed sphere. Now let us, for technical reasons we shall consider later, set the earth rotating upon its axis and orbiting around the sun. Would the earth fly apart? By comparison, consider the great speed at which the whole outer sphere of the fixed stars would move! Simple experience cannot decide; it is like floating along in a ship on a calm sea. If we sight another ship, can we be sure which ship is moving—ours or the other? Falling objects are carried along with the earth; gravity is a natural inclination which may exist on other planets; the earth is a sphere and rotation is the natural movement of a sphere; stability is more appropriate to nobler bodies; and, finally, the sun as the source of heat and light ought to rest in the center of the universe—here Copernicus has Hermes as a witness. In short, we have heard these arguments before!

Many problems remain, some purely observational. A rotating earth accounts for the diurnal revolution of the stars, but an orbiting earth should cause the terrestrial observer to notice a slight shift of each of the fixed stars relative to the pole of the stellar sphere. This phenomenon, called the stellar parallax, was simply not apparent and remained undetected until 1838 (see Figure 13.1). Copernicus therefore is forced to

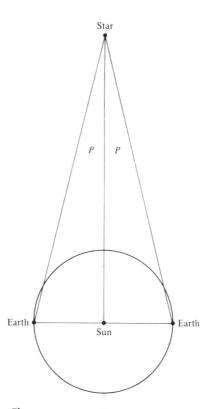

Figure 13.1 Stellar parallax "*P*".

expand the sphere of the fixed stars so that the entire orbit of the earth becomes as a geometrical point relative to the enclosing stellar globe. What a price to pay! The size of the universe becomes unthinkable; it is not infinite, but the human imagination is certainly stretched, along with the poor cosmic balloon, to conceive it. Is this rational? Has the theory been falsified?

Now let us see what possible benefits we may derive by exchanging the position of the earth for that of the sun. Copernicus could finally give an explanation (not simply a representation) of retrogression. Retrogression is an optical illusion. Take Mars for example; its orbit is passed by a faster-moving earth, and against the background of the fixed stars the planet *seems* to stop and retrogress (see Figure 13.2). The inferior planets pass the slower-moving earth in the same manner. Thus we no longer require epicycles. In fact, the deferent of the inferior planet in the Ptolemaic system actually represents the earth's orbit, and its epicycle is really its own orbit around the sun. So this is why the inferior planets had their epicycle centers always on a line with the sun in the Ptolemaic system. We can also see that the year of about 365 days is the

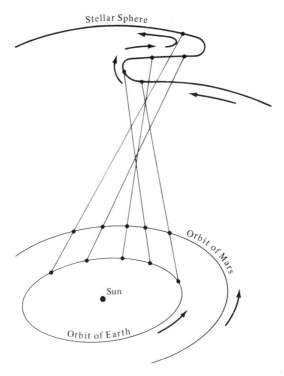

Figure 13.2 A simplified representation of retrogression for a superior planet (Mars).

earth's orbital time, and therefore the planets' synodic period is actually a combination of the earth's orbit and their own. Using this information, we may compute the true orbital times of Venus and Mercury, which are 225 days and 88 days, respectively. It is also easy to see that their positions from the sun can now be fixed. The periods of the superior planets can be computed also from the orbit of the earth. Furthermore, if we take the size of the earth's orbit as an astronomical unit, we can compute the *relative* distances of all the planets from the sun.

Copernicus has indeed presented us with a profound harmony, and this harmony is possible only when the earth is placed in motion about the sun. The Pythagoreans would have been proud. So would Plato. The search for harmony, exemplified by mathematical constructions, had demonstrated that the apparent disunity of the phenomena as perceived by the senses was illusionary. And now there were only *six* moving spheres in the universe! Six, as Rheticus pointed out, was a divine Pythagorean number.

What about complexity? Shockingly, the *Revolutionibus* shows little improvement over the *Almagest!* In order to account for the varying speed of the sun and the inequality of the seasons in the old system, Copernicus, keeping everything perfectly circular, had to make the earth's orbit eccentric. Thus the center of the earth's orbit is not the sun at all but a geometric point removed from the sun. Actually, the earth's eccentric point ends up revolving around a point which in turn revolves around the sun! And since the other planets have their orbits computed in conjunction with the earth's orbit, the Copernican system is not really heliocentric at all. The earth is not "just another planet." It still holds a unique position in the cosmos.

Although Copernicus eliminated the epicycles which accounted for retrogression, he still had to deal with the fact that the planets moved at varying speeds in their orbits and were observed to be brighter or darker, deviating north and south from the ecliptic. The moon also presented problems. In the end Copernicus was forced to reintroduce epicycles and eccentrics just like Ptolemy, although he was able to dispose of the equant. Disposing of the equant was indeed a triumph for Copernicus, but his system was hardly less complex.

Copernicus, after all, never did solve the physical problem of the earth's motion—he was only a mathematician, not a physicist, and he still seemed to hold to the Aristotelian division of the sciences. Nor does he address the theological issues involved. In fact, he smells of mystical Pythagoreanism and even Hermeticism! By the science of his day we might sincerely ask if an astronomer could be expected to discern cosmic reality? The other "revolutionary," Martin Luther, may well have expressed the common opinion of laymen when he caught wind of the Copernican system. Luther is reported as saying: "That fool will upset the whole science of astronomy!"

The reaction of sixteenth-century astronomers to the new "hypothesis" was more complex. The first news of the system reaching the learned world did not come directly from the pen of Copernicus himself. Rather it came from a little work entitled *Narratio Prima (First Report)* printed in 1540. The author was Georg Joachim, called Rheticus, an Austrian. Possessed of tremendous enthusiasm but somewhat emotionally unstable, Rheticus had sought out Copernicus in 1539 to learn of the older man's theories, and he persuaded the reluctant canon to share the new ideas with the world. The *Narratio Prima* was less a "first report" than an enthusiastic plea for the new system, which, according to Rheticus, had "lifted a fog" from astronomy.

Even more than Copernicus, Rheticus stressed the unity and harmony of the system and its mathematical ordering of the planets, which resulted from the correlation of the magnitudes of their orbits with the roughly sun-centered periods (again, to be sure, the center of the earth's orbit). Besides the elimination of the equant, this harmony commended the system, and Rheticus went even further in the comparison with Platonism and Pythagoreanism, concluding his treatise with a passage on the harmony of the soul.

While astronomers might well (and did) applaud freedom from the equant and view the Copernican theory as a new and improved scheme to calculate such things as the calendar, on the whole they did not follow Rheticus in his overzealous acceptance of the conceptual basis. Rheticus had surrendered the task of editing the *Revolutionibus* to the Protestant reformer Andreas Osiander, who added an unsigned introduction. Undoubtedly, Osiander wrote, the learned will take offense at a moving earth and at the confusion such a proposition would cause in other areas of the liberal arts (earthly physics, for example). However, the duty of the astronomer is to devise "hypotheses" by which the motions of the spheres may be calculated correctly from the principles of geometry. Such hypotheses need not be true or even probable; it is sufficient that they agree with observation. In short, astronomy does not speak of certainties—the reality of the system—but provides methods devised by the imagination to facilitate calculation. Although Copernicus had also used the word "hypothesis," which may have suggested "saving the phenomena," it is more likely that he believed his system was the true description of the universe. Osiander, on the other hand, cautioned that accepting such ideas as true would result in one's departing from the study of astronomy a greater fool than when one entered it!

Osiander's position—that the system was simply a new calculating device—was a variation of the theme most scholars, Protestant and Catholic, seemed to take in the decades following the death of Copernicus and the publication of the *Revolutionibus* in 1543. In some

Catholic universities the treatise was read and, in general, was considered yet another speculation or a convenient fiction. After all, the book had been dedicated to Pope Paul III, and Copernicus had been urged to publish it by some high Church officials, including a cardinal.

Luther's reaction was shared in part by his disciple Philip Melanchthon. At the University of Wittenberg an informal circle of scholars gathered around Melanchthon as he launched a vigorous campaign for educational reform. Thus Melanchthon and his followers, while rejecting any notion of the earth's motion, praised those parts of the theory that would result in better calculations. The earth's motion could be interpreted as probable but must be rejected, since it is contrary to Scripture; on the other hand, there may be valuable observations and computational methods in the theory which recommend it for study. Therefore, the theory as a calculating device was split from its cosmological implications, which were generally passed over in silence. Rheticus was simply ignored.

One of the leading figures at Wittenberg, Erasmus Reinhold, actually used the Copernican planetary mechanisms to prepare a new set of astronomical tables. Reinhold's *Prutenic Tables,* so named after his patron Albrecht, Duke of Prussia, were completed in 1551 and were superior to all others. In fact, Pope Gregory XIII's calendar reform of 1582 was based upon the *Prutenic Tables.* Nevertheless, while praising the elimination of the equant, Reinhold ignored the cosmological innovations, even though the mechanisms he used in his calculations were based upon them. Thus the fledgling "revolution" survived; the old theory of Aristarchus was left dangling. Only later would attention be drawn to it and the recognition arise that the moving earth was a fundamental presupposition behind the improved calculating devices.

Trouble erupted in an entirely different quarter. Indeed, Copernicus had severely stretched the enclosing celestial sphere, but it was still a finite orb and the heavenly spheres still carried the planets. Giordano Bruno, the one-time Dominican and later Hermetic philosopher, smashed the spheres with a vengeance. According to Bruno the universe was infinite; there was no center of the world, no boundaries; there was only a single space which we may call the Void and in which infinite planets revolve around infinite suns. God is infinite and omnipresent, and creation must be an unfolding of His power. So, asked Bruno, how can we restrict that Power to a finite world? Rather, the universe, as the Hermeticists believed, was alive, constantly in the flux of movement and change, and united by the infinite presence of the Creator. Of course the earth moves; of course the universe is infinite—any limitations upon the powers of God were, for Bruno, absurd.

Bruno was not philosophizing "according to the imagination," nor was he constructing a hypothesis to help calculate the motions of the planets, nor was he maintaining Pythagorean silence. His doctrine was a direct assertion of reality. Aristotle's stratified universe was rejected for a basically homogeneous geometrical continuum—an abstract geometrical void without end. His principle of unity was God, not the Christian God of Dante's onion universe, but the God of the magicians—the *spiritus mundi*. Bruno appears to be more of a heretical pantheist than a champion of the new astronomy. It is no wonder, then, that he was tried in Rome by the Inquisition and burned at the stake in 1600. His was a religious revolution, and he was not burned because of his astronomy. Strictly speaking, he was not even a Copernican!

Neither was the Danish astronomer Tycho Brahe. Copernicus had derived his system mainly as a paper-and-pencil exercise; he had made observations, yet they were hardly systematic, nor was his astronomy dependent upon them. Tycho Brahe was an obsessive observer. Granted the island of Hveen by the King of Denmark in 1576, Tycho gathered the best instruments of his day and measured the heavens. Like none since the Arabs he studied the skies night after night, meticulously recording his observations. He was a true empiricist. Therefore he rejected the Copernican system!

Tycho objected to a moving earth for the usual physical reasons, but he marshaled some very interesting astronomical problems as well. Before the telescope, the stars, like the planets, were conceived to be discs, not simply luminous points. The Arabs had endeavored to measure these discs, and so did Tycho. If the parallax of the stars is imperceptible because of their great distance, and yet we are able to measure their diameters up to, say, 120″, imagine how large they must be! Why, some would fill the orbit of the earth! Surely that was unthinkable. And there were other problems—for both Copernicus *and* Ptolemy. The movements of Mars, upon close observation, revealed difficulties. Its epicycle in Ptolemy's system must vary in size, or the orbit of the earth must vary. The moon and other planets presented similar problems.

Thus Tycho rejected both Ptolemy and Copernicus. Actually he combined them. He made the earth the center of the orbits of the sun and moon, setting the fixed stars revolving again. Around the sun he placed the orbits of the five planets, making Venus and Mercury encircle the sun alone and the superior planets encircle the sun and earth. Hence all the planets are in some way connected to the solar motion, and there is no need for epicycles. However, Tycho found that the orbits of Mars and the sun intersected! How could that be? Would not the solid spheres shatter?

No, because there are no solid orbs! Like Bruno, but for very different reasons, Tycho rejected the solid spheres. One night in 1577, while out fishing, he saw a comet. Rushing back to his observatory, he tried to calculate its path. He was astonished: the comet was moving *outside* the orbit of Venus! The comet was a true celestial body, not some vapor exhaled by the earth. Further, it was moving on an irregular path—perhaps an oval thought Tycho—*through* the spheres. How could anything move through solid spheres? That was easy: there were no such things!

Years earlier, in November of 1572, Tycho had been startled to see a new star (a nova) in the constellation of Cassiopeia. Since it presented no parallax, it had to be a *new* fixed star. Then, in March of 1574, it disappeared. A divine sign placed in the heavens by God? Many medieval scientists would have thought so. But we are now in the world of the Renaissance, and at least part of that world (not the magi) leans more toward natural causation than supernatural. So another ancient belief has been shattered: the sphere of the fixed stars also admits change. Bruno had posited celestrial changeability from a mystical metaphysics; Tycho did so from empirical observation.

Tycho died in Prague in October of 1601. His new patron had been the strange, erratic Holy Roman Emperor Rudolf II. During the last year of his life he worked with a brilliant "assistant" named Johannes Kepler. It was actually this Kepler who made the Copernican revolution, and in Kepler, more so than in Copernicus, we begin to see clearly the sixteenth-century idea of "revolution." His writings reveal the inner workings of his mind; we are privy to his secret motivations, his false starts, confusions, mistakes, and triumphs. We follow a twisting road through unknown country, only to discover a map of the highway at the end of our journey.

Renaissance Neoplatonism ran deep in Kepler. The glory of the heavens spoke to him of an underlying unity and simplicity, manifesting the mind of the Creator. He was convinced that the heavens operated by the archetypal laws of geometry, forms which existed in the mind of God and could be grasped by human reason. Mathematically computing the movements of the planets allowed him to participate in the divine thoughts. Yet Kepler also brought along a strong *physical* intuition, perhaps the first mathematical astronomer to do so. The metaphysical flashed through the physical. What makes the planets move? What force holds the celestial unity together? Kepler was even more of a Pythagorean than Copernicus. Mathematics sewed together the material fabric of nature, and its patterns presented spiritual ideas woven into the cloth by the Divine Mind.

As a student at the University of Tubingen Kepler was inspired by the writings of Cusa and also came into contact with Copernicus. He was converted to Copernican astronomy for "mathematical reasons," for its ability to describe those harmonious connections, and for its appeal to his esthetic sense of mathematical unity. Then there was the nobility of the sun—a doctrine to be found in both Pythagoras and Hermes.

Kepler became a teacher of mathematics at the Protestant seminary in Gratz. Tradition has it that while giving a lecture he was suddenly struck with a grand inspiration of cosmic unity, one he was to carry with him to the end. He had been searching for some numerical relationship between the planets themselves, their orbits, and their periods of revolution. Now it hit him that the answer might be found in geometry, specifically the five regular solids. The sphere which enclosed the universe was related mathematically to the cube, so Kepler placed the cube between Saturn and Jupiter. Next comes the tetrahedron between Jupiter and Mars, followed by the dodecahedron, the icosahedron, and the octohedron, all interspaced consecutively between the planets. But calculating the thickness of these figures presented problems, and Kepler required more accurate observations. Here was a man who strongly believed that his *a priori* reasoning had to accord with the numerical data of exact observation. Nothing less would reveal the mind of God.

Here, too, his physical intuition came into play. There had to be a single motive force which governed planetary motions. This force, similar to the laws of light in optics, propagated from the sun and varied in strength over distance. It spread out on a plane and therefore was directly proportional to distance, thus connecting the periods of the planets with their distances from the sun. Kepler called it *anima motrix*, a kind of "moving spirit." Since it radiated from the sun, holding the planets in their orbits, the planes of these orbits *had to intersect at the sun.* For physical reasons the Copernican eccentric would not do. For physical *and* mystical reasons the center of the universe had to be the sun. But how to compute the planetary eccentricities from the sun? Kepler needed better observations, and the man who had them was Tycho Brahe.

When Kepler arrived at Prague in 1600, Tycho put him to work on Mars. The proximity of Mars to the earth afforded more precise observations, yet it also revealed the full eccentricity of the planet's orbit. On his deathbed Tycho had begged Kepler to follow his system over that of Copernicus. But the problem of Mars defeated the ingenuity of both. For nearly ten years Kepler fought a losing battle with Mars; he tried everything, including the discarded equant point. Finally he achieved an accuracy of 8' between observation and computation. This would have satisfied Ptolemy and Copernicus. Not Kepler! The archetypal patterns of the Divine Mind had to be exact. God had also given to us the genius of Tycho Brahe—his observations—and we must make full use of this gift.

Now a number of things dawned on Kepler. Surely it was absurd to believe that a physical force from the sun would move a nonmaterial geometric point. The epicycles whose centers were such points made no *physical* sense! The same was true of the equant. Also Tycho had shown conclusively that the sun's orbit expanded and contracted, yet Copernicus had assumed a uniform circular orbit for the earth. Why should the earth's orbit be any different from the others? It, too must be nonuniform, unless referred to an equant. Equants would not do. Epicycles were physically impossible. The harmony of the universe seemed totally elusive. Kepler would not believe it. The thought may have come to him, as it later would to Einstein: The Lord is subtle, but He is not malicious!

Let us imagine the earth's orbit as if we viewed it from Mars. Its velocity, too, varies through its orbit, as we know from observing the sun. How does the change of velocity vary with the distance of any planet, including the earth, from the source of the *anima motrix*, the sun? Kepler plunged into a maze of mathematical considerations. Slowly a radical new pattern began to emerge. Had not the great Archimedes divided a circle into infinitely many triangles in order to find the ratio of the circumference to the diameter? Let us divide the orbit into areas bounded by rays of the sun. Perhaps we may be able to find the *time* it takes for a planet to pass over the arc described by the end points of this area. And here Kepler discovered a pattern. The relationship between varying velocities and distances was to be found in the comparison of areas—equal areas are described in equal times! To have this law, however, he was forced to surrender the most ancient tradition of astronomy, uniform circular motion!

The problem once again was Mars. Computations of its assumed circular orbit according to the new law did not agree with observation, and Kepler's celestial dynamics prohibited the use of epicycles. All along this had been the key element in his thought, this *anima motrix* which seemed to spring from his solar Neoplatonism. Only one road led out of the impasse: *the orbit of Mars must not be circular!* At first he tried an oval. The margin of error was reduced but still did not fully agree with observations. Finally, as Kepler himself wrote: "I awoke as if from sleep." The orbit was an ellipse with the sun located at one focus (Figure 13.3). And with this "awakening" the second law fell into place: In lines describing

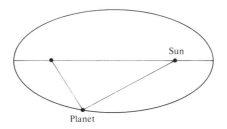

Figure 13.3

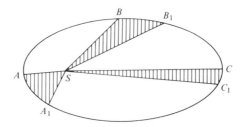

Figure 13.4   Areas SAA_1, SBB_1, and SCC_1 are equal, and the planet moves through the arcs in equal times.

equal areas from the planet to the sun, the planet will sweep the corresponding àrcs in equal times. Thus the varying velocities are accounted for. In a elliptical orbit the second law shows that the planets move faster when they are nearer the focus of the sun and more slowly near the empty focus (Figure 13.4).

At last the harmony of the universe had been revealed to Kepler in all its simplicity–indeed, a mathematical simplicity which had escaped even Copernicus. Having been forced to choose between perfect circular motion of the celestial bodies (the basic presupposition of all astronomical science) and a kind of Pythagorean mathematical harmony (which originated from fundamentally religious and mystical feelings), Kepler adopted the latter. In 1609 his "discoveries" were published in a book entitled *Astronomia Nova.*

To Kepler the inverse distance law which sprang from his concept of solar force was equivalent in principle to the second law of areas (although actually it is not). It suggested a force which *acted upon* the planets, weakening further out from the sun. Implicit in this idea is a basically *inert* planet. We may, if we wish, substitute the word "force" for *anima,* wrote Kepler in 1621. And so with a simple change of names we begin to journey into a new world, for the motive species of the sun has finally become mechanical! Behind, retreating into the clouds of memory, is the Hermetic cosmos of *living* force or spirit; ahead, appearing now upon the horizon, is a new mechanical universe of matter in motion.

Kepler had first assumed that the sun rotated, creating a whirlpool or vortex which carried the planets. Then he came across William Gilbert's *De Magnete,* published in England in 1600. Magnetic attraction, claimed Gilbert, is a kind of incorporeal force or virtue with one of its properties being rotation. Gilbert's magnetism, as much as Kepler's *anima motrix,* tended toward the occult forces of Renaissance Naturalism. On the other hand, Gilbert made the earth a huge magnet and assigned its diurnal rotation to that inherent magnetic property. The earth revolved because of its "magnetic soul."

If the earth is a magnet and one of the planets, why not assume that the planets are also magnets and the sun is a magnet as well? And so Kepler did. The sun has one pole on its surface and the other at its center. The poles of the planetary magnets are situated at either end of their axes. As the planets swing around the sun, they present their "friendly" poles or their "unfriendly" poles. Hence they are attracted and repelled according to the properties of magnetism, and this explains their behavior apropos to Kepler's laws.

Notice that there is no reason to suppose any relationship between this celestial magnetism and the fall of bodies on the earth. The force of gravity is qualitatively distinct from the "force" which holds the planets in their orbits. While Kepler destroyed the wall separating geometric heavenly circular motion from curved motion in general—as Tycho knocked away celestial perfection—there still remains a solid mechanistic barrier.

Kepler, however, was a true Pythagorean, and his search for harmony could not be complete until he found a pure numerical law which bound the planets. According to the Pythagoreans, numerical harmonies were to be found in the musical intervals. There must be, thought Kepler, such a musical harmony of the planets, dependent upon some pure numerical relationship. In the course of searching for this "music of the spheres," Kepler decided that the interval must depend in some way upon a relationship between the mean revolution of the planet and its mean distance from the sun. In 1619 he published *The Harmony of the World*, which contained this musical score (the music is intellectual, of course, not sensate). In the process of his search for this musical harmony Kepler "discovered" his Third Law!

The Third Law is quite different from any previous astronomical laws. In contemplating the average periods of revolution and the average distances of the planets, Kepler found what must have been a deeply moving relationship. Taking the earth's orbital time, 1 year, as a standard, and its *mean* distance as 1 astronomical unit, let us look again at Mars. In comparison, Mars' orbital period is 1.88 years (T) and its mean distance is 1.524 (R). Squaring T, we get $T^2 = 3.54$, and the cube of R gives us $R^3 = 3.54$. Amazing! It works for all the other planets! The squares of the orbital periods or the times of revolution for each planet are proportional to their mean distances from the sun cubed. T^2/R^3 gives a constant for all the planets, which is unity.

The Pythagorean vision had finally been realized. God was the archetypal architect whose immense worldly castle embodied the beautiful mathematical simplicity of the Divine Mind. Beneath apparent chaos existed a divine order and regularity which could be grasped by the human mind, a mind which thought in the same patterns as the Creator.

All this time the ghostly visitor from the sixteenth century has been peering over our shoulder, watching the revolution unfold. A shake of the head, a sigh . . . the spirit moves back into the mists.

"What faith!" we hear the phantom whisper. "To think that the frail human mind could actually conceive the thoughts of God operating behind the phenomena!"

"Faith?" we call into the corridors of time. "But the mathematics, the new discoveries, the detailed observations?"

Silence.

"Sleepwalkers," Arthur Koestler labeled the great discoverers. Perhaps. Certainly they were walking, their steps guided by faith and observation. If they were asleep, their slumber was surely not the sibling of death. It was alive, flowing and restless, a sleep filled with wonderful and vivid dreams.

SUGGESTIONS FOR FURTHER READING

BUTTERFIELD, HERBERT. *The Origins of Modern Science 1300–1800*, rev. ed. New York: The Free Press, 1957.

CASPER, MAX. *Kepler*, trans. Doris Hellman. London: Abelard-Schuman, 1959.

COHEN, I. BERNARD. *The Birth of a New Physics*. Garden City, N. Y.: Doubleday, 1960.

GINGERICH, OWEN, ed. *The Nature of Scientific Discovery: A Symposium Commemorating the 500th Anniversary of the Birth of Nicolaus Copernicus*. Washington, D. C.: Smithsonian Institution Press, 1975.

HALL, A. R. *The Scientific Revolution 1500–1800: The Formation of the Modern Scientific Attitude*, 2d ed. Boston: Beacon Press, 1962.

KOESTLER, ARTHUR. *The Sleepwalkers: A History of Man's Changing Vision of the Universe*. New York: Macmillan, 1959.

KOYRÉ, ALEXANDRE. *The Astronomical Revolution; Copernicus-Kepler-Borelli*, trans. R. E. W. Maddison. Ithaca, N. Y.: Cornell University Press, 1973.

WESTMAN, ROBERT S., ed. *The Copernican Achievement*. Berkeley: University of California Press, 1975.

14

The Language of Nature
Galileo

Does the earth *really* move? Or, was the Copernican system just another hypothesis, like the visions the fourteenth-century scholastics created "according to the imagination"? We must still explain how local motion is possible as we experience it on a moving earth. We must still deal with Aristotelian physics. Here, too, we have the example of the scholastics, yet again we are left wondering if by their intricate criticisms they meant to expound the true physical world or simply to create new "worlds on paper." Nonetheless, the great Aristotelian bridge to nature was weatherbeaten; the winds and spray of criticism had eroded its once-strong foundations. The planks had rotted and the cables were frayed. Perhaps it was beyond repair.

The Aristotelian bridge was a language, a system of symbols and the rules for operating with them, meant to carry us over from the island of the human mind to the great continent of nature. The scholastics, despite all their criticisms, had still used that bridge. It had buckled beneath their feet, and now, two hundred years later, it would no longer support the weight of Copernicus. The time had come to construct a new bridge, a new language. Such was the task of Galileo Galilei.

Galileo himself has become a symbol. Few would deny his profound influence upon the course of Western science. It was Galileo who finally changed the language of physics from causes and qualities to quantity, measurement, and description. It was Galileo who proposed that experiments—designed to answer specific questions—be the deciding tests of hypotheses. He defended the Copernican system publicly, loudly, becoming its greatest propagandist. Yet, a great deal of controversy swirls about Galileo.

Did his experiments really assume the tremendous importance he seems to imply? Or were they propaganda tools, ideal "thought-experiments," which when performed in the physical world could hardly be conclusive? Were his observations with the new telescope exaggerated, again part and parcel of his polemics for the Copernican system? Was he a Platonic mathematical idealist, attempting to mold nature into a preconceived metaphysical pattern? Was he the working scientist of our elementary textbooks, allowing himself to be guided by observation and experiment, searching for mathematical relationships only after the relevant data had been gathered?

The most important proponent of Galileo's Platonism was Alexandre Koyré. Aristotle sought to explain motion and change as we experience these things here on earth by rationalizing the evidence of common-sense perception. But Galileo begins with a faith, a belief, that matter must be mathematical, that motion obeys mathematical laws which we do not actually see by sheer unaided observation. For him, as for Plato, matter is only an *approximation* of eternal geometrical forms. Galileo's moving bodies are not physical, said Koyré, they are ideal bodies moving in mathematical Euclidean space. Free fall in a vacuum, perfectly smooth inclined planes, compound motion of bodies on the earth—we can only imagine these things in the abstract. Science is a series of ideal statements about the world, yet such statements in the language of mathematics have an ontological veracity which overshadows common sense. Galileo signifies the triumph of Plato over Aristotle—"Plato's revenge!"

Others have questioned Koyré's assertions. Galileo, like Aristotle, believed that science must give us knowledge of the real sensate world. Nature itself is rational, not the chaotic receptacle of the *Timaeus* about which we can merely construct "likely stories." Mathematics is simply a better tool for the investigation of that rational structure. We need not torture ourselves over *why* mathematics works so well, over the metaphysical implications of our ideal descriptions. That is not the job of science. Mathematics for Galileo was an instrument, says Stillman Drake, but the instrument need not be part of the finished product.[1] Galileo's use of mathematics need not imply metaphysical Platonism. Whatever demonstrations he made had to be supported by experience, controlled experience—experiment. And Galileo, better than anyone, knew that experiment could be only approximate—that abstraction must not be confused with material nature.

---

[1]Stillman Drake, *Galileo At Work* (Chicago: University of Chicago Press, 1978), p. 365.

No matter how one decides to view the problem, one thing is certain: Galileo changed both the language and the tasks of physical science. With Galileo metaphysical questions are pushed into the background, and pure sensate experience is restricted and even given up for formal descriptive mechanisms. Biological animism, still present in *both* Plato and Aristotle, is finally read out of *physical* science. Perhaps it was, ironically, a metaphysical decision, or as the philosopher Edmund Husserl said, an "idealized cloak of thought" thrown over the fluid lifeworld of human experience. Nevertheless, Galileo's physics divided the world into levels—the level of initial subjective perception and the realm of measurable quantities which we have come to label "objectivity." Whether objectivity is subjectivity in another guise remains an open question. And it is possible that all the debates over Galileo's Platonism arise from this unanswered riddle.

Galileo Galilei was born in 1564 near Pisa. His father Vincenzio was a cloth merchant and talented musician, hence he had a good knowledge of mathematics. In 1581 Galileo was enrolled at the University of Pisa to study medicine. The course in the medical faculty was, of course, based upon the works of Aristotle and Galen. Imagine the young Galileo, already showing some predilection for mathematics, attending the tedious lectures on Aristotelian natural science. He listens and takes notes, attempting to follow the tortuous discussions of the scholastics. The requirement that science conform to actual experience was not, as we have seen, a major consideration in such study; the discussions were over logical relations in the text.

In his spare time he begins to study Euclid, then Archimedes. What clarity! In many ways mathematics is an art; its symmetry and precision compel the student not only to assert the necessity of its conclusions, but also to feel an esthetic pleasure just as visual art stirs the emotions. Galileo loved art, too; he read poetry and found joy in music to the end of his life. His pleasure came from the form rather than the content. When asked by the Florentine Academy to lecture on the topology of Dante's Hell in the *Divine Comedy,* he geometrized its structure using Archimedean conic sections and proportiions of the German artist Albrecht Dürer! Even in poetry, form and measurement appealed to him over content.

As a student Galileo gained the dubious reputation of contradicting his professors. The speeds of falling bodies in the same medium are proportional to their weights—a body twice as heavy will fall twice as fast. So says Aristotle. Yet, Galileo protests, I once saw hailstones, some twice as large as others, falling together and striking the earth nearly at the same time! Aristotle's physics *does seem* to contradict experience. Now here was Archimedes, who applied his mathematics to physics, an ancient mathematician who used his demonstrations to discover things about the physical world. Geometry at play . . . ?

Galileo quit his medical studies and took up tutoring mathematics. In 1589 he became Professor of Mathematics at Pisa. There he stayed until 1592, when he left to take a position at Padua. During these years he worked on the problem of motion; later he would write that to be ignorant of motion was to be ignorant of nature. He rejected Aristotle's laws of motion, yet there was nothing new in that. The problem was to find something which worked better. By Galileo's time the Spaniard Domenico de Soto had applied the Mean Speed Theorem to account for the observed acceleration of falling bodies. Perhaps Galileo knew of it second hand. Perhaps, too, the fourteenth-century scholastics were becoming for him dreary Aristotelian logic-choppers, bookish and sterile, no matter what their theories. Armed with his Archimedes, he was determined to step out into the world and begin anew.

His first treatise on the subject, *De motu (On Motion)*, was that initial, halting step. Was there a qualitative distinction between heavy and light bodies, as Aristotle maintained? Did not Archimedes prove that some bodies which are heavy in air will be light in water? Consider a piece of wood; it falls in air, yet when released under water it rises. Therefore a body's weight—its heaviness or lightness—must be relative to the medium in which the body is found. All bodies are actually heavy; they seem light only in a medium denser than their own material. If we subtract the medium, then all bodies will have a specific weight, and it is this specific weight which determines their velocity in free fall. In a vacuum two bodies of the same material, no matter what their gross size, will fall at the same rate and not instantaneously.

We have seen such arguments dating back to Philoponus. And in order to explain acceleration and the violent motion of projectiles, Galileo even adopts a kind of impetus physics, not unlike Buridan. Throwing a heavy body or supporting it by a tower *impresses* a force which, as the body falls, slowly drains away. This impressed force actually acts against the downward force of natural motion (gravity), and thus the observed acceleration of falling bodies is a kind of evaporation—the overcoming of impressed force by natural motion. Uniform acceleration, then, is an *accidental* feature of free fall. And, like those of the impetus school, Galileo holds that the speed is proportional to the distance traversed.

So far Galileo is merely repeating earlier criticisms of Aristotle, whether he knew of them or not. The significant difference is that he experiments! First, there is the colorful story, reported by his biographer Viviani, of his dropping objects of the same composition but different weights off the Leaning Tower of Pisa. If the story is true, it would only have had the affect of proving Aristotle's proportionalities wrong—the heavier body would still strike the ground a little before the lighter. As we have seen, many others had questioned Aristotle's formulas before.

Next Galileo rolled balls down inclined planes, the inclined plane being a limiting case of vertical fall, just as Jordanus had shown it was a limiting case of a body's weight. Significantly, Galileo attempted to compute ratios of speeds in relation to the distances traversed on the planes for bodies of different densities. He failed. It was one thing to criticize Aristotle, but it was quite another to find the proper law of free fall. Impetus physics did not seem to work either.

How important were these experiments? Was it, instead, Galileo's inability to mathematize impetus physics which gave him the clue that uniform acceleration might be a natural consequence of free fall, as, for example, Koyré maintains? Probably it was a little of both. Certainly he is thinking about his planes in the abstract: the plane is considered incorporeal, the balls perfectly smooth, friction and air resistance neglected. Impetus physics was still a semiqualitative causal description of free fall. Galileo's problem was how to match his experiments with a new level of mathematical abstraction—Archimedean bodies moving in Euclidean space. After he had moved to Padua, Galileo abandoned impetus physics for good.

Motion must be considered as a state of being to be described; no longer are we interested in what causes motion. Uniform acceleration, then, can no longer be considered as accidental. It, too, is a natural fact which must be mathematically tamed. The goal now is to completely dissect motion, ignoring weight, medium, friction, and the other bare facts of perception, searching for the laws which govern the relationship of abstract entities. This approach makes it easier to establish the laws of motion for the moving earth of Copernicus. Let us digress for a moment and glance at Galileo's Copernicanism. The two, local motion and the moving earth, go hand in hand.

Hearing of a Dutch lensgrinder's spyglass, Galileo in 1609 experimented with lenses and constructed a crude telescope of his own. For some years he had been a wary Copernican. Now, gazing into the heavens with his telescope, he became convinced. It was the same year Kepler published his *Astronomia Nova*. But while Kepler's book was a massive, forbidding volume, Galileo's publication of 1610—*Sidereus Nuncius (The Starry Messenger)*—was a charming little tract for the educated layman. More a chronicle of his observations than an astronomical treatise, it had implications no less revolutionary than Kepler's labyrinthine researches.

The moon, Galileo claimed, was not unlike the earth—full of mountains, craters, and valleys. He even measured the height of a lunar mountain and found it to be about four miles. Implication: the Aristotelian distinction between the celestial and terrestrial realms was apparently without substance. The Milky Way proved to be a massive collection of stars, and while the planets were magnified in size by the telescope, the stars themselves were not perceptively increased. They were luminous points, and the lack of a parallax no longer seemed all that odd. Galileo was most surprised to discover four tiny planets orbiting Jupiter. A major planet had satellites like the earth. This fact ought to "quiet our doubts" about having the moon circle a moving earth in the Copernican system—*four* moons kept pace with Jupiter! Later in his *Letters on Sunspots* Galileo found that Venus went through a complete cycle of phases like the earth's moon—a phenomenon difficult to account for in the Ptolemaic system but making perfect sense if the orbit encircled the sun inside that of the earth. Sunspots, of course, were another argument against the incorruptibility of the heavens.

Now these were observations, and opponents could always question (as we might) the reliability of Galileo's telescope. Some Aristotelians claimed that the instrument itself produced illusions; others decided that the moon was really covered by a perfectly smooth and invisible sphere. The sunspots might actually be tiny planets close to the sun. Further, none of these observations conclusively proved Copernicus; they could easily be accommodated by Tycho's system. Thus it was perfectly rational and scientific for men like Cardinal Bellarmine to accept Galileo's observations but deny them any conclusive proof of the Copernican system. Strictly speaking, they "proved" nothing until one looked at them through *Copernican telescopes*! How to get people to see the heavens this way? Galileo knew: propagandize!

Galileo had to convince people that the Copernican system was the *only possible* explanation for what they saw in the heavens as well as the motion they experienced upon the moving earth. Aristotelians and their "worlds on paper" were the enemy. He could rail at them for not getting out of their books to study nature directly, but underneath runs the question of *how* we are to observe. We are to have an entirely new experience of the world. The old facts must be made to suggest new theories, and in order to do this Galileo must change the very language of physical science. So he writes in *The Assayer:*

> Philosophy is written in this grand book, the universe, which stands continually open to our gaze. But the book cannot be understood unless one first learns to comprehend the language and read the letters in which it is composed. It is written in the language of mathematics, and its characters are triangles, circles, and other geometric figures without which it is humanly impossible to understand a single word of it; without these, one wanders about in a dark labyrinth.[2]

[2]Galileo, *The Assayer* (1623), in Stillman Drake, *Discoveries and Opinions of Galileo* (New York: Doubleday, 1957), pp. 237–238.

The die is cast. How may we explain what our senses obviously show: that bodies fall at right angles to the earth, a "fact" hardly possible, it seems, if the earth moved. We must first adopt the new language.

Aristotle had always taken into account the natural context of motion. His distinctions between heavenly and terrestrial, natural and forced, were based upon the context of our experience. Ah yes, says Galileo, writing in his *Dialogues Concerning the Two Chief World Systems— Ptolemaic and Copernican*, but it is also true that we ourselves are part of the natural context. Imagine yourself on a ship, and let us say that your eyes are fixed on the sail or the top of the mast. The ship is moving rather quickly. Must you constantly adjust your eyes to keep pace with the moving sail? No. By sight alone you cannot detect the motion of the ship and all its contents (including you). Now, what if a ball is dropped from the top of the mast? The character Simplicio, who represents the Aristotelian point of view, gives the standard answer: the ball will land somewhat behind the mast. Surely no Aristotelian has ever performed this experiment, answers Galileo. The ball falls to the base, and we need not even experiment to show it!

Let us say that the ship is moving uniformly (nonaccelerated) on a calm sea, and let us, for the moment, neglect wind, waves, and other *extraneous* things. Could a person on the ship say with certainty that the ship was actually moving or at rest? No matter what experiments we performed on the ship—dropping bodies or whatever—we would not be able to predict uniform motion. Only changes in our *rate* of motion would be perceptible. Thus there is a form of motion which is invisible— the frame of reference which the observer shares.

Therefore the falling body is really a *compound* of two motions—one we see and one we do not. We see only its vertical motion. We do not see its horizontal motion because we share in it. And since the revolving earth is the frame of reference shared by all observers, it is no wonder that we experience only vertical fall. If we assume a uniform rotation of the earth about its center, we may demonstrate *geometrically* that the falling body moves through equal arcs in equal times, as does any point upon the earth. Hence we see only motion that we do not share. We must extrapolate or abstract from common experience in order to find the reality behind our initial perception.

Notice Galileo's clever procedure. From the onset he has assumed a moving earth. In order to convince us that our perceptions cannot judge *against* its uniform rotation he concocts an ideal situation—"a painted ship on a painted sea." He calls upon the evidence of experiment, although he already knows the outcome. And he caps the whole thing off with a mathematical demonstration. Any frame of reference which preserves a uniform rate of motion or state of rest is therefore *relative* to the observer in that frame; that is, an observer within the system cannot physically detect its uniform motion or rest. The implication is that only a privileged observer, say someone at a fixed point in space, could decide between the two states. Common sense gives way to mathematical abstraction. But this is no Platonic likely story; the physical principles inherent in natural experience, once abstracted, are fully true and real.

The outraged Aristotelian might well wonder how the ball keeps pace with the mast, or how it preserves the rotation of the earth when dropped from a tower? Galileo already knows the answer from his earlier work on motion, and now he formulates the principle in a Copernican universe. It is *inertia*! No one really knows what forces cause the fall of bodies or the rotation of the earth, says Galileo in the *Dialogue*. No matter! Let us *imagine* a ball on a perfectly horizontal plane, excluding all resistance. If no outside forces acted upon the ball, it would remain forever at rest. Now give a push. Remember: there is no resistance, nor is there up and down—and let us add that the space or plane itself is boundless. Would not the ball preserve its uniform motion forever?

We may conceive another "thought experiment." Galileo, in fact, already has one in mind, a situation he imagined while writing about his sunspots. A ship on a tranquil sea, if given impetus to movement, will continue to circle the globe forever if all *extrinsic* forces are removed. A ship at rest in such a situation will remain at rest. As purely abstract states of being, the ship at rest or the ship in uniform motion are perfectly equal to anyone on board. Thus balls dropped from the mast preserve this "state" and fall in the same place whether or not the ship is at rest or in motion. Again we are persuaded.

The earth's inertial motion is circular, and since gravity draws a body to its center (the point about which it rotates), the actual preserved or inertial state of such a body is also circular. Galileo's principle of inertia therefore, unlike that of Descartes or Newton, must be called "circular inertia." In the absence of any extrinsic force a body in motion or in rest will preserve that state. Yet, Gaileo's universe is still finite; at least he makes no positive assertion of its infinity. Thus there can be no infinite straight line. Nor does Galileo seriously entertain the possibility of Kepler's ellipses. It seems, then, that circularity was one tradition with which he did not break. Neither does he speculate on the force which holds the planets in their orbits. Why should he? His task is to explain local motion and how it is possible in our evident experience on a moving earth. On the other hand, space has lost all its qualitative distinctions; it has become like a blackboard upon which we sketch Euclidean figures. Once space has been abstracted, can the infinite void be far behind?

The poor Aristotelian natural philosopher has been handled roughly indeed by Galileo. Even Galileo's own mouthpiece, Salviati, is amazed at how "the sheer force of the intellect" overcomes common sense. Galileo no longer stands upon common scientific ground with the Aristotelians. And what is worse, he raises the old ghost of Democritus and the atomists! We read in *The Assayer* that only those qualities—size, weight, etc.—which are measurable are the real scientific properties of bodies. Tastes, odors, colors and so forth are products of the imagination! Galileo, in short, wishes to reduce and restrict our scientific experience. If we consider the gross experience of change, we are forced to admit that Aristotle and the scholastics explained much more— remember Oresme? Such are the wages of mathematical rigor!

Galileo had argued his case too convincingly, although he never asserted that he had *irrefutable* proof of the Copernican theory; rather, the preponderance of evidence taken together seemed to indicate that the earth moved and Aristotelian natural philosophy was mistaken. Galileo himself was a good Catholic and was certain that any conflicts between his science and his religion must lie in interpretation. But the philosophers, smarting from his attacks for many years, were not adverse to using theology and the considerable power of the Church in order to find the weak spots in his armor. To Galileo such questions were beyond the scientific pale.

Galileo's visit to Rome in 1611, after the publication of his *Starry Messenger*, was in general a great success; he found supporters among the Jesuit astronomers of the Roman College, and even Cardinal Bellarmine, one of Bruno's judges, showed his esteem for Galileo. Nevertheless, his foes could and did make use of the complications which might arise from the theological issues.

Part of the problem was the challenge of the Reformation and the heated question of Biblical interpretation which divided Catholic and Protestant. Any Catholic reinterpretation of Scripture—say, to accommodate Copernicanism—could become ammunition for the Protestants in their propaganda war with Catholicism. Galileo probably did not believe that reconciliation was even necessary. Scripture, he held, did not favor one astronomy over another, nor did it in general pronounce on anything scientific. But theology was too interwoven with cosmology, as we know, and so complications were inevitable.

Copernicanism *as a hypothesis* had not been condemned. Cardinal Bellarmine said that as long as the motion of the earth was treated as a hypothesis, not actually true, all was well. The first clash with Church authority precipitated by Galileo's Aristotelian enemies came in 1616 and resulted in Bellarmine's prohibition against holding the Copernican system as true. What actually occurred when Galileo met with the Inquisition and the Cardinal has been debated. Certainly attempting to reconcile Copernicus with Scripture was prohibited, yet this did not mean that Galileo was not free to discuss a "hypothesis," weighing the arguments on both sides. If he was not allowed to "teach" the system, then he would be required not even to describe it. Galileo left Rome with an affidavit given to him by Bellarmine saying that he must no longer hold or defend Copernicanism, which would mean that simple discussion of the two sides was perfectly legal.

Thus the *Dialogue,* published in 1632, was supposed to be an impartial, hypothetical discussion in accordance with Bellarmine's warnings. Galileo even received the *imprimatur* of the Inquisition for the book. But alas, the power of his own polemics may have betrayed him, for his scientific enemies could certainly make a good case that he was actually demonstrating (or trying to) the reality of the system. Worse, Galileo's opponents successfully spread the rumor that the character Simplicio was in fact a clownish caricature of Pope Urban VIII, who had been friendly toward Galileo. An old document from 1616, an unsigned notary's memorandum recounting the single meeting with the Inquisition, was dug up and presented to the Pope. It included the words "nor teach in any way," and so in August of 1632 Galileo was summoned to Rome to stand trial. There he was condemned by the Inquisition in 1633; he was forced to abjure and curse his theories, forbidden to publish his books, and condemned to perpetual house arrest. The tragedy was that Galileo actually wanted to protect the Church from upholding erroneous scientific beliefs, and some Cardinals and other Church officials who had no scientific stake in the matter actually supported him. His "true" persecutors were the philosophers, the "scientists" of his day. These were the culprits who forced the Copernican issue with the Church. Galileo, the great martyr for science, was in reality a victim of the alliance of the Church and the scientists of his day!

The Condemnation turned his attention back to the physics of motion, and his final book, *Discourses and Demonstrations Concerning Two New Sciences,* was published in Holland. It was the culmination of his work on motion, establishing at last a new and successful law for the free fall of bodies, the flight of projectiles, and the mathematical demonstrations he had been seeking. Geometry, once confined to the heavens, was finally brought down to the earth.

Motion is a state of being in which a body finds itself, and this state in free fall is exemplified by acceleration. The accelerations acquired by bodies falling from the same vertical height are identical. A body falling from rest through a vertical displacement and a body descending by an inclined plane from the same height acquire equal velocities even though the path along the plane is longer. It is not the distance traversed but the time which is the measure of free fall—in our notation, $v = at$. Let us see how Galileo reached such startling conclusions.

Consider two bodies of different specific weights, say lead and ebony, falling in a very dense medium, says Galileo in *The Two Sciences*. The difference in their velocities will be perceptible. Is this difference dependent upon the ratio of their specific weights? If the medium becomes rarified, we should expect this ratio to hold. But it does not. In the rare medium the difference is nearly imperceptible, and in rarer mediums becomes less, so that we should expect in vacua all bodies will fall *at the same* rate. In a vacuum, then, bodies will fall from rest together, whatever their weights or composition. And in this ideal situation we see that uniform acceleration—a body receiving equal increments of velocity through equal intervals of time—is the constant dynamic principle of free fall.

We can already hear the Aristotelian protest: "Let us return to the real world! If we allowed the fall to continue for some time, should we not expect the heavier body to outrun the lighter? And since the medium is the same, should this not be due to their specific weights?"

Galileo again shows that what actually happens conflicts with Aristotelian common sense. The medium resists motion, and the resisting force is proportional to the speed of the body. As a body continues to accelerate and gain more speed, the resistance of the medium becomes greater, reducing the rate of acceleration. The falling body in this way generates a resistance of the medium, which is a force the body experiences as "a diminution of weight." A lighter body will have its rate of acceleration diminished faster, and the perceived difference of speeds is due *only* to the increasing resistance of the medium upon this *rate*. In fact, given a long enough time, the retardation becomes so great that it balances the rate of gain of speed, and free fall becomes uniform—the body reaches its terminal velocity.

Notice that for Galileo weight or heaviness is still a unique property of a body. Gravity, therefore, is not something acting upon the body. Also, according to Galileo, the tendency of heavy bodies to fall toward the center of the earth is still their "natural motion." Yet, having established the dynamic principle of free fall—acceleration—and having swept away the "accidental" conditions of our experience, we may now study this motion mathematically—an abstract description which holds true for *all* bodies. Let us glance once more at our inclined plane.

The inclined plane is a limiting case of free fall. Galileo wants to show that speed is proportional to time, since uniform acceleration says that equal increments of speed are gained in successive intervals of time. For any inclination of the plane (from the same height) he found that the distances traversed were proportional to the square of the times. This relationship implies only that speed is proportional to time. Further, we are asked to imagine that a rolling ball exhibits the same behavior as a freely falling ball, since the experiment is designed to prove the relationship for free fall. But what of it? We already know how misleading unaided, common sense experience can be.

The experiment, then, reveals the significance of the Mean Speed Theorem, which Galileo proves geometrically in a way similar to Oresme. The time in which any space is traversed by a body starting from rest and uniformly accelerated is equal to the time in which that same space would be traversed by the same body moving at a uniform speed whose value is the mean of the highest speed and the speed just before acceleration began. And we have already seen in Chapter 11 how we may derive the formula $S = \frac{1}{2} at^2$. Since the distances described by a body falling from rest are to each other as the squares of the time intervals (Galileo's next theorem for uniform acceleration), it immediately follows that, taking any equal time intervals from rest and assigning a numerical sequence from 1, squaring these intervals gives 1, 4, 9, 16, 25, . . . , and the differences between them are the odd numbers 1, 3, 5, 7, ... Hence the distances traversed in these equal time intervals bear to each other the same ratio as the series of odd numbers 1, 3, 5, 7, ... And the experiments "always" agree with these ratios! The same body takes more time to "fall" (roll) along the inclined plane than to fall from the vertical height. Bodies acquire identical accelerations for the same vertical displacement, and the inclined plane, no matter what its inclination (and hence length), is simply a limiting example of this rule. Therefore the time for movement on the plane is greater, which makes up for the limiting factor of the plane. The terminal velocities, then, for the same body falling along the plane or directly from the vertical, are equal, and the ratios hold (see Figure 14.1).

The Aristotelian Simplicio gives in. Although he would have liked to be present at the demonstrations, he is still satisfied and accepts their validity. What is it that persuades him? Most likely Galileo did perfect his experiments and found that the results "roughly" agreed with the ratios he sought. Surely they were also idealized. Yet in the *Discourses*

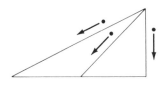

Figure 14.1

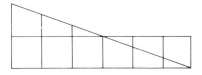

Figure 14.2

Simplicio is made to say, "If I were again beginning my studies, I should follow the advice of Plato and start with mathematics."[3]

There are other marvels to behold. Consider the pendulum. If the strings are of equal length, bobs of any material pulled aside through whatever number of degrees will take equal times to pass through these arcs. With cords of different lengths the period varies as the square root of the cord's length, and the lengths are to each other as the squares of the times. So the pendulum, a very common example—Galileo mentions his observations of swinging church lamps—also exemplifies the times-squared law. What Aristotelian would have believed it? A pendulum is analogous to a falling body!

The analysis of projectiles brings in once again the concept of inertia. A shell fired from a cannon will have two independent motions: the uniform horizontal motion and the acceleration of free fall. Again neglecting the retardation of the medium, the horizontal motion is preserved while gravity bends the projectile downward. Galileo proceeds with geometric proofs, demonstrating mathematically how the two motions are combined to give a curved trajectory, a parabola. He is able to find the elevation of the cannon which will give the projectile maximum range (it is 45° as shown by Niccolo Tartaglia in 1537). Gone are causes, impetus, Aristotle's "air machine," and the like. We must, says Galileo, "cut loose" from accidental properties in order to grasp nature. Such properties, we are tempted to add, were the very materials of the Aristotelian bridge.

So changed the language of nature. What about the grammar of that language? Indeed, lurking behind the beautiful mathematical demonstrations is the awful shadow of Zeno. Look at Galileo's figure for the Mean Speed Theorem (Figure 14.2). The figure is divided into parallel lines representing increments of speed gained through time. Each line, therefore, represents an "atom" or indivisible increment. May we not, however, divide the figure further—to infinity? Can an infinite series, then, have a sum and yet be made up of indivisibles such as atoms? Are we not again facing the perplexing problem of instantaneous speed?

[3]Galileo, *Discourses*, trans. Henry Crew and Alfonso de Salvio (New York: Dover, 1954), pp. 90–91.

Galileo had to admit that the paradox of the infinite and the indivisible were by nature incomprehensible to "our finite minds." Yet he stressed that such concepts as equality, greater than, and less than had no place in comparing infinities or infinity to the finite. Consider the infinite class of all positive integers. Now imagine an infinite class of all perfect squares—4, 9, 16, . . . Up to 100 we have ten such squares. As we continue, we encounter squares less frequently. Yet as an infinite class, the number of such squares may be placed into a one-to-one correspondence with the integers . . . to infinity! What a strange situation! It is much easier, says Galileo, to think of the infinite in terms of a continuum, the indivisibles or "atoms" merging into a unity like a fluid. Galileo's student Cavalieri used indivisibles in geometric proofs. Kepler had already considered a sphere as being made up of an infinite number of infinitesimal cones, vertices at the center, and bases making up the surface. He was working on the problem of determining the volumes of wine barrels! Many other mathematicians, in different ways, would wrestle with the same or similar problems.

What are we to make of it? In the end the language itself was far from clear. The geometrical thinking of the ancient Greeks still dominated physical thought, carrying with it all the inherent paradoxes of antiquity. But faith in that language to truly represent physical nature had seized hold—a faith which overshadowed its logical inconsistencies. And this faith rests at the heart of the Galilean symbol.

SUGGESTIONS FOR FURTHER READING

Burtt, Edwin Arthur. *The Metaphysical Foundations of Modern Physical Science.* Atlantic Highlands, New Jersey: Humanities Press, 1951.

Drake, Stillman. *Galileo At Work: His Scientific Biography.* Chicago: University of Chicago Press, 1978.

Koyré, Alexandre. *Metaphysics and Measurement: Essays in Scientific Revolution.* London: Chapman and Hall, 1968.

————. *Galileo Studies,* trans. John Mepham. Atlantic Highlands, New Jersey; Humanities Press, 1978.

McMullin, Ernan, ed. *Galileo: Man of Science.* New York: Basic Books, 1967.

15

~~~~~~

From Magic to Mechanism

It was St. John's Day, June 24, 1527. Copernicus was still quietly reworking Ptolemy; Galileo had not yet been born. At the University of Basel a bonfire had been set to celebrate the day. Suddenly, up stepped Philippus Aurelius Theophrastus Bombastus von Hohenheim (thankfully he is known simply as Paracelsus). Before the horrified professors he threw Avicenna's *Canon of Medicine* into the roaring blaze. "Your Galen, your Avicenna, and all their followers know less than the buckles on my shoes!" he shouted as the flames consumed the ancients. "All the old writers put together are less than the hairs of my beard!" Four hundred years later the nineteenth-century anarchist Michael Bakunin would proclaim that destruction was also a creative act. No one knew this better than Paracelsus!

Here was Paracelsus—mystical philosopher, alchemist, surgeon—crying out for experiment and observation to replace the authority of the ancients. And he did more than cry out. He roamed Europe actually doing experiments—healing, observing, ceaselessly working at his alchemical furnace, attacking authority. His own curious blend of experiment and mysticism viewed the universe as filled with incorporeal spiritual forces. He was the perfect embodiment of the magician-scientist.

We, the survivors, would like to see the scientific revolution as primarily the work of people like Galileo. From the vantage point of our "modern" science we seize upon such lines of descent. Yet alongside this tradition there existed a complex kaleidoscope of change, sometimes separate and sometimes intermingled with the main stream. Organic nature, as important as physical nature for science, presented an interwoven tapestry of diverse fields and problems. The word "biology" was unknown until the nineteenth century. Alchemy itself was hardly a single scientific discipline like our modern chemistry; it was a philosophy of the universe. Medicine was an academic field, while surgery was still a simple craft. But all were changing.

So here was Paracelsus burning medical books. Medicine is not to be learned from books, he claimed; patients are not cured by theories. The physician must be guided by experiment. Not only Galileo, but Paracelsus too proclaimed that the method of science was experiment. The alchemical laboratory was to replace the scholar's study. But this alchemy is different from that of the Middle Ages. Its goal is not to transmute metals, for alchemy is more than a method. It is the key to life and nature. All changes in the body should be understood through chemicals, and diseases are to treated with chemicals. Man is a unique microcosm uniting vegetable, mineral, animal, and even celestial forces. The human body is a chemical factory, as is all nature.

This chemical machine, however, runs upon incorporeal spiritual essences. The outward chemical functions of material bodies are determined by vital spirits. The forms of all organized bodies may be attributed to an incorporeal essence which Paracelsus called the *Archeus*. These invisible virtues cannot be grasped by logic; rather they must be uncovered through intuition and experiment. Medicine is a constant seeking, the hard and tedious work of the laboratory.

The chemical doctrines of Paracelsus form a maze of spiritual principles and forces which sometimes defy logical categories. What interests us was his rejection of Galenic medicine. The humors were dropped, along with the four elements. The new basic principles of matter were mercury (vaporous and fluid), salt (inert), and sulphur (combustible). Yet these principles were not really chemicals; they were spiritual entities, and they differed in quality with each individual body. For every individual their union was determined by an organizing force, the *Archeus*, and even individual organs had their own identities, their *Archeus*. Thus Paracelsus was led to consider diseases as specific species like any natural thing, extrinsic to the body. Disease was not an imbalance of humors; it was an attack, and its essence was chemical. So remedies had to be specific, too, and they could be discovered only through chemical experiment.

This chemical medicine, this "iatrochemistry," was a vague collection of the fantastic and the practical. The emphasis upon the laboratory resulted in an increased empirical knowledge of chemical properties and remedies. Paracelsus himself produced the anesthetic we call "ether."

But most important, his was one of the first serious attacks on the four elements and qualities. Mystic, experimenter, mountebank, even drunk, Paracelsus declared his independence from authority and stressed the beneficial goals of knowledge for humanity.

Similar in approach was Gilbert's treatise on the magnet. He overturned many fables through his experiment and observation. Static electricity and magnetism were identified as different phenomena, yet magnetism for Gilbert was like chemistry for Paracelsus—an incorporeal force which was no less than the primal "energy" of nature. Magnetism was the soul of the earth and responsible for its rotation as the vital principle of matter. No doubt it was this idea which "attracted" Kepler!

For both Paracelsus and Gilbert nature was animate through and through, and Hermetic and Neoplatonic occult virtues guided their experiments and observations. The tradition they represented has been called Renaissance Naturalism, and its basic principle was the dynamic concept of nature. Recalling Albert the Great, though, we know that there was an impressive body of naturalistic and rational alchemical literature during the Middle Ages. Surprisingly, the mystical and religious aspects of alchemy actually increased during the sixteenth and seventeenth centuries. Why?

The two centuries were consumed by unparalleled religious strife. The wars of the Reformation touched nearly every country in Europe, and the universality of the Church was washed away. Conflict created dogma on both sides, and religion became mired in national politics. Many turned away in disgust—"a plague on both your houses!" Then there was the excessive logicism of the schools, not only far removed from nature, but reducing religion to sterile disputes and rhetorical exercises. To the seekers for deep religious feeling, as well as the Galileos who sought the reality of nature, Aristotle and the schools seemed to be the great corrupters and dissemblers. Both aversions, at times, joined hands in a single person.

One such individual was J. B. van Helmont of Brussels, born in 1579. Disgusted by the sophistry of the schools as well as the dogmas of theology, van Helmont found in Paracelsan alchemy a method to pierce the facade of reason and the false vanities of learning. The door of nature, he said, can be opened only by the "art of fire"—chemistry. And with the handling of material nature the mind is able to penetrate to the essence of things, which is God's command.

Van Helmont was a reformer of Paracelsus. He did away with both the four elements and the three principles. There are only two principal elements, water and air. Since air undergoes no chemical changes, the material matrix of all things is water. But van Helmont was a very different sort of Thales, for he used an experiment to prove the primacy of water. He weighed an amount of earth and planted in it a weighed willow. For five years he did nothing but add water. The tree, which initially weighed five pounds, after five years was found to weigh 169. Conclusion: 164 pounds of willow came directly out of water! Fossils, too, show that minerals develop from water, van Helmont thought. So if water is the common matrix, what differentiates things?

Water is the general matrix of matter, but material informed is animated by "seeds" which produce all objects and life alike. Now what would this "animated" water look like? How would we recognize it? It cannot be plain old lifeless water, nor water vapor, nor air. It must conform to the spirituality of God; it must be living, volatile water that is "signed" by the active semina. It is, says van Helmont, found by releasing the spiritual seed from its material indwelling, and it is different for each substance. It is the point of union for matter and spirit. It is gas!

So the identification of gas as something distinct from water vapor or air comes from a mystical doctrine. What is more, van Helmont identified gases roughly equivalent to carbon dioxide, nitrous oxide, and carbon monoxide. When bodies are burned by the alchemical art of fire, this animated, volatile vapor is released. Yet gas is produced not only by burning; it comes also from fermentation and putrefaction, like the odors which arise from decay or from the fermentation of grapes into wine. Digestion, for example, is a species of fermentation which takes place in the stomach. The digestive element is not heat (how do *cold* fish digest?) but a strong acid, which causes fermentation and separation of food into its good and harmful parts. Significantly, van Helmont also visualized respiration as an exchange of "gas" in the lungs.

Like Paracelus, van Helmont localized disease and saw it as something external. Fever, he said, was not a disease but the sign of disease. His most notorious cure was the so-called weapon salve: When a person was wounded, the wound was bandaged, and the remedy—the salve—applied to the weapon. This doctrine was an old one based upon the concept of sympathy. To van Helmont the weapon salve worked because of magnetic attraction: particles of the remedy mixed with blood on the weapon were attracted to the wound. Spirit was interwoven in the blood, and like joins to like. This was also similar to the old doctrine of signatures, which held that certain plants cured certain organs because of similarities in names: liverwort cured the liver, feverwort relieved fever.

The Renaissance alchemists and Naturalists present us with a duality. The practical, experimental side sought useful knowledge in the laboratory. The religious, mystical side saw in iatrochemistry a profound philosophy of the universe and the fulfillment of religious needs. The two cannot always be separated. Robert Fludd believed that the weapon salve indicated the fundamental sympathy of the microcosm and the macrocosm. Fludd saw in iatrochemistry more than a new method for medicine; indeed for him it was the key to God. The study of chemistry was the study of the "finger of God" revealed in nature.

And there is more! The desire for a sweeping reform of education, and indeed society itself, captured the minds of many who stood aghast before the destruction of the religious wars. The practical benefits of alchemical medicine and even Hermeticism were generalized into programs of reform and spiritual enlightenment. From Germany came a manifesto issued by a so-called "Fraternity of the Rosy Cross" in 1614. These Rosicrucians, as they are known, called for a closer association of philosophers of the occult arts in order to reform the whole world. Popular utopian literature appeared, mixing the call for scientific research with religious reform and occultism. Experiment, human welfare, mastery of nature, spirituality—again and again these words are heard from the alchemists. In this mood we must turn to Francis Bacon, so long hailed as the prophet of scientific progress.

Bacon's program was a call for a reform of education. *The New Organon*, published in 1620, was to be the "new tool" of the new learning. It is the reverence for authority, the endless repetitions of the same things, Bacon claimed, which hinder human progress. The senses may well deceive us, but logic—word-play—creates its own fantasies of the imagination and substitutes them for reality. Bacon decries "four idols" which have hindered the progress of science and knowledge: (1) idols of the tribe, which arise from human nature; (2) idols of the cave, from individual predisposition and education; (3) idols of the market, from social relations; and (4) idols of the theater, from inherited systems. forming individual acts by fixed laws.

Bacon's program was a call for a reform of educaton. *The New Organon*, published in 1620, was to be the "new tool" of the new learning. It is the reverence for authority, the endless repetitions of the same things, Bacon claimed, which hinder human progress. The senses may well deceive us, but logic—word-play—creates its own fantasies of the imagination and substitutes them for reality. Bacon decries "four idols" which have hindered the progress of science and knowledge: (1) idols of the tribe, which arise from human nature; (2) idols of the cave, from individual predisposition and education; (3) idols of the market, from social relations; and (4) idols of the theater, from inherited systems.

The Renaissance had a dark side—a kind of pessimism and despair about "the decay of nature." Some believed that the world had grown old, the ancients would never be surpassed, and the physical nature of humans themselves had degenerated. Bacon, in effect, sought to cut through such gloom. If the senses deceive, if education has failed, or if the ancients hold us in thrall, we must seek new pathways. Only experiment is able to shed light upon the true causes of things, and from well-planned and executed experiments we slowly proceed from particulars to the general. Science in this way is a universal natural history of experiments. Most important, if science is to benefit society, furthering human progress, it must be a collective endeavor of all individuals sharing knowledge gained through collaborative experiments. Here Bacon makes his departure from elite occult societies like the Rosicrucians.

Bacon's hero was not the magician but the artisan, who builds from a collective knowledge of physical facts. His scientific community was akin to the guild and would later be embodied in the Royal Society. Further, he made a clear distinction between science and religion. Natural philosophy treats of visible nature, physical fact. Rather than being an assault against religion, it is the "surest medicine" against superstition. Here Bacon was not alone. In France, Martin Mersenne spoke out against the mingling of physical science with the supernatural. Later, the mechanical philosophy would belittle the "occult virtues" and the spirits of the magicians. Nonetheless, many of the magical categories would find their way into the mechanical philosophy (active principles and passive matter, for example), and the mechanical reduction of physical phenomena to matter in motion would open the dangerous doors to materialism, such as in the system of Thomas Hobbes. Indeed, while Bacon's program of reform *sounds* secular, and many mechanical philosophers did deprecate magical explanations, the influence of the magicians was still to be felt—even in Newton. And both joined hands in their protests against the authority, indeed the veneration, of the ancients.

Yet, passing into the realm of the organic sciences, those we today label the biological, the picture becomes more complex. The age of exploration revealed just how great the diversity of plant and animal life was across the world. Here Aristotle's classification scheme was a benefit rather than a hindrance, for Aristotle *was* a great biologist. If the revolution in astronomy and physics was basically a change in thinking about old experience, the change in the natural history of organic life during the Renaissance was the recognition of its diversity. Massive encyclopedic natural histories exemplified the mind-boggling complexity of living things. Konrad Gesner, a Swiss professor of Greek, compiled a tremendous encyclopedia of animals, basing his classification scheme upon Aristotelian principles. Others would alter that scheme, and not always for the better. For example, G. Rondelet's study of marine life in the sixteenth century placed whales and seals in the category of the fishes; another writer, Pierre Belon, included beavers and otters. Not until the eighteenth century would a satisfactory classification scheme be developed, and it would be tied to the question of how to determine the species.

Further surprises awaited naturalists when they, like Galileo, peered into another new optical device, the microscope. In the seventeenth century Robert Hooke's *Micrographia* described, like a newly discovered America, the minute structures of insects, plants, and other formerly invisible marvels. Inspecting the fine structure of cork, Hooke saw very regular tiny pores, like miniature honeycombs. He called them cells, yet he interpreted them to be channels or pipes which conveyed the natural juices of plants, analogous to the veins and arteries through which the blood flows in animals. Another researcher, Jan Swammerdam, showed that insects possessed organs and reproduced like other forms of life; insects, at least, did not arise by spontaneous generation. With a more powerful microscope Anthony van Leeuwenhoek discovered an even smaller world, a world of "little eels" and thousands of other creatures in a drop of water. These little "animalcules," as he called them, also seemed to possess body parts and organs. To Leeuwenhoek the microscopic world became thus a kind of nature "writ small," and it even provided a vehicle for satire in Swift's *Gulliver's Travels*.

Leeuwenhoek, with some embarrassment, also reported seeing the tiny spermatozoa swimming in semen. If these spermatozoa were like his little aminalcules, Leeuwenhoek thought that they might well contain the entire embryo in miniature, *preformed*. Since there seemed to be no lower limit to organic size, there was no reason to doubt that all the organs of the embryo might be preformed in the sperm (or later in the egg). Opposed to this preformation was the doctrine of epigenesis, held by, among others, William Harvey, which maintained that the embryo developed out of homogeneous matter, with one organ forming successively after another. Yet, without some outside (occult?) principle, how could the epigeneticists explain the amazing complexity and obvious blueprint of the organism's physiology? We shall hear more of this debate.

Physiology and anatomy generally belonged to the province of the medical schools, and here the physicians labored beneath the great weight of Galen. For example, consider a typical dissection at the school of Padua. The professor sits on a elevated chair with the corpus of Galen; the surgeon, a craftsman, does the actual work, while the physician, a scholar, reads from the text. If differences arose, usually they were not considered to be between Galen and observation but between the texts, since medieval Galenism was actually a mixture of the original Galen with Byzantine and Arabic sources. However, the Renaissance revival of Greek classics yielded many of the important works of Galen in the master's own language, and the first printed Greek edition of Galen appeared in 1525 in Venice. The initial response of many physicians was to search the new texts for what Galen "really meant." When Galen disagreed with observation, many undertook his defense—some even claimed that since Galen's time the human body had changed!

During the Middle Ages anatomical diagrams—illustrations of Galenic anatomy—were rare. Even after Galen's major works had been edited, translated, and printed in the sixteenth century, no illustrations accompanied them. Anatomy was still like philology or discussions of what Aristotle "really meant," and the basis of a medical student's education was the reading of Galen. Dissection had to be done quickly (without refrigeration), and the bodies of executed criminals were the main source of cadavers. The slow emancipation from Galen began with Andreas Vesalius.

Vesalius became a professor at Padua and he did his own dissections. He published a new anatomy based upon this work in 1543, and much as Copernicus in that same year "revised" Ptolemy, Vesalius "corrected" Galen in his *Fabrica*. Still, the main tenets of Galenic physiology guided his hand: food is cooked, respiration cools the blood, and so forth. While he presented a detailed study of the cardiovascular system, it was fundamentally Galen's, and he searched long and hard for the pores in the septum (never to find them). Nevertheless, he was certain that the pulmonary vein brought air to the lungs and carried off vapors. But in many other things his own observations altered the old accepted anatomy.

Earlier, in 1538, Vesalius had prepared illustrations of anatomy in an effort to render Galen into visual form. Now in the *Fabrica* he was determined to give precise illustrations for the structures he himself had handled. The result was a massive printed folio containing scores of corrections of Galen—showing how Galen had substituted animal anatomy for human—and all illustrated. Having done his major work, Vesalius was appointed physician to the court of the Emperor Charles V and spent the rest of his life practicing medicine. He had shaken the foundations of Galen's authority, as Galen himself might have done had the Greek physician been alive in the sixteenth century.

Those who followed Vesalius, motivated by his new techniques and detailed study, continued the criticism: The pulmonary vein did convey blood from the lungs to the left heart; the vapors which supposedly returned to the lungs were nonsense, since valves in the pulmonary vein prevented such an exchange. Andrea Cesalpino even called this process "circulation," yet he still considered the pores to exist in the septum. Then, in 1599, a young Englishman came to Padua to study medicine. His name was William Harvey.

Poor Harvey—how he would have protested the label "revolutionary"! His respect for Galen and Aristotle, especially the latter, was immense. In addition to his medical education at Padua he studied Aristotelian natural philosophy under Cesare Cremonini, one of those who upheld Aristotle against Galileo. Yet, it was William Harvey who proposed the theory of the complete circulation of the blood, and he did so by means of mechanistic thinking.

Harvey's medical professor at Padua, Fabricius, had been the first to describe the valves in all the veins, and Fabricius believed that these valves prevented the blood from falling into the lower extremities of the body (Galen said that this was prevented by "attraction" of body parts). Harvey also learned of the "lesser circulation" between the lungs and heart. After he returned to England, he made another observation which may have provided an important clue. The diastole of the heart had long been considered an active movement, akin to drawing blood into the heart. Harvey, observing a slowly beating dying heart, saw that the systole was a muscular contraction which drove the blood into the arteries, and the diastole was simply the rest phase. The pulse, therefore, must be caused by the systole, since the arteries expand as the blood spurts through. Obviously the blood does not ebb and flow as Galen held; rather it is driven through the body as water is pushed by a pump through pipes. So what are those valves in the veins actually there for?

It seems so absurdly simple, especially as we read Harvey's *De motu cordis et sanguinis (On the Motion of the Heart and Blood)* published in 1628. Everything seems to point toward general circulation of the blood. When we see the veins as part of a mechanical hydraulic system, it immediately becomes clear that the valves are there to direct the flow of blood back to the heart. Harvey himself later told Robert Boyle that this was one of the keys to his theory. Severing veins and arteries in a live serpent, Harvey found that the veins empty and blood gushes from the arteries. He made a mathematical calculation of the amount of blood passed into the body in a minute and found it to be about ten pounds. Surely such an amount of blood cannot arise from chyle converted in the liver! It must return! The heart drives the bright vital blood into the body and exhausted blood into the lungs. Hence the circulation is one large mechanical system of pipes connected to a pump, the heart. Obviously! But is it really so obvious?

How, then, does blood pass from the veins to the arteries in the lungs? Harvey did not use a microscope—the discovery of the capillaries was made by Marcello Malpighi four years after Harvey's death. Yet the system hinges on them. Thus Harvey had to postulate "invisible" connecting ducts in the lungs, in a sense no different from Galen's perforated septum!

Harvey's biological ideas can hardly be called pure mechanism. The air must contain some vital form, some incorporeal pneuma, which vitalizes the blood. In his study of animals Harvey observed that the pulsating blood was the first act of life, hence it is akin to the macrocosmic cycle of generation and corruption—carrying the vital spirit throughout the body, it returns to be renewed. The primacy of the heart in circulation reflects the position of the sun as the heart of the world. Circulation is analogous to the eternal cycle of life and death.

Yet we are now into the seventeenth century, the century of the mechanical philosophy. At least in one mechanical system, the vitalism of the magicians and Harvey was eliminated from the physical world. This system was the mechanical philosophy of the Frenchman René Descartes. It was perhaps premature, for surely Descartes' picture of nature, bereft of animating (occult) principles, was forced to fall back upon all sorts of imagined mechanisms to fill the void. Nonetheless, his influence was tremendous, especially in France.

Poor physical nature! How bleak and lifeless it was to become in his hands! Descartes, born to an old noble family, was a sickly youth and was allowed to spend his mornings in bed. There he lay, thinking. Mathematics delighted him with its certainty and clarity. The classics, he perceived, were barren. If only philosophy could establish the kind of certainty he found in mathematics! A few basic principles built up a world of demonstrations. Here in embryo was the sort of thinking which would lead Descartes to his mechanical philosophy.

It is important to realize that Descartes himself was also a creative mathematician. Along with Pierre Fermat he was the founder of analytical geometry, the application of algebra to geometry. Let us begin by glancing briefly at this portentous invention.

Like many great ideas in science, the concept of analytical geometry is surprisingly simple. Imagine space as a kind of grid, and for convenience lay down two axes on the grid at right angles (actually Descartes began with his figures). We have in effect created a "map" of homogenous space. Two axes give two dimensions, three a solid, and we may even have more—n dimensions mapped out by n axes. Any point in relation to the axes, say two, x and y, may be defined by the coordinates of x and y. So we may wander about the map, finding every point in every figure in space defined by the coordinates x and y (x, y, and z for a solid). And—significantly—these points or coordinates may be expressed in relationship to each other by an *equation*. An equation of *any* degree of complexity may now be interpreted geometrically, or *any* geometric figure may be studied algebraically. In short, our geometrical intuition may be expanded to the numerical or variable relationships of algebra. Thus we attain a new level of abstraction—for, operating with equations, we do not necessarily require visual representations. Physical geometry, hence physical space, can be treated purely intellectually through the manipulation of symbols by logical rules—at higher and higher levels, until . . . .

What is the essence of matter? asked the mathematician Descartes. Why, it is *extension*—spacial extension like our coordinate map. So wherever there is space there is extension and thus matter. Space without matter is unthinkable; a vacuum cannot exist. Matter fills the universe, stretching indefinitely (only God is infinite), and it is in motion: colliding, rebounding, sticking together. Now motion itself is not intrinsic to matter, rather it is a state in which matter finds itself. It is a state given to poor lifeless and inert matter by God. Since all space is full—a plenum—motion, perception, change in general must be by *contact*.

Galileo himself in his discussion of the cohesion of material bodies had suggested atomism, and he had postulated infinitesimal vacua between particles. He had also noted the commonly known fact that a syphon would not draw water over a height of thirty-four feet, indicating to him the maximum extension of a column of water to support its own weight. Yet closing one end of the syphon and inserting the open end into a container of water shows that the water still stands. Evangelista Torricelli substituted mercury for water and so constructed the first mercury barometer. The column of mercury stood lower than the water, and since its specific weight (density) was greater, it was assumed that both experiments indicated the weight of the atmosphere. What if the weight of the atmosphere was less, say atop a mountain? Blaise Pascal carried out this experiment in central France and found that, as expected, the mercury column dropped. Later, after the invention of the air pump (by von Guericke), Robert Boyle enclosed a barometer in the receiver of the pump and evacuated the air. The column fell, suggesting to Boyle that it was not weight alone of the atmosphere, but the pressure maintained by the weight. Air was like an elastic fluid—each particle a little spring—compressed by the atmospheric weight, Boyle thought. These experiments, coming after Descartes, suggested atomism along with the existence of vacua.

But this was not the world of Descartes! Although he divided his extension into roughly three types of particles—earthy matter, aether, and light—he would not allow vacua to exist between them. Theoretically, at least, matter was infinitely divisible, like a line on the coordinate map. Gaps between atoms might require some sort of action at a distance, and this sort of occult virtue Descartes would not allow.

In the beginning God gave a certain quantity of motion to the universe. This quantity, according to Descartes, remains constant—a concept similar to our conservation of momentum (mv). A particle in motion preserves that state—inertia—but the primal form of inertia is rectilinear motion. Bodies act upon one another by impact alone, and the universe is completely full. The constant collision of matter in a plenum sets up indefinite whirlpools of continually rotating particles. Thus Descartes pictured the universe as an indefinite sum of vortices. For example, the whirlpool of the earth's vortex exerts a centrifugal pressure, yet at some point the pressure of the next vortex balances this outward pressure, and so on indefinitely throughout the universe. Gravity is the pressure of the earthly vortex by which the larger particles are forced downward by the pressure of the more subtle second element. There is nothing mysterious about gravity; it is simply matter in motion! The same is true for magnetism. This formerly occult force is due to screw-shaped particles which fit into the gaps of magnets and iron, drawing the two together. And since there are two magnetic poles, there are two kinds of screws—right-handed and left-handed.

We shall see more of the Cartesian mechanical system later. Let us, for now, eschew the details and glance at the overall philosophy. Descartes sought to rebuild the principles of philosophy from the bottom, reconstructing Western thought upon firm foundations as clear and as evident as those of mathematics. I may well doubt everything, he wrote, yet I cannot doubt one thing: the fact that I am doubting! Therefore Descartes concluded: I am a being whose essence consists in thought. But the rest of the world is extension—mechanical, inert matter given a fixed quantity of motion by God. Even animals are automata, machines without souls. Life itself is simply a mechanical process, and the entire biological world a machine. In the brain the so-called animal spirits—the former vital principle—are simply subtle particles separated from the blood and passed through the nerves. Only humans have a soul, that incorporeal mind which is able to grasp the rules of geometry and reason.

Much as Bacon had taken experiment out of the hands of the magicians, Descartes "cleansed" theoretical science of occultism, mind, and even God. Not necessarily mathematical techniques, but mathematical ways of thinking prevailed in his science and philosophy both. The ineffable, intuitive, and spontaneous activities of mind were driven out of his system. God's only activity was the First Cause of motion; His role in science and the mechanistic universe had become as pale and as fleeting as it had been with Aristotle. An impenetrable wall was erected by Descartes between mind and matter, and ever since then psychology and philosophy both have labored beneath the hanging sword of Cartesian dualism. In such a universe mind can be only an alien presence, and perhaps this is so even of life itself.

Yet the day of the magicians was not finished. What the mechanical philosophy lacked, their concepts could provide, although in the end they would not have been able to recognize their descendants. As representatives of a particular world-view, however, the wizards and magicians were slowly withdrawing from the world of science—retreating into the murky waters of spiritualism. The world of science and the world of religion were slowly, ever so slowly, becoming compartmentalized. It probably had more to do with European society itself rather than science, for the words "God" and "spirit" were still to be heard from the greatest scientists. Yet, like the Cartesian dualism of mind and matter, secular pursuits were drawing away from theological ones, and eventually science would be dragged along. Not that science had disproved anything about the spiritual world, but these subjects simply had no place in physical science. This was still far in the future; for now we shall take from the wizards what we can and leave them, biding their time, awaiting the next scientific revolution.

SUGGESTIONS FOR FURTHER READING

BONELLI, M. L. R., and SHEA, WILLIAM R., eds. *Reason, Experiment and Mysticism in the Scientific Revolution.* New York; Science History Publications, 1975.

DEBUS, ALLEN G. *The English Paracelsians.* New York: Franklin Watts, 1966.

———. *The Chemical Philosophy: Paracelsian Science and Medicine in the Sixteenth and Seventeenth Centuries,* 2 vols. New York: Science History Publications, 1977.

DIJKSTERHUIS, E. J. *The Mechanization of the World Picture.* Oxford: Oxford University Press, 1961.

HALL, A. R. *From Galileo to Newton.* New York: Dover, 1981.

PAGEL, WALTER. *William Harvey's Biological Ideas: Selected Aspects and Historical Background.* Basel: Karger, 1967.

———. *Joan Baptista Van Helmont: Reformer of Science.* Cambridge: Cambridge University Press, 1982.

ROSSI, PAOLO. *Francis Bacon: From Magic to Science,* trans. Sacha Rabenovitch. London: Routledge and Kegan Paul, 1968.

16

~~~~~~

"Such a Wonderful Uniformity"

In June of 1661, a new student from rural Lincolnshire was on his way to Trinity College, Cambridge. His father had died some three months before the boy's birth on December 25, 1642 (actually January 4, 1643, but Protestant England had refused to accept the "popish" Gregorian calendar), and his mother had remarried an older clergyman, leaving the child in the care of his grandmother. He had grown up a solitary, dreamy child, finding solace in building small machines, constructing sundials, dreaming of perpetual motion, and charting the equinoxes. He had the tendency to lose himself completely in such lonely activities; sometimes he even forgot his meals. The servants thought him a bit odd, even silly. He in turn treated them none too kindly. It must have seemed quite natural for such a strange lad to be off to the university that June. He was good for little else. His name was Isaac Newton.

The Cambridge that Newton attended in 1661 was still governed as to intellectual content and standards of behavior by the Elizabethan Statutes of 1571. While the application and enforcement of the Statutes had been altered, the prescribed curriculum at both Cambridge and Oxford was still immersed in the scholasticism of the thirteenth and fourteenth centuries. An enterprising professor might attempt to insert some of the new natural philosophy, or a good tutor might steer his ward toward the new learning, yet such activities were rare, and even the official curriculum was often neglected. Some of the masters themselves disdained the curriculum and the medieval formalisms which hung about the university like a mildewy blanket. Disdain led to neglect, neglect to apathy and laxity—on the whole students were left to fend for themselves.

Like the rest of English society during Newton's youth, Cambridge had passed through the upheaval of civil war and the establishment of the Puritan Commonwealth by Oliver Cromwell. While Oxford had militarily resisted parliamentary arms, Cambridge had been occupied by Cromwell's forces without much damage. But many Royalist scholars quit the university; others lost their positions to Puritans for refusing to pledge allegiance to the Oath and Covenant decreed by Parliament in 1644. In January of 1649 King Charles I was executed, and in 1653 Cromwell was installed as Lord Protector. In 1660, the year before Newton entered Cambridge, the son of the executed king, Charles II, returned to the throne of England. The Restoration brought further purges to Cambridge but no real acts of vengeance. Despite the turmoil, the official intellectual rigidity remained basically unaltered; if anything, the political upheavals deepened the apathy about the place.

Such benign neglect was an open invitation for young Newton to follow his own inclinations. Soon he discovered the new mechanical philosophy; he studied Cartesian geometry without formal instruction; in short, he had no intellectual debris to clear away in his private intellectual world.

While the universities provided little stimulus to the study of the new natural philosophy, Puritan values and the desire for educational reform did tend to foster an interest in experimental and mechanical science as a means of both glorifying God and benefiting society. The emphasis was upon a life of public service; a passion for hard work dedicated to self- and community improvement; an interest in nature as the "second book of God's revelation," separate from Scripture, whose order and hierarchy was a visual manifestation of the divine creative act; a Baconian emphasis on experiment and command of nature; education as a "fitting of men's minds for the business of this life." Such values were not exclusively Puritan, for they could be found in the Hermeticism of the European continent, which was also devoted to humanitarian projects, experiment, and alchemy, all of which found its way to England. Some of the reformers brought the new mechanical philosophy into their alchemical studies, rationalizing the old mystical chemical philosophy.

As Newton pondered the mechanical philosophy in the midst of medieval Cambridge, outside his study the organized life of science itself was also changing. The new king, Charles II, had given the appellation "royal" to a spontaneous gathering of English natural philosophers in London. This group, tracing its roots back to the 1640s, had originally been an informal discussion circle. In 1662 they took the name Royal Society. Early in the century a similar scientific academy, The Academy of the Lynx, had flourished in Rome—Galileo had been a member. In France, Colbert, Louis XIV's finance minister, established the *Academie royale des sciences* in 1666. While many of the new scientists were university trained, the societies offered things the universities, still formally medieval, did not.

The president of the new society of "learned men" who met at Gresham College was John Wilkins, and Robert Boyle was one of its members; but most important was the German master of languages Henry Oldenburg, who became secretary of the Royal Society. In this capacity Oldenburg handled the Society's correspondence, including foreign correspondence, which he translated. Thus it was Oldenburg who conceived the momentous scheme of collecting and publishing scientific correspondence in a journal dedicated to scientific communication. Guided by Oldenburg, scientific research emerged into the public arena through *The Philosophical Transactions*, the first issue appearing in March of 1665. Two months earlier in Paris, the *Académie* published the *Journal des Scavans*. The scientific journal, one of the most important forms of communication among scientists in the modern world, had been born.

The "business" of the Royal Society, as stated in its first charter, was to enhance the knowledge of nature and of useful arts including manufacturing, mechanics, and other experimental inventions. From agriculture to improvements in navigation the Society was to fulfill Bacon's dream of promoting experiments for their social usefulness, as well as for virtue and wisdom. Robert Hooke became Curator of Experiments. A polymath, rich and inventive, Hooke steered the Society toward this Baconian vision. The Society became a kind of clearing house, a forum. If it did not actually finance projects, it encouraged them, and, free from theology, its somewhat chaotic nature offered a degree of freedom and a sounding board for innovative ideas. The Society also served to bring the new science before the public; it became a passageway through which science emerged from the shadows, the "secrecy of the Pythagoreans," into the general consciousness of the educated layman.

Such were the circumstances during the 1660s as Newton began to ponder the new science. Consider the intellectual dilemmas. Newton was heir to the mechanical philosophy, the new astronomy, Baconian experimentation, English Puritanism, even alchemy. The mechanical philosophy alone raised many problems. Descartes had given a visual picture of motion, yet one which, in the end, must be imagined to exist behind natural motions. Galileo, Descartes complained, had built "without foundations"; he had merely described motion and not explained it. However, reducing the mechanism of motion to the impact of bodies in a plenum, we are actually led to wonder how Galileo's uniform acceleration is at all possible. By the same explanation circular motion becomes like a static balance of opposing vortex pressures. How may this be conceived in an *infinite* universe? And Kepler's laws of planetary motion fare no better. In short, how can mathematical abstractions be made to agree with Cartesian visual mechanisms? Descartes had ignored the problem.

There was one problem, however, Descartes could not ignore: impact. His entire system was based upon the impact of bodies. Impact, or percussion, was a difficult subject; Galileo had wrestled with it and had little success. Descartes' science of motion demanded he deal with it. Since God had initially infused the universe with movement, and all resultant motions must be by contact, it follows that "the quanitity of motion" in any impact must be preserved. What is "quanitity of motion"? Is it not the size of the body multiplied by its velocity? If so, then two unequal bodies, unequal in size or velocity, would not necessarily conserve their individual quantities upon impact. Only the *sum* of motion would be equal after impact. Therefore, is *something* transfered from one to another?

Descartes was led, probably with some distaste, to considerations of what we would label "force." If, for example, a small body strikes a large body at rest, the large body endeavors to remain in its present state and the smaller must rebound with the same velocity, changing direction only. The large body resists impact. Resists? Endeavors? How can passive, inert matter resist anything? In the case of two equal bodies moving at different speeds, the quicker body transfers excess velocity to the slower. Transfer? Then there is also the tendency of matter to recede from its center in the swirling celestial vortices. How may all of this be reconciled with Descartes' first principle, the passivity of matter to motion? How does matter resist, endeavor, and so forth while remaining inert?

To admit force of any kind acting upon a body was to open the door to occult principles, even if force was considered mathematically. Torricelli was a good example. Suppose, he said, that I try all day to push down a wall. Probably I will fail. But if the entire day's force can be gathered in one instant of time, the wall will crumble "like Jericho." If impact is spread over time, the force is diminished. If the force of impact is instantaneous (again *that* word), the strength becomes greater. Thus a falling body gathers increments of momentum like water poured into a jug: a falling body of, say, ten pounds gathers the strength of a static body of a hundred pounds if it is allowed to fall far enough without resistance. Now what is this strength? Why, it is a subtle *spirit* of force which flows into the body! And we can measure it as momentum—the size of a body multiplied by velocity in a given time. But a subtle spirit of force? Occultism!

And if this were not all, Descartes' conservation of motion faced another problem. In a fully occupied universe, a universe of moving vortices, all motion must be seen only in relation to other moving bodies. So
if all motion is actually relative, the "quantity of motion" in any impact
must vary according to the frame of reference. Christian Huygens, a
Dutchman, realized that in the Cartesian plenum a body moves or is at
rest in relation to another body; and in the case where a smaller body in
motion rebounds from a larger at rest with its velocity intact, we may
shift the frame of reference, considering the smaller body at rest, and
find that the larger body puts it into motion, losing as much motion as it
gives the smaller body, with both moving off together after impact. Also
from a new frame of reference, the motion lost by a large body moving a
small may be visualized as motion imparted to the large by impact of the
small. Obviously the Cartesian conservation of motion for the impact of
bodies was in error, yielding different results for different frames of reference.

Following the lead of Torricelli, Huygens concluded that in a given
frame of reference any isolated system of bodies, however many you
wish, may be considered as a single body concentrated at their common
center of gravity, and this center of gravity, before and after impact, suf-
fers no change. From the perspective of the center of gravity each body
changes instantaneously the direction of its motion on impact, yet both
depart from impact with their original motions unchanged, and henceforth there is no dynamic action or reference to force. Huygens found
another quantity which did remain constant in the impact of perfectly
hard bodies: the magnitude of each body multiplied by the square of its
velocity—mv^2. The sum of the two quantities before impact always
equals the sum of the two quantities after, and this quantity remains
constant within each frame of reference for the impact of bodies. For
Huygens this quantity was merely a number which served as a substitute for the erroneous Cartesian quantity of motion; it was not a measure of force, since Huygens' basic principles were still Cartesian. Yet G.
W. Leibniz would take this simple formula and label it "living force."

The problem, as we are beginning to see, is simply that the mechanical philosophy of matter in motion was finding it difficult to account for
that motion without active principles—force. On the other hand, such
principles were deemed occult. In his consideration of circular motion,
Huygens did use the dreaded word; in fact he coined the phrase "centrifugal force." Here, too, he attempted to escape the occult connota-
tions of the word and to conceive it in terms of statics. Huygens knew
that a rotating body, say a weight twirled round on a string, has the tendency to pull away from the center of rotation. Such a force, he said,
arises from the inertial tendency of the body to follow a straight line
away from the curved path. However, Huygens viewed this force in

terms of weight, a kind of balancing of weights as in statics, where centrifugal force and weight are two sides of the same coin. He believed that weight was actually caused by a deficiency of centrifugal force: when a heavy body falls, an equal quantity of subtle matter (in the Cartesian plenum) moves away from the earth. Relating centrifugal force to weight in rotary motion, Huygens demonstrated that if a body moves in a given circle with a velocity equal to that which it would gain in falling from rest through half the radius of the circle, its centrifugal force will equal its weight. He was even able to deduce a mathematical formula for it: Centrifugal force equals the magnitude of the body multiplied by its velocity squared over its radius—mv^2/r. And he found other treasures to mine from it.

Imagine a pendulum, only with its bob rotating in a circle. At a certain angle from the vertical (45°), centrifugal force exactly equals the weight of the bob. This "conical pendulum" is thus analogous to a simple pendulum and hence to free fall: The radius of the described circle is equal to the vertical height of the cone, and the period of oscillation varies as the square root of the vertical length. Huygens derived a formula for this period, $T = 2\pi \sqrt{L/g}$. The unknown g is, of course, the acceleration produced by gravity. Being able to measure the other variables, Huygens found that g equaled a little more than 32 ft/sec^2 at the latitude of the Netherlands. In short, Huygens was able to find a very exact constant for gravity near the surface of the earth with his conical pendulum. He also used his knowledge to design clocks of great precision and (in 1674) a watch with a balance spring (over which he became involved in a priority dispute with Hooke).

Yet Huygens was a Cartesian to the end. Centrifugal force was nothing more than the movement of rotation, and bodies fall by lack of centrifugal motion (let's not use force any more), since weight is simply a complementary phenomenon to this motion. However, we still have not accounted for uniform acceleration in fall. And acceleration goes through a whole range of velocities, changing at any instant. The problem, then, is *rate* of change. How to compute it? How to account for it? The same holds for Kepler's laws; we have come no closer to an understanding of them within the mechanical philosophy of matter in motion. It is almost like cubism in modern art; in the painting the forms and colors present a scene seemingly out of focus, awaiting the invisible imagination of the viewer to give it meaning.

In Cambridge that imagination grappled with the new science and its problems. But there were other interesting philosophies to consider. One was alchemy (not that again!). The fact is, Newton conducted his own alchemical experiments, albeit from the mechanical and corpuscular perspective of Robert Boyle. Still, Newton read in the most esoteric literature of Hermeticism, believing that beneath the cryptic language lay profound truths. And there was another influence at work upon him, part of the climate of Cambridge itself.

Isaac Barrow, who held the first Lucasian chair of mathematics at the university (which shows the university curriculum was changing), was also interested in alchemy. Descartes, so Barrow believed, had been overzealous in reducing everything to dead matter in motion. Unlike Descartes, who actually began with *a priori* concepts, the Hermetic philosophers were experimenters. They were the true Baconians, consulting nature first rather than the sanctity of their own minds. And they had found evidence for a vital spirit in nature, an immaterial soul so desperately absent in Cartesian extension. Why should there not be something incorporeal to account for motion?

Then there was Henry Moore, who had exchanged letters with Descartes and to whom Newton referred in his early notebooks. Moore found much to admire in Descartes' philosophy, yet upon sober reflection Moore decided that he could not accept the identification of matter with extended space. It is true, Moore believed, that spirit must be separate from matter, but spirit *acts* upon matter—it *must* act upon matter! Space is infinite, as the ancient atomists held, and at first appearance is a void. But space is actually filled with spirit, in fact *is* spirit. If only God is rightly called infinite, and space is infinite, then the two must be coeval! God is everywhere present in space, and as the "soul of the world" God moves it, gives it life, pervades it. God, we should like to say, listening to Moore, is an immaterial *force!*

Newton was a religious man. His scientific work was only a small part of his life's studies, which included Biblical chronology and alchemy. God's active and creative role in the "wonderful uniformity" of the world was a fact uppermost in his mind. Moore had indeed picked at a sore spot in Cartesian mechanism: Eliminating active spirit from the world could lead (as it did with Hobbes) to atheism. In short, the concept of force called out to Newton from the other side of the scientific revolution.

But only called! Newton was still a mathematical philosopher and a follower of mechanism. His approach to physical mechanics was similar to Galileo's. Beginning with idealized mathematics, which have little connection to physical situations, Newton deduces consequences, compares observations and experiments to the data, and corrects or alters his idealized system when necessary. Only after this process is a physical system of the world announced, based upon the necessary mathematical and observational conclusions. I. Bernard Cohen labels this procedure the "Newtonian style," a kind of initial freedom of thought which begins in the abstract unburdened by any hypotheses of physical reality ("I frame no hypotheses" is Newton's famous phrase).[1] And so Newton was led, as he pondered the mechanical philosophy, to accept attractive force as a fact of the physical world, providing the missing vitalistic piece to mechanism.

Was it the overwhelming power of his mathematical demonstrations? Many would like to think so today. Imagine, however, the other influences that swirled about him like the thick smoke of his alchemical furnace. Indeed, like the fire of that furnace, Newton shared with Renaissance naturalism a doctrine often labeled the *prisca sapientia*. The most moral and brilliant ancients had in the past proclaimed Nature's truth— for example, Pythagoras knew the law of universal gravity—but had hidden their wisdom in veiled form in order to protect it from the vulgar. Using his reason and patient Puritan diligence to strip away the myths and allegories of the ancients, Newton could discover the gold of wisdom he believed lay hidden in and encrusted by mystical obscurity. Evidently Newton saw himself as not only among the Calvinist elect, but also a member of that special "brotherhood" of the ancient esoteric "elect." And, while he held to the rigorous standards of mathematics and experiment, bringing these into his alchemy, the universe still appeared to him to be a kind of aperture through which he might catch a fleeting glimpse of ultimate and divine truth. The gate he had to open was mechanical—nature's mechanical contrivance which served to bar those who did not belong to the elect. True knowledge was sacred; approaching it was a kind of moral act, a search for redemption. John Maynard Keynes went so far as to call Newton "the last of the magicians." Perhaps it was an overstatement, yet we must not forget those voices in the furnace, in the mystical literature, in his religion, kindred voices of the elect calling. Does he not hear the whisperings of the Hermetic philosophers, their words of vitality and active forces? Is he not appalled by the poor Cartesian God, a God as far removed from the world as the do-nothing deity of Aristotle? These influences, too, would converge to form the great unifying concept—gravity, the force of attraction across the universe.

So how he did he do it? In his old age, looking back to the decade of the 1660s, Newton recalled that during the years 1665–1666 he experienced the most creative period of his life. The terrible plague had once again struck England, and Cambridge was deserted by its students. Newton himself had retired to the country, where in the undisturbed quiet he contemplated mathematics and physical science. It was during

---

[1] I. Bernard Cohen, *The Newtonian Revolution* (Cambridge: Cambridge University Press, 1980), p. 109.

that time, he reminisced, that he discovered his method of fluxions (the calculus), his theory of colors (the heterogenous quality of white light), and began to think of gravity as extending to the moon. It was then, too, that he deduced from Kepler's Third Law the inverse-square rule—that the *force* which keeps the planets in their orbits must be inversely proportional to the squares of their distances from the center (the sun). And finally he deduced that gravity was indeed the force that kept the moon in its orbit. He did so by mathematically comparing this force with the force of gravity at the earth's surface, finding the two agreed "pretty nearly." Yet none of this (except the theory of light) reached the public until 1687!

It could be that the system was long in the perfecting, but it is more likely that Newton was reading back into the past the idea of attraction which he did not accept until much later. Let us look closer. Before 1666 Newton had been investigating mathematically the Cartesian rules of motion and impact. To him, as to Huygens, force was only a quantitative description of impact. As he continued, he also began to consider circular motion. Again the analogy of the lever—a static balance of opposing forces—seemed to dominate his thinking. Newton imagined a rotating ball striking four sides of a square. Next, he increased the square to a polygon of infinite sides and deduced Huygens' law for centrifugal force. Owing to rectilinear inertia, the ball's tendency is to follow the tangent away from the curved path. Thus Newton imagined the ball striking an infinite number of similar bodies as it is deflected to the curve. Now (for the purposes of mathematical abstraction) consider the planets to be revolving in circular orbits. It was an easy step for Newton to substitute his formula into Kepler's Third Law (D^3/T^2), since Kepler's mean distance D was actually Newton's radius R of the circle. Thus Newton was able to show that the tendency to recede decreases in proportion to the square of the radius—the famous inverse-square law, $1/R^2$, which was to become the cornerstone of universal gravity.

In 1666 it was not universal gravity! Nonetheless, he could still compare the force required to keep the moon in its orbit with the force of acceleration on the earth. In Newton's mind this comparison actually was between the acceleration of gravity on the earth and the centrifugal tendency of the moon *to recede*. If the moon is 60 earth radii from the earth's center, this *tendency* by the inverse-square law ought to be $1/60^2$ or 1/3600 of the acceleration of gravity on the earth's surface. Newton, because he was using an inaccurate value for the radius of the earth, found it to be 1/4000. It was "pretty nearly" a good correlation, and indeed it does suggest that gravity extends to the orbit of the moon. Did Newton not announce it because he desired the correlation to be exact? Of course he was also assuming that gravity on the earth acts *as if* all the earth's matter were concentrated at its geometric center—an assumption he still could not prove. So did he simply hesitate?

Probably not! Newton was thinking in terms of the moon's tendency to recede, its centrifugal force *away* from the earth. He still considered a balancing vortex. Indeed, the mathematics did show some connection between Galileo's natural acceleration and this tendency to recede, hence some relationship to Kepler's Third Law. But falling bodies suggest *attraction*. Centrifugal force is quite the opposite! Further, attraction carries along all those occult suggestions of sympathy, whereas the tendency to recede may be explained mechanistically (as Huygens did), even with its inherent difficulties. In short, we have the perplexing situation of two conceptually different things—attraction and the tendency to recede—exhibiting "pretty nearly" the same values! No wonder Newton put the whole thing aside!

Now the mercurial figure of Hooke enters the story. In 1672 Newton, by this time Lucasian Professor of Mathematics at Cambridge, was elected to the Royal Society. His election was secured by his reflecting telescope. That same year he sent Oldenburg his theory of colors. Within a week Robert Hooke wrote a critique of the theory, and Newton, stung by the criticism, replied angrily. He then withdrew into isolation, remaining in contact with the world of science only through Oldenburg and a few friends. Oldenburg died in 1677, and Newton's nemesis Hooke became Secretary of the Royal Society.

Hooke wrote to Newton in 1679, asking him to resume correspondence. Newton's reply, as we may well imagine, expressed little enthusiasm, yet it did contain the description of a body falling from a tower to the center of the earth. Newton's diagram showed a spiral. Hooke, jumping on the error, replied that the body would describe an ellipse. Back in the 1660s Hooke had demonstrated something similar with a conical pendulum, a pendulum set revolving in a circle. With the proper combination of "deflecting" shoves Hooke demonstrated that the bob would trace an ellipse just like Kepler's planets. What was this deflecting force? In his letter Hooke said that the orbit arises from the inertial velocity of the body following its tangent and *attraction* to the center! In another letter Hooke surmised that this attraction decreases in proportion to the square of the distance! It is attraction, not the tendency to recede, which gives the orbit by this rule.

And here was the revelation. Hooke had, as the saying goes, turned the problem upside down and thus set it right side up. Rather than being an equilibrium like centrifugal tendencies and Cartesian vortices, the orbits result from a single force and inertia. The motion of the planets was, in fact, a question of dynamics. The moon, so to speak, is constantly falling. Like an apple falling from a tree (another famous legend which is probably the invention of historians), the moon is constantly being pulled toward the earth by gravity. This attraction *acts upon* the moon and at every instant produces a change in the rate of its motion. Yet the moon's inertial tendency to continue in a straight line resists the acceleration-producing force of gravity. Thus we begin to see the significance of *rectilinear* inertia. Gravity and inertia combined give a uniform orbital velocity which varies according to the square of the distance from the central point of attraction—a point that in the case of the planets is the sun. From these principles Kepler's elliptical orbits (somewhat corrected) and the area law follow. Inertia and gravity! How incredibly simple!

Newton was the mathematician that Hooke was not, and so he fell to computing Kepler's laws with the new insight of attraction. Hooke's last letter remained unanswered. Finally, in August of 1684, Edmond Halley visited Newton in Cambridge. Halley blurted the question: "What path would a body, which orbits another attracting it, follow if that force acted by the inverse square of the distance between them?" Newton replied: "An ellipse!" How did he know? Why, he had computed it! Unfortunately he could not find the paper, so after Halley left, Newton set to work. The final result was entitled: *The Mathematical Principles of Natural Philosophy* (commonly referred to as the *Principia*), published in 1687. It is one of the great works of Western science.

The central concept is now attraction, and in Book I Newton coins a new word for this force, *centripetal* force, in conscious contrast to centrifugal. It is that force which "seeks the center." This force which acts upon the planets is nothing else but gravity. Gravity extends across the universe; it is universal. It accounts for the fall of bodies upon the earth and the movement of the planets by Kepler's laws. By now Newton was finally convinced that gravity as an attractive force really exists. Was it the power of his mathematical demonstrations?

During his years of silence Newton had been actively pursuing his alchemical researches. He had come to the conclusion that active forces were at work between particles of matter. What more confirmation did he need? Force was as real a thing as matter in motion. Call it occult if you must (Newton's critics did, of course). Newton, at least in the *Principia*, would frame no hypotheses about its ultimate nature. But once again the universe is alive (in the figurative sense), governed by a single force—a "wonderful uniformity."

The *Principia* begins with the fundamental laws of motion. The first is inertia: A body maintains its state of rest or uniform motion in a straight line unless compelled to change by forces impressed upon it. By force, not impact alone! The second law says: Change of motion is proportional to motive force impressed and is made in the direction of the right line in which that force is impressed. Here force assumes direction as well, what we refer to as a vectorial quantity. The third law says: To every action there is always opposed an equal reaction; that is, as you are attracted by the earth, you also attract the earth, or as the horse pulls the cart, the cart pulls the horse.

What interests us is the second law, for Newton has defined force as that which changes *motion* rather than that which produces or causes acceleration, a change in the *rate* of motion. Gravity is a force pertaining to all bodies, Newton continues, which is proportional to their quantity of matter and inversely proportional to the square of the distance between them. Our own contemporary rendering of the second law says force is *measured* by a change in the *rate* of velocity, or $F = M$ (mass) times a (acceleration). Gravity in our notation is $F = GM_1M_2/d^2$ (G being the universal gravitational constant). Newton in his geometrical demonstrations of Kepler's laws implied these relationships, which are commonly found in modern physics textbooks. In order to grasp their significance we must go a bit beyond what Newton actually wrote in the *Principia*. First, what is "M"? Why, *mass* of course! Mass?

Let us be a little ahistorical and say simply that force is that which produces acceleration. If we have two bodies and discover that one requires a greater force to change its state than the other, we have in essence a measurable quantity between them. One body *resists* the force in greater quantity than the other. Now inertia says that a body at rest or in uniform motion requires an outside force to change that state. Hence the resistance any body offers to force says something about the body itself. This "something" is called *inertial mass*. The inertia or resistance of an object to force provides us with an exact operational procedure to measure "quantity of matter"—inertial mass.

Next we consider a particular force, the force of gravity or the attraction between two bodies. Let us say one of these is the earth. The odd thing is that the earth is also attracted by a falling body, say a feather, although the inertial mass of the earth is so much greater that, roughly speaking, the force of the feather's attraction does not exist. Rather, the force of the earth's attraction is what pulls down the feather and all bodies, including the moon. Therefore, the mass of a body near the surface of the earth is called the gravitational mass, since it, too, can be measured, in this case by the acceleration due to gravity. Roughly speaking, this gravitational mass is weight, the acceleration due to gravity times the mass of the object. Thus an object of greater mass, say iron, weighs more on a scale than a feather. But there is a catch!

If we drop our two different weights in a vacuum at the earth's surface, it would seem that the heavier should fall faster than the lighter. Was Galileo wrong? No, because we have forgotten that gravity is also a force, and the resistance of a body to *any* force is proportional to its inertial mass. Thus iron is attracted more strongly than a feather, owing to its gravitational mass (weight), but it resists that attraction just as strongly, owing to its inertial mass! Hence, without the resistance of a medium like air, the two fall at the same rate in the earth's gravitational field. And for inertial mass to balance gravitational mass, the two dissimilar concepts give an equal measurement! How strange. Here are two distinct concepts, and yet we discover that the two measurable definitions of mass are actually equal. Is it coincidence? Or . . . a clue . . . to something deeper?

When mass is conceived of as measured by forces acting upon bodies, the science of statics now becomes merely a special case of dynamics, a case in which the forces acting upon a given body are in equilibrium. This was the key to solving the motion of the planets. Newton was no longer a prisoner of the static lever analogy, as had been Descartes and Huygens. Instead, the entire planetary system could now be treated dynamically, which in turn allowed Newton to easily derive Kepler's laws.

Newton demonstrated Kepler's law of areas geometrically. The planet is first imagined to move inertially in a straight line. At regular intervals it receives a momentary impulse toward the sun. This creates a new motion, which is simply a combination of inertial motion and the impulse, which in the case of the planets is the attraction of the sun's gravity. Next we construct triangles at each impulse and to the sun. Since the impulses are at regular intervals, the areas of the triangles are all equal, and we have equal areas swept out in equal times (Figure 16.1).

Newton next tells us to increase the impulses and hence our triangles. Ultimately the path of the body (the planet) becomes a curve. The centripetal force which draws the planet acts continually like an infinite number of instantaneous impulses; so even in the limit of the curve any described areas are swept out in equal times—Kepler's law.

Kepler's third law follows when we imagine the moon to be falling at every instant by the acceleration of gravity. Gravity produces a constant change of the rate of the moon's velocity as it endeavors to follow its inertial path. Newton again demonstrated geometrically that the two combined—representing the orbit dynamically—yielded the inverse-square law and Kepler's third law. It was here, too, that Newton implied $F = ma$: Force produces a change of the rate of velocity. Finally, Newton demonstrated that this force yielded an ellipse and that homogenous spheres do attract each other as though their entire mass were concentrated at their centers. It was all simply amazing! The *Principia* was an almost unbelievable exercise of mathematical power.

And there was more: The tides were now shown to be caused by the attractions of the sun and moon; the precession of the equinoxes was demonstrated to be due to oscillations of the earth's axis caused by its

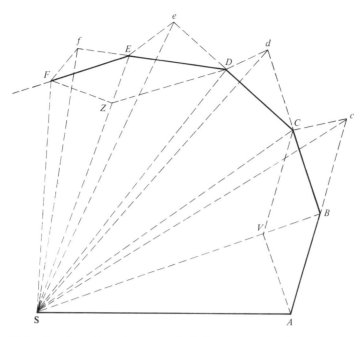

Figure 16.1 Centripetal impulses are at *A, B, C, D, E, F.* From the *Principia*, Book I, Section II. Translated by Andrew Motte and revised by Florian Cajori (Berkeley: University of California Press, 1960).

equatorial bulge, like a wobbling top; even the comets were shown to have gigantic orbits governed by the gravitational law—another reason to doubt the Cartesian vortices. Newton also surmised that the small irregularities of the planets were probably due to the mutual gravitational attraction between them, although exactly computing these perturbations, the mutual attraction of three bodies, defied his resourcefulness. Strictly speaking, Kepler's laws could be considered only approximations because of the "three-body problem," and since some of the inequalities seemed continuous, not periodic, Newton believed that God occasionally intervened to tinker with the celestial machine. These final two problems caused ineradicable difficulties for Newton's mechanistic universe.

Accurate tables of lunar motions were of practical importance in navigation, since they provided a means to determine longitude at sea. However, the moon is attracted strongly by two bodies, the earth and the sun, pulling it at different angles. Its movements therefore are more erratic than those of the planets, which for all practical purposes can be assumed to be attracted only by the sun. While we are able to write differential equations for three mutually attracting bodies—the simultaneous pull on the moon by the earth and the sun—it is impossible to solve

them directly, for the problem lends itself only to successive approximations. Newton once said that his lunar theory was the only problem that
ever made his head ache, and it is easy to understand why: Approximations were not the absolute values he longed for. In 1747, some twenty
years after Newton's death, Alexis Clairaut even went so far as to announce to the French Academy of Sciences that Newton's law of gravitation was in error, basing his startling opinion on the lunar problem (two
years later, after a simple mathematical error was discovered, Clairaut
reversed himself).

And as for Newton's prospect of requiring God's intervention, consider Leibniz's protests. Is it not strange, Leibniz wondered, that God,
the most perfect Being, would create a machine which he must wind up
from time to time? Does this not violate the mechanistic universe, since
it can be maintained in good working order only by "extraordinary concourse"? Without precise mathematical values and with his belief in the
activity of God in the world, Newton could not expunge charges of occultism from his law of gravity.

All this time we have been speaking of Newton's mathematical
powers, yet a glance at the *Principia* reveals (at least on the surface) the
standard geometry which dominated the science of his time. Within it,
however, was a powerful new mathematics whose concepts he had invented twenty years before. Indeed it had been anticipated by many,
and even invented independently by Leibniz. Eventually it would become the language of physics. It was the calculus.

To grasp the significance of the calculus we must again depart a little
from Newton's original methods. Newton was interested in problems
which demanded considerations of the instantaneous—for example, the
velocity of a body at an instant. This is quite different from average
speed, how far a body has traveled in a given time, or S (distance)/T
(time). In Newton's problems, to know the average velocity is not
enough, for an object falling to earth is accelerating, its velocity
changing from instant to instant. As we leave the earth, the rate of acceleration itself changes by the inverse-square law. Kepler's laws demand
that we consider instantaneous velocities. It is now time to face up to
these mathematical monsters.

Ah, but here comes Zeno. What is the velocity at an instant? How
much time elapses in an instant of time? Why, no time! How much distance? Why, no distance! So, mathematically, velocity at an instant is
0/0, zero distance divided by zero time. But that is meaningless! How do
we mathematically "freeze" velocity and yet escape the paradoxes of
Zeno?

Now we know from Galileo that, during free fall, distance is
changing with time, or $S = \frac{1}{2}at^2$, and we know from Huygens that $a = 32$ near the earth's surface, so we may write $S = 16t^2$. It is obvious that S

is changing as t is changing and that the change of distance is dependent upon the change of time. Leibniz called this relationship between two variables, where one is dependent upon the independent change of another, a function. Later in the eighteenth century the mathematician Leonhard Euler would develop the function concept. For now let us say that S is changing by $16t^2$ and we want to know the rate of that change, the velocity S/t, at an instant. But, as we saw, it is 0/0! What to do?

Let us perform some inspired nonsense (mathematicians are good at that!). If $S = 16t^2$, in order to keep everything flowing we shall, after a given time value, assign a small, unspecified increase to time, expressing this increase in symbols as Δt. All this means is that after, say, a second, we have increased time by a small amount Δt. Obviously S also increases as a function of time, ΔS. Now we may write our equation:

$$S + \Delta S = 16(t + \Delta t)^2$$

Expanding the equation by the binomial theorem (which Newton generalized) and multiplying through, we get

$$S + \Delta S = 16t^2 + 32 \, \Delta t(t) + 16 \, \Delta t^2$$

and, since $S = 16t^2$, we may subtract it from both sides:

$$\Delta S = 32 \, \Delta t(t) + 16 \, \Delta t^2$$

Since we wish to find the velocity, we divide by Δt and get

$$\Delta S/\Delta t = 32t + 16 \, \Delta t$$

But this is still only the average speed in the small increase of time, Δt, after the given value. The problem is that, after all this work, if we make Δt equal to zero we end up again with 0/0!

Here comes the nonsense: Let us say that Δt gets smaller and smaller, so small that it approaches the instant of time, 0 time elapsed, "infinitely close." But it *does not become* zero. For all practical purposes, however, we may "blot it out." So as it comes infinitely close to the instant, $\Delta S/\Delta t$ follows, until it is infinitely close to $32t$. Therefore, $32t$ is taken to be the value of velocity at an instant! After one second of fall near the earth the speed is 32 feet per second at this instant, at two seconds it is 64, at three 96. In calculus the $32t$ is called the differential coefficient; it measures the rate at which S is changing with respect to t. The method is called the differential calculus. Newton called it his method of fluxions and used the symbol $\dot{y}$ to denote this rate of change. Leibniz used dS/dt, which is our notation today. There were some differences between the two approaches, and later an angry priority dispute erupted between Newton and Leibniz over the calculus.

The differential calculus can become a great deal more complicated. For example, velocity is the rate of change of distance with respect to time, but acceleration is the rate of change of velocity with respect to time. To compute instantaneous acceleration we take a second derivative, dV/dt, in which V is actually the first dS/dt. This second derivative is actually the rate of a rate and is denoted today by d^2S/dt^2. The integral calculus is used for finding the areas under curves (like the Greek exhaustion) and volumes of curved solids (like Kepler's wine casks). Roughly speaking, it is the reverse of differentiation, since the curve actually represents the limit of our function. We know the equation of the curve (from Descartes), so we work back to find its function. In this case the area which was approximated by the Greek exhaustion (as in Archimedes) has its unwieldy summations of triangles replaced by a simple method which finds the area exactly. But it all hinges on an assumption.

What are these vanishing quantities which approach zero and may be *blotted out*, as Newton himself phrased it? In the first edition of the *Principia* Newton spoke of ultimate ratios as limits approached by quantities without limit, to which they come closer than any "given difference" . . . and "have diminished indefinitely." He also referred to them as the ultimate ratio of "evanescent quantities" with which or at the moment they vanish. It was evident that some of his geometrical proofs in the *Principia* made use of the calculus. Leibniz, who worked more extensively with the integral, spoke of his infinitesimals as quantities which can be made as small as we please, so that the error is less than any given number. In short, there is no error! What logical right do we have to speak like this?

It seems as if, like true Pythagoreans, we are dealing with mathematical mysticism. Bishop George Berkeley certainly saw it this way when he criticized the calculus in 1734. According to Berkeley the whole thing rested upon defiance of the law of contradiction. What are these quantities, he asked, which are neither finite, nor infinitely small (since they vanish) nor yet nothing (which they should be if they did vanish)? Berkeley knew mysticism when he saw it; they were "the ghosts of departed quantities"! What scientist, using the calculus, had the right to complain about the mysteries of religion? Voltaire put it bluntly when he quipped that calculus was the art of measuring something whose existence could not be conceived! Yet it worked, and scientists less concerned with metaphysics than applications used and improved it.

The rational Greeks would probably have shaken their heads in dismay. A cynic might have said that Newton used "ghosts" to prove occult sympathies (gravity) and labeled it all science. The scholastics might have smiled and called it "according to the imagination." Nonetheless, physical nature, if looked at in this way, did seem to conform to these "mysteries." Why it did is, perhaps, the greatest mystery of all.

Remarkably, Newton was not only an abstract (mystical?) mathematician and physicist, he was a good Baconian too. His experimental side was most evident in his work on light, the *Optics*, a very different work than the *Principia*. Again the story must begin with Descartes.

Descartes declared in 1637 that light was an instantaneous transmission of pressure through a luminous medium. Yet, true to his mechanical philosophy, when Descartes explained the laws of reflection and refraction he used the mechanical analogy of a tennis ball rebounding off a surface or passing through it. Pressure, said Descartes, being a tendency to motion, may for all practical purposes be treated by the laws of motion, which for him, of course, were the laws of impact. So the law of reflection easily follows: the tennis ball (light) strikes a perfectly hard surface and rebounds at its original speed, and geometrically the angle of incidence equals the angle of reflection. Refraction, however, is more difficult.

Passing from one medium to another the light ray bends like a stick in water. And motion in two different media suggests not only a change of direction like the stick but also an alteration of velocity. Again Descartes called upon his tennis ball; the ball breaks through the surface as if it were passing through cloth. Instead of losing velocity, Descartes claimed that light *gains* speed in a denser medium, as if the ball received an extra push as it passed through the cloth. Using this assumption and a rather complicated geometrical procedure that related velocities to angles of incidence and refraction, Descartes did prove that the sine of the angle of incidence is proportional to the sine of the angle of refraction, sine i/sine $r = n$. Again, with the use of questionable physical relationships we achieve a mathematical law which works.

Perhaps Descartes lifted the sine law from the Dutchman Willebrord Snel, or perhaps the mathematics led him past the physical difficulties. Pierre Fermat derived the same sine law from two different assumptions: Light travels slower in a denser medium, and the sum of time light takes to pass from one point to another through a refracting medium should be the least possible, the latter being an idea vaguely similar to the old precept of the economy of nature. Fermat's proof was derived from a mathematical concept called the maxima/minima of the calculus, yet for all of this he found the same sine law as Descartes!

What about color? For Descartes color could also be explained mechanically. Since his luminous aether is made up of tiny spheres, and light is actually a pressure imparting motion to these spheres, color results from different tendencies of motion. The spheres spin on their axes; a faster spin gives red, a slower blue, and all variations in between are dependent upon rates of rotation. Color, then, is a modification of the medium, or better, a modification of primitive white light as it passes through different media. Again we have invisible, *a priori* mechanisms, and again Newton challenged them—this time with experiment.

Newton took a prism and projected a narrow beam of light through it onto a wall across his room. If color was but a modification of white light, we should expect the projected beam to give a round spot, since all modified rays would be equally refrangible (refracted at equal angles). But Newton found a spectrum of colors about five times as long as it was wide. Had the prism separated the light into different rays corresponding to a specific color and angle of refrangibility? Newton thought so, and to test this startling conclusion he set up behind the first prism a board with a small hole and across the room another board with a small hole aligned with the first. Behind the second board he set another prism. If white light really consisted of various rays having their own color and refrangibility, we should expect a single ray of the spectrum, a particular color passing through the boards, to be refracted by an equal angle in the second prism. This was the crucial experiment, and this was exactly what Newton found. White light was in reality a heterogeneous mixture of rays, each with its own color and degree of refrangibility. Color was no modification of light; it was the essence of light! Who would have believed it?

The existence of colored bodies suggests, therefore, that reflection also must somehow separate white light. Newton pressed a lens against a flat sheet of glass, forming a thin film of air between. He found that white light gave rise to a series of colored rings, which varied with the thickness of the film. The rings indicated that light was either reflected or transmitted according to the thickness of the film, and Newton also discovered that the same thickness did not transmit two different colors. The differences in colored bodies, he concluded, were due to the thickness of their transparent particles. In physical terms light was composed of corpuscles moving at tremendous speeds. Different sizes of corpuscles create different colors, since they are refracted or reflected differently by varying densities. He also concluded, like Descartes, that light travels faster in denser media because of stronger attraction.

But physically explaining his rings posed a problem. The rings suggest a periodic quality more akin to waves than to particles. In fact, monochromatic light (of a single color) actually produced rings of light and dark, suggesting some sort of interference—periodic waves reinforcing or canceling each other. Other phenomena—crossing beams which did not interfere, polarity, and diffraction—also suggested waves. Newton hypothesized (a great deal in the *Opticks*) that the rings were due to light rays' possessing "fits of easy transmission" and "fits of easy reflection."

In short, the emission theory of particles was insufficient to account for the periodic phenomena.

Robert Hooke, who had criticized Newton's theory of light, suggested that light was a kind of impulse through the aether, like a wave spreading over the surface of water. Huygens developed a complicated wave theory based upon the idea of a wave front reinforced by wavelets caused by rapid motions communicated to particles of the aether. His reinforcing wavelets accounted for the sharp images in projection, since we should expect waves to continue spreading behind some obstacle as they do in water. Particles or waves? For the next century Newton's tremendous influence led most physicists to accept the former.

Newton died in 1727. He had been knighted, he had served as warden of the Mint, and he had become president of the Royal Society. He was lionized in England; as the century wore on his influence would spread to the continent, and eventually he would outshine Descartes, even in France. Yet his science would also be transformed, so much so that by the end of the eighteenth century many of the principles of the physical world view might have left him aghast. In fields such as chemistry the immediate benefits derived from his theories were problematic at best. And despite his avowed "I frame no hypotheses," he did do so, especially in his Queries appended to the *Opticks,* reaching a total of 31 in later English editions.

What was this gravity, this universal law of the cosmos? Was it not action at a distance, with all the occult overtones of that concept? Newton was certain that forces existed, that they were real, and that he had shown conclusively the effects of gravity in the *Principia.* Its causes, however, remained unknown. Unknown causes would no longer satisfy the mechanical philosophers. So in later editions of the *Opticks* Newton added new Queries (the first edition of 1704 had 16) in which he attempted to explain gravity by the aether. In the end, the mechanical hypothesis of matter in motion was still deemed necessary, certainly by those who accepted the fact of gravity and perhaps by Newton himself. Yet Newton the religious thinker could not help but see in the "wonderful Uniformity" of nature the hand of God, moving things by His will within His boundless "Sensorium," absolute space.

Absolute space? In the *Principia* Newton thought of space as absolute and immobile, the frame of reference for all relative inertial movements.[2] Did it not seem obvious? And the same was true for time—

[2]Newton admitted that in practice there could be no observable distinction between rest and uniform (inertial) motion. In principle, however, they must be treated as if they are absolute. How? The resistance of bodies to acceleration must be interpreted in relation to absolute space. Such inertial forces may also be seen in rotating systems in the form of centrifugal forces. Rotate a bucket filled with water, said Newton. At first the surface of the

absolute mathematical time ran simultaneously throughout the universe without reference to anything external. In an exchange of letters with Newton's friend Samuel Clarke, Leibniz complained that absolute space was unobservable, and absolute time without reference to the world was unimaginable. But these concepts were not unimaginable in Newton's mathematical dynamics; in fact, they follow naturally. Like force, his demonstrations led him to accept as simply given what seemed to be evident conclusions. The difference is that, unlike his contemporary mechanical philosophers who sought the very essence of nature, and so were forced to imagine all sorts of invisible mechanisms, Newton was aware that his mathematical principles only suggested truths about nature. In this he was closer to the scholastics; the ultimate reality of nature lay beyond the human gaze. Unlike them, however, he was willing to accept the physical necessity revealed by his mathematical demonstrations. Nature *was* open to exact quantitative description, even though its ultimate reality—the inner workings of the clock, as Einstein phrased it—could never fully be attained by science.

In his old age Newton said that he had been like a boy playing upon the seashore; he had found some pretty pebbles, but the great ocean of truth lay beyond, beckoning and undiscovered. The ocean still called, for behind the scientist stands the human being.

SUGGESTIONS FOR FURTHER READING

BOYER, CARL B. *The History of the Calculus and Its Conceptual Development.* New York: Dover, 1959.

CHRISTIANSON, GALE E. *In the Presence of the Creator: Isaac Newton and His Times.* New York: The Free Press, 1984.

COHEN, I. BERNARD. *The Newtonian Revolution.* Cambridge: Cambridge University Press, 1980.

DOBBS, BETTY JO TEETER. *The Foundations of Newton's Alchemy or "The Hunting of the Green Lyon."* Cambridge: Cambridge University Press, 1975.

HERIVAL, JOHN. *The Background to Newton's "Principia": A Study of Newton's Dynamical Researches in the Years 1664–1684.* Oxford: Clarendon Press, 1965.

KLINE, MORRIS. *Mathematics: The Loss of Certainty.* New York: Oxford University Press, 1980.

MANUEL, FRANK E. *A Portrait of Isaac Newton.* Cambridge, Mass.: Harvard University Press, 1968.

SABRA, A. I. *Theories of Light: From Descartes to Newton.* London: Oldbourne, 1967.

---

water is flat. Then it increasingly becomes concave, until at last the water is rotating at the same rate as the bucket. What causes concavity? Surely it is not the rotation of the water relative to the bucket, for concavity is greatest when the relative rotation of the water and bucket is zero. It must be due to absolute space. The flattening of the earth is also due to its rotation, surely not relative, hence absolute. Absolute rotation requires absolute space.

WESTFALL, RICHARD S. *Force in Newton's Physics: The Science of Dynamics in the Seventeenth Century.* New York: Elsevier Publishing Co., 1971.

————. *The Construction of Modern Science.* Cambridge: Cambridge University Press, 1977.

————. *Never At Rest: A Biography of Isaac Newton.* Cambridge: Cambridge University Press, 1980.

17

$\infty\infty\infty\infty\infty\infty\infty\infty\infty\infty\infty\infty\infty\infty\infty\infty$

Skeptical Chemists

The chemist bends over the open crucible, watching the metal inside a s it is slowly heated. The metal begins to lose its luster; gradually it changes into a dull, ashy powder. The process is called calcination, and the reduced metal, the ashy poweder, is called a calx. Obviously there has been some kind of *qualitative* change. And this is but a single example of countless others the chemist has witnessed. Two different solutions, for example, when mixed, change color, perhaps even form a precipitate. The alchemists knew of all sorts of similar reactions; even the unlettered peasant could perceive the rusting of iron tools as a fundamental qualitative change in the nature of the metal. Some reactions are reversible. Paracelsus had written that "dead metals" (calces) may be restored to the original metal by "soot" (charcoal). Chemical reactions seem preeminently qualitative. The affinities of some substances for others—the "occult" properties of attraction or repulsion—seem to cry out as we watch the reactions. Surely there are unseen virtues in brute matter.

Mechanical philosophers scoffed: Unseen virtues no longer served as good scientific explanation. But what did? Pierre Gassendi saw in Torricelli's experiments proof of the existence of a vacuum and hence evidence for the existence of Epicurean atoms in the void. While Gassendi believed that the qualities of matter could be reduced to size, weight, and shape of atoms, the practicing chemist might well wonder how atoms could explain the complex qualitative reactions observed in the laboratory. Chemistry was a kind of two-headed monster: One head was the mechanical philosophy, the other the practical experience of the crucible. The two heads scowled at each other.

Robert Boyle was skeptical (his famous book, published in 1661, was entitled *The Sceptical Chymist*). He attacked both the Aristotelian elements and the Paracelsan principles, showing by experiment that neither could be extracted from *all* bodies. And could Boyle experiment! He was responsible for developing many qualitative identification tests used in chemical analysis. He knew at first hand the extreme complexity of chemical reactions. His recorded experiments filled many dense volumes. But such lists of chemical analyses tend to become tedious. Everybody praised Boyle, said Samuel Johnson, but nobody read him!

People did read Boyle, however, in his role as mechanical philosopher. Boyle believed that gross matter was composed of insensible particles which form stable corpuscles of a second order, giving rise to the varying qualities of mixed bodies. Matter was conserved in reactions. Yet his laboratory work seemed to indicate an almost infinite range of transformations. Hence there was no reason for Boyle to suppose that nature could not produce an *unlimited* number of possible products by altering even the smallest particles of matter. Thus Boyle the mechanical, corpuscular philosopher accepted the possibility of transmutation, and Boyle the skeptical chemist realized the inherent difficulties in any elemental theory. At the same time he hoped for the day when a truly scientific chemistry would be able to establish a reliable program of particles in motion. But without waiting for theory he plunged into his experiments.

Air, as we have seen, was made up of particles with springs, which accounted for Boyle's so-called Law relating volume and pressure. Yet air apparently reacted chemically—fire could not exist without air, and animals died in exhausted containers. Boyle also roasted metals and discovered the ashy calx. Weighing it, he found that it weighed *more* than the original metal! Roasting metals in a closed vessel still resulted in a calx of greater weight (unfortunately, he broke the seal before he weighed the contents). Thus Boyle believed that the evident chemical properties of air were actually due to various particles of vapors floating in the springy, elastic air. The calx in his closed vessel had weighed more because fire particles in the air had penetrated the glass and united with the calx. Fire had weight, so did light. Air was a kind of universal solvent. But in France the chemist Jean Rey had written in 1630 that the observed increase in weight of the calx comes from its mixture with air. Two quite different theories explained an identical chemical phenomenon.

Like Boyle, Newton also sought to make chemistry a science based upon mechanical principles. The most primitive particles of primary matter are solid, massy, impenetrable, of different sizes and figures, formed by God in the Beginning—so wrote Newton in the famous Query 31 of the *Opticks*. This was Newton the mechanical philosopher, and it appears that these primary particles—call them atoms—were truly uncuttable. But now from Newton's alchemy comes another "hypothesis": It is probable not only that particles of matter have inertial mass (inertia, too, seems to endow matter with a certain active principle) but that they are moved and cohere by active forces, virtues, or affinities. Here, finally, was something to measure: Attractive and repulsive forces between particles pass into one another just as positive quantities pass into negative quantities in algebra.

"How do I *measure* these forces in the laboratory?" asks the chemist.

"What *are* these active principles?" asks the mechanical philosopher.

So Newton was forced to fall back upon the aether. It was a kind of elastic medium which communicated forces by a differential density. It was like the Stoic pneuma—yet unlike; the aether itself was composed of particles which endeavored to recede from each other (elastic). Thus there is gross, passive matter (but not quite passive, because of inertia) and there is the aether, which carries the active forces. Are there, then, *two* specific kinds of matter? Newton said elsewhere that there was one "Catholic Matter!" And how do we account for repulsions (a force) between the aether particles?

Nevertheless, many Newtonians believed that the problems of chemistry could be reduced to the physics of forces—that is, the study of affinity. The Croatian Jesuit, R. O. Boscovich, dropped the aether and simply drew force curves between unchangeable, dimensionless points. These points, he suggested, acquired a force in relation to each other, and the same particle could exert repulsion on one and attraction on another at the same time, or no action at all. Hermann Boerhaave of Leyden strove mightily to reconcile the divisibility of matter with Newton's atomism. If the "attractive virtue" in any part of matter resists all power of separation, said Boerhaave, then we have an atom. Fire was a weightless fluid composed of minute spherical particles which penetrate the porous nature of substances. All of this was, of course, a kind of physics "writ small." It was also physics "writ invisible." Unlike the motion of gross bodies, there was a wide chasm between Newtonian theory and what chemists actually observed.

The first truly workable chemical theory came from a different quarter—alchemy (of course!). It had occurred to J. J. Becher of Speyer that Genesis had spoken of creation only in terms of organic materials, earths, and therefore metals must be a byproduct. Becher held that the principles of compound bodies were three sorts of earths: the vitreous (like Paracelsan salt), a kind of mercury or fluid, and *terra pinguis*. Now this *terra pinguis* was a "fatty earth" found particularly in organic substances. It was moist and oily, and it gave bodies their qualities of odor, taste, and . . . combustibility! But it was Becher's follower, Georg Ernst Stahl, who converted this *terra pinguis* into a grand unifying theory. He called it *phlogiston.*

Phlogiston became the Renaissance Man of chemistry—it could do all things. When wood burns and turns to ashes, when metals are reduced to their calces, when metals rust, it is the principle of phlogiston which is released. Phlogiston not only was the agent of combustion but it also gave color and solidity. Its escape from a burning body stirred up particles and produced heat. Phlogiston was found in great quantities in organic materials, hence wood burns better than metal.

Stahl was particularly interested in the combustion and restoration of metals. If the calx of the reduced metal was once again impregnated with phlogiston, the metal was restored! Thus a very mysterious reversible chemical process was explained by a single principle. Also, air acts as a medium to carry off phlogiston in burning. Air does not become overburdened with phlogiston, because plants are especially good at absorbing it (in wood, for example). From plants it passes to animals. In short, phlogiston becomes a kind of chemical cycle of the biosphere. And almost as an afterthought we now know why combustion is impossible in a vacuum: there is no air to carry off the phlogiston! There seemed to be nothing which phlogiston was unable to explain.

Except . . . some chemists noticed that the calx (metal minus its phlogiston) weighed *more* than the original metal. How could this be if the metal actually lost something? Those chemists who accepted the phlogiston theory could, of course, always dispute such findings. Others, accepting this nasty little fact, explained that phlogiston possessed levity; when united with metals it made them lighter. Gravity certainly did not like that! Well, then, they replied, phlogiston has a *specific gravity* less than that of air. Phlogiston proved to be remarkably flexible, and it could explain many qualitative reactions better than the mechanical philosophy.

During the eighteenth century another field of chemistry was opened which also challenged the ingenuity of chemists, the study of airs (gases) or what is called pneumatic chemistry. In 1727 the Anglican clergyman Stephen Hales described to the Royal Society a new method for liberating "airs" from organic and inorganic materials. The apparatus he used was called a pneumatic trough. The main idea was simply to bubble up the gas through a flask of water to insure its purity. With this device Hales found that the properties of his airs varied; however, they all obeyed Boyle's law, so he naturally believed that they all must be a basic form of common air, differing only in their impurities.

But a professor of chemistry at Edinburgh, Joseph Black, found that gases took part in reactions and differed chemically. In 1756 Black showed that certain substances—magnesium carbonate and limestone—lost a similar gas upon reduction and that new substances were produced. Since this gas seemed to be fixed in solids, Black called it "fixed air." It possessed different properties from common air: It was a component of "air" exhaled in respiration, and it could also be obtained by passing common air over charcoal. Therefore Black reasoned that is must be dispersed through the atmosphere as well as fixed in solids.

Black also studied heat and made the distinction between sensible heat (temperature) and quantity of heat. Different substances of the same weight required different quantities of heat to raise their temperatures an equal number of degrees. Each substance seemed to have its own *specific heat*. As ice melted in water, the temperature of the water dropped off in proportion to the weight of ice melted, yet the temperature of the ice itself remained constant. There must be, thought Black, a certain part of heat which is *latent*, insensible but evidently transferred from water to ice by a given proportion. Whatever the physical nature of heat, Black had found a way to quantify it, demonstrating that it is conserved.

There were also some curious peculiarities about the evaporation of fluids. William Cullen, Black's teacher, discovered that not only water but many liquids cooled on evaporation, something which seemed to indicate that evaporation was not the dissolving of liquid in air, as was commonly believed, but rather was the combination of liquid with fire particles to create vapor. Further evidence for this idea came from the discovery that liquids could be evaporated in a vacuum, and it was noted that water in a vacuum boils at a lower temperature. Yet the physical properties of elasticity and expansionibility (an element's expanding to completely fill its container) were held to be the essential qualities of air as an element, and it was this air which made evaporated substances elastic. Only now, such experiments seemed to indicate that elasticity was caused by fire, and perhaps air itself might be a liquid combined with fire.

In France, Anne Robert Jacques Turgot, later director general of finances, expressed these ideas in an article he contributed to the great French *Encyclopédie*. Expansionibility, he declared, was a property not only of air but of all substances in a "vaporous" (his word) state. Heat overcame the attractive forces of matter and separated its parts; fire, not Boyle's mechanical springs, caused expansion. Thus Turgot distinguished this vaporization (the change of liquid to a gaseous state by heat) from evaporation, which happens, for example, in the case of ice that dissolves into the air without first melting. Evaporation occurred on the surface and required air; vaporization was caused by fire particles attaching themselves to matter. Hales' air, then, fixed in matter, was air minus fire. Vaporization indicated that matter could exist in several physical states and pass through these states by change of temperature. Therefore, the "aeriform" quality as a physical state was not confined solely to a specific element, the implication being that "air" was not an element but also a physical state.

Pneumatic chemistry was proving amenable to measurement. Perhaps its methods and techniques could be used to isolate and measure the elusive phlogiston. In London Henry Cavendish noticed the release of a gas when he dropped zinc, iron, or tin into an acid. Each gas burned with a blue flame and each had the same specific gravity. A gas of identical properties had been released by different metals. What would it mean? Cavendish was looking for phlogiston and now it seemed that he had found it. But it had weight!

In the course of his readings, Cavendish discovered reports of a strange dew generated when his inflammable air, as he called the gas, was fired with common air. Using the newly discovered Leyden jar, he passed sparks of electricity through his phlogiston (inflammable air) and common air. Next he tried another newly discovered gas called dephlogisticated air. In each case he found the dew! For ten years he carefully weighed and measured his airs, noting their proportions in combination. By 1783 he had reached a conclusion. Common water, once thought to be an element, was really a compound of two gases: phlogiston and dephlogisticated air! Further, two volumes of phlogiston always united with one volume of dephlogisticated air to form an equal weight of water. So departed water from the elements. From Thales to van Helmont it had lived a long and honored elemental life. No more.

But wait! What was this "dephlogisticated air"? The answer came from another clergyman (this time a Dissenter who rejected Anglican dogma) and teacher of languages named Joseph Priestley. Chemistry for him was simply a fascinating hobby. He experimented with many gases, following whatever whim took his fancy. He succeeded in isolating nitrous air from metals and nitric acid, and he noticed that this nitrous air diminished a volume of common air by about one-fifth. Priestley saw that the proportion of this gas in common air evidently had some effect upon its "goodness" for respiration. So the question naturally dawned upon him: What gave common air its goodness? What made it fit for breathing?

Priestley had found that heating the calces of metals, specifically the red powder of reduced mercury, gave off another air, which, he reasoned, had been taken up from the atmosphere upon reduction. The chemist Carl Wilhelm Scheele had named this gas "fire air," since a candle placed in it burned very brightly. Priestley prepared two vessels over water, one with common air and the other with "fire air." In each vessel he placed a mouse. What he saw was astonishing. The mouse in the common air expired in a short time, but the other mouse in the fire air lived a long time afterward. Priestley decided to inhale the fire air himself and found that his breathing became light and easy (although he cautioned that breathing too much would cause us to "live too fast"). His experiments with metals seemed to indicate that fire air was a part of the atmosphere. What else could it be but that part which gave the atmosphere, common air, its goodness?

The phlogiston theory taught that common air took up phlogiston during combustion. Priestley believed that he had found the mechanism. This fire air must really be dephlogisticated air, a gas originally deprived of phlogiston but eminently capable of absorbing it. Thus common air becomes unfit for respiration when it is full of phlogiston, and dephlogisticated air makes common air fit for respiration and combustion, since it is able to carry off phlogiston. How neat! The facts of the laboratory beautifully fit the phlogiston theory. But did they? No one (except perhaps Cavendish) had identified phlogiston for certain. Yet that need not bother us: Who had identified the cause of gravity? The problem, rather, was the confusion of all these "airs" and their roles in chemical reactions. Who could make sense of it all? In France, waiting in the wings, was the man who could and did, Antoine Laurent Lavoisier.

Lavoisier, the son of a wealthy lawyer, was sent to the Collège Mazarin, where he studied law, mathematics, physics, botany, and, yes, chemistry. Here finally was someone grounded in physical science and possessing a thorough knowledge of qualitative chemistry. Lavoisier the physicist believed in conservation—that is, in all chemical reactions nothing is either destroyed or created. Lavoisier the chemist realized that quantitatively establishing *what* actually was conserved— the elements—was a difficult task. Hypothetical atoms were all very fine, but they must not be allowed to lead us astray from our practical chemical experience. Thus, if we should find in our experiments a substance which cannot be divided further, for all practical purposes this must be an element! For example, heat was evidently conserved; it entered into reactions and it could be quantified (Black). So for now let us treat heat as an elemental substance—*caloric*, Lavoisier named it. It could change a liquid into a gas. Thus gas was a base element combined with caloric, in this case Black's latent heat. Fire was free heat. Latent heat added to water produced steam.

There are no transmutations in chemistry, only modifications. But such modifications are qualitative, and new substances may be formed which have different properties from the originals. By careful weighing, Lavoisier demonstrated that the grey material found in a heated pelican of water came not from the transmutation of water to earth, as van Helmont had believed, but from the glass of the pelican itself. Substances reacted chemically and yet were conserved.

And this was the clue to the puzzle. The gases found in the atmosphere by the English chemists were not "transmutations" of pure air, nor separate vapors; they were actually distinct substances which reacted chemically to give the atmosphere its gross qualities. The atmosphere *was* a chemical mixture of gases which played a role in chemical reactions like combustion and respiration. Phlogiston was superfluous!

In November of 1772 Lavoisier deposited a sealed note with the French Academy of Sciences. In it he had written that when phosphorus and sulfur are burned, they *gain* weight, and this weight comes from a huge quantity of air which unites with these substances. Then Lavoisier made the leap. Is it not probable that the gain in weight of all calces is due to the same cause? The fact that calces did actually gain weight had only now been shown conclusively. Lavoisier's sealed note was meant to establish his priority, not in facts, but in theory.

Lavoisier knew that reducing the calx of lead (red lead) restored the metal and also resulted in an effervescence. Tests showed that when the red lead was reduced with charcoal, the resultant gas was none other than Black's fixed air. Now if the calx of a metal fixes some part of the atmosphere, and reduction of the calx gives off a gas, evidently the two processes are related. But was it some component of the atmosphere that was required for combustion? Perhaps fixed air?

In October of 1774 Priestley visited Paris and dined with Lavoisier. The precise significance of the visit has been debated by historians. Did Priestley tell him of the properties of his newly discovered dephlogisticated air—how brilliantly the candle burned in it and how well it supported respiration? *Respiration!* Fixed air did not support respiration! And what if respiration was a form of combustion . . . ?

Whatever the case, Lavoisier went to work. Heating mercury calx with charcoal produced fixed air, but heating the calx *without* charcoal produced a different sort of gas. It supported respiration and combustion! Evidently the charcoal had misled him: The air given off by the calx had been taken up by charcoal, which in turn resulted in fixed air. Earlier, like Boyle, Lavoisier had heated metals in closed containers, only unlike Boyle he had found that there was no gain in weight until the vessel was opened and air rushed in. In a closed bell jar he heated mercury with a given quantity of air. The results showed that with the formation of the calx the air had lost about one sixth of its volume and that the remaining air was "azote," lifeless—nitrogen. By 1778 the pieces had finally fallen into place; the part of the air which supported combustion was Priestley's dephlogisticated air or Scheele's fire air.

Priestley had not known his own riches. It was not the facts Lavoisier changed but the way of looking at them. Air is not an element! It is a compound of distinct gases, one of which is responsible for combustion and fixes with calces (in our terms "oxides"). This vital air

Lavoisier renamed. It was not air minus its phlogiston; rather it was a specific gas all its own—*oxygen!* Actually, since Lavoisier believed that gases were produced by the addition of his imponderable caloric, oxygen gas was the base, oxygen, containing caloric.

Oxygen was that part of the atmosphere which supported combustion, the oxidation of metals, and respiration. Thus respiration could now be viewed as a slow combustion, the oxygen combining with carbon in the lungs (Lavoisier believed) producing carbon dioxide (formerly fixed air) and heat. With the help of the mathematician Pierre Simon Laplace, Lavoisier performed experiments on his assistant and noted that oxygen intake was increased as work was done. So respiration, too, was a mechanical chemical process.

Water was a chemical compound of oxygen and inflammable air, or, as Lavoisier renamed it, *hydrogen.* When metals reacted with acids, oxygen was taken up and hydrogen released (Lavoisier believed that all acids contained oxygen, which later Humphrey Davy proved incorrect). Lavoisier had called into existence a new chemical world, much as Galileo did for physics, by changing its language.

Lavoisier carefully uncovered the inconsistencies of the phlogiston theory in a polemic of 1783. Phlogiston simply failed to meet the criteria of quantitative exactitude. Notably, the concepts of mass and weight adopted by physicists following Newton clashed with "negative weight" some chemists believed was the property of phlogiston. Also, phlogiston seemed to hark back to Aristotle's element fire, which had the principle of levity. A principle, a substance, sometimes having negative weight, sometimes positive, or even weightless, phlogiston was too vague and self-contradictory, changing its qualities and properties like a chameleon according to its experimental environment. It was a term, a word, which had been reified into existence, and worse, it was a throwback to the imprecise language of the alchemists.

True to the Enlightenment, Lavoisier saw language as not only a collection of symbols but also an important component of clear and precise reasoning. Mathematics was the perfect prototype: Algebra, claimed Abbé Condillac, was a model of precision, for its symbols carried no ambiguities and its conclusions followed strictly from its premises—it was the best of all scientific languages. Lavoisier, too, believed that the logic of chemistry was essentially dependent upon its language, a language which did not admit the inconsistencies and confusions of a term such as phlogiston.

Phlogiston did not leave the stage all at once and without protest. Priestley remained unconvinced to the end of his life, and others, long after Lavoisier, used the word. But Lavoisier not only presented a new theory, he followed through with his program and changed the very language of chemistry. In 1789, his *Traité élémentaire de chimie* introduced the new terms—oxygen, hydrogen, caloric, and so forth—and a new generation of chemists grew up speaking the language. Phlogiston finally disappeared as a chemical fact.

In 1789 there was another revolution in France, not one of scientific abstractions but political and social. It had begun perhaps with ideas, as we shall see in the next chapter, many connected to science; soon, however, it passed into blood, passion, and war. Lavoisier faithfully served the new regime in France, and his chemical work on gunpowder provided the French armies of the Republic and later of Napoleon's Empire with a superiority over their enemies, revealing the new practical importance of chemistry in the wider world. In fact, the beginning industrial revolution furnished an outside stimulus to chemistry by requiring improved methods of manufacture in such things as textiles and metallurgy. But Lavoisier had been a tax farmer under the old regime, an activity much despised by the revolutionaries, and he had also earned the enmity of the revolutionary Jean Paul Marat by criticizing Marat's studies of fire. Thus Lavoisier was caught up in the flood and executed in 1794.

The language lived, yet it would not go unaltered. For some, the caloric was like the ghost of a departed quality—the spirit of phlogiston. Benjamin Thompson, an American by birth, later made Count Rumford by the Elector of Bavaria, was working in a foundry in Munich in 1798. Rumford noticed that the mechanical work used to bore a cannon caused water to boil; mechanical work produced heat. Perhaps both were convertible. Some mechanical philosophers imagined heat to be particles of matter in motion, and so heat need not be an imponderable fluid, caloric, to obey the laws of conservation. Yet such speculation was premature, for imponderable fluids were scientifically respectable in the eighteenth century, as we shall see in the next chapter.

A more pressing problem was the elements of matter. What were they? Could they be quantified? How did they combine in compounds? It is now the early nineteenth century, and a mostly self-educated Quaker named John Dalton is studying gases in the atmosphere. He muses: "How are the gases held together in the common atmosphere?" Dalton wanders about England, taking samples of the atmosphere and making meterological observations. "The composition of the atmosphere is the same wherever I go!" he exclaims. In France the chemist Gay-Lussac ascended in a balloon and collected samples, finding the composition of the atmosphere nearly the same as at ground level.

If the atmosphere is made up of different gases, all having different densities, how is it that the heavier gases do not settle to the bottom and the lighter rise to the top, forming layers?

Perhaps, reasoned Dalton, the air was a purely mechanical mixture of gases. Dalton also noted that the pressure of dry air was increased when water vapor was added by the *same* amount of vapor pressure at a constant temperature. This seemed to indicate that the total pressure of the mixture was really the sum of the separate or "partial pressures" of the constituent gases. Now, if the composition of the atmosphere was constant; if the various gases were of different densities; if their pressures at constant temperature were independent of each other and, added together, give the sum of the atmospheric pressure; was it not logical to assume that the *particles of each gas* differed in number and weight? It seemed evident that particles of the same element only exerted repulsion (Dalton had read Newton) upon one another, thus keeping the atmosphere in a steady proportion.

Joseph Louis Proust, teaching chemistry in Spain, held that when elements combined to form compounds, the elements united in definite proportions. Dalton's friend William Henry had shown that the solubility of gases in liquids was directly proportional to their pressure. All of this suggested to Dalton that Newton's atoms differed in weight from element to element. Ultimate particles differing in weight? Dalton believed that this was something entirely new. Yet if every particle of a certain gas was like every other particle of that gas, and if the composition of compounds was constant, why then the *atoms* of each element must weigh the same!

No one, of course, can see an atom, but we all know that mathematics is able to visualize many unseen things. Take hydrogen, said Dalton. It is the lightest element. Therefore the weight of a hydrogen atom may be fixed arbitrarily as one. By gross weight hydrogen unites with oxygen in a ratio of one to seven. Is it not reasonable to assume that one atom of hydrogen unites with one atom of oxygen, and thus the *relative* weight of the oxygen atom is seven? Relative atomic weights! At last chemists had something to measure. The dream of atomists from Democritus to Newton had come to pass: Chemical measurement could be made to infer atoms.

Chemical reactions now revealed to Dalton a remarkable mathematical simplicity. He constructed little models of his atoms, little wooden spheres, and used different symbols to represent them: Hydrogen was ⊙, oxygen ○, silver Ⓢ, nitrogen ①, carbon ●, and so on. Hence he could visualize what was happening in reactions. Take a simple example: One carbon atom unites with one oxygen atom and forms a gas. The ratio of carbon to oxygen is 3 to 4. Carbon also unites with oxygen to form another by the ratio 3 to 8 (carbon dioxide). Two distinctly *qualitative* gases could yet be related mathematically—8 was a perfect multiple of 4. Other compounds seemed to follow this rule, and Dalton's models showed the simplicity of it all. In two compounds of the same elements there was evidently a simple ratio of weights—a law of Multiple Proportions. At last the essence of matter itself had been tamed by mathematics. The Pythagoreans would have applauded.

But any applause was premature. Atoms were to prove troublesome; after all, they had only been inferred from gross combinations. What were the true elements? Were Dalton's simple ratios correct? And Dalton had used the word atom quite freely—an atom of hydrogen, an atom of water, an atom of gas. It was the first decade of the nineteenth century, and chemists clearly had their tasks set out for them. Nonetheless, the two heads of chemistry had finally become one, for the working physical hypothesis of atomic weights and proportions seemed to account for the qualitative changes in the crucible. And—to the dismay of all alchemists—the ultimate particles of matter, whatever they turned out to be, could not be transmuted. Elemental atoms were truly uncuttable.

Or . . . they were immutable for the next hundred years! There were still doubters all the way into the twentieth century. No one had proved conclusively that atoms even existed. But if they did, surely they were uncuttable. Surely the ultimate elements, if found, were stable and unalterable. Surely?

SUGGESTIONS FOR FURTHER READING

BOAS, MARIE. *Robert Boyle and Seventeenth-century Chemistry.* Cambridge: Cambridge University Press, 1958.

GUERLAC, HENRY. *Lavoisier—The Crucial Year: The Background and Origin of His First Experiments on Combustion.* Ithaca, N.Y.: Cornell University Press, 1961.

IHDE, AARON J. *The Development of Modern Chemistry.* New York: Harper and Row, 1964.

PARTINGTON, J. R. *A History of Chemistry,* Vols. 2–3. London: Macmillan, 1961–1962.

THACKRAY, ARNOLD. *Atoms and Powers: An Essay on Newtonian Matter-Theory and the Development of Chemistry.* Cambridge, Mass.: Harvard University Press, 1970.

————. *John Dalton: Critical Assessments of His Life and Science.* Cambridge, Mass.: Harvard University Press, 1972.

18

∞∞∞∞∞∞∞∞∞∞∞∞∞∞∞∞∞∞∞∞∞∞∞∞∞∞∞∞∞∞∞∞

Optimism, or
Cultivating the Garden

Such was the revolution in physical science. Ernst Mach said that Newton's principles thoroughly established the necessary assumptions to treat any problem in mechanics. The field had been ploughed and fenced, and whatever problems arose were of a formal character and did not involve questions of principle. The following generations cultivated the garden, but its furrows were deep and well marked. The new physical laws which grew were like seedlings in the fertile soil.

If only this were the case! Actually, the development of physical science after Newton was much more complex. Perhaps Newton's science was the center of the garden; yet, there were other traditions—Descartes, Leibniz, Huygens, not to mention other scientific gardens in other fields of the human endeavor to understand the world. And for the first time the world beyond the garden was beginning to take note of its crops—science was about to enter the social consciousness. Also there were different ways of looking at Newton: the mathematical, mechanical Newton of the *Principia* or the speculative, experimental, and even religious Newton of the *Opticks*.

But oh what an exciting harvest! New areas of mechanics opened—fluid dynamics, the physics of elastic media, dynamical systems analysis—all requiring new techniques. The calculus was expanded into partial differentials and variations; probability theory was developed and applied to such things as social statistics. The logical foundations of the calculus led to a maze of speculation. Electricity became a new subject for physics and, like caloric, was first perceived as an imponderable fluid. Physical science was, in short, a kaleidoscope of assumptions, methods, principles—even metaphysics. And what it all meant . . . now that was a real question!

Consider the conceptual problems raised by force and inertia. Matter, as Newton suggested, was essentially passive, and gravity, as Newton clearly protested in a letter to his friend Bentley, was not an inherent principle of matter. Yet inertia says that matter *resists* change of state, suggesting that at least inertia is a passive force inherent in matter. The mathematician and *philosophe* Jean d'Alembert was ready to accept the abstract mathematics of gravity, but where no impact was observed causing change of motion, d'Alembert would describe it mathematically without recourse to force. Cartesianism was still strong in Frenchman d'Alembert.

Then there was the debate over Leibniz's *vis viva* (living force), which continued into the eighteenth century. The roots of the controversy are actually found in metaphysics. Had God created a self-sufficient world—the best of all possible worlds—or did he have to tinker with the machine? If matter was solid and inert, attraction according to Leibniz was God's tinkering, a "miraculous thing." Would a perfect God create an imperfect world? And is not inertia a contradiction? Force *is* the essence of matter, Leibniz believed, but the force we see and measure in mechanics is actually a phenomenal derivative of a primitive metaphysical force. The phenomenal forces are two: dead force, which is the tendency of bodies to move, and living force, which arises from infinitesimal and successive impulses of dead force like the instantaneous fall (or rise) of a body. Now the force which is conserved in the universe is this *vis viva* or living force, and it is measured by mv^2 (our kinetic energy is $\frac{1}{2}mv^2$). The conservation of *vis viva* makes for a well-oiled, self-sufficient universe. And since living force is comprised of infinitesimal increments of dead force, all change must be continuous. There are no *jumps* in nature. Thus all bodies must be elastic, because the collision of perfectly hard bodies would mean instantaneous rebound and the loss of living force.

The debate carried on into the eighteenth century. The Swiss mathematician Johann Bernoulli adopted the conservation of living force in his discussion of the collision of elastic bodies. Like Leibniz, Bernoulli viewed dead force as a tendency to motion and living force as successive motion over a distance in time. Unlike that of Leibniz, Bernoulli's living force was no phenomenal derivative of a metaphysical force; it was inherent in matter, conserved in the universe, and fully real. Perhaps Newton in the quiet of his study saw God's activity in the world in terms of force. The conservation of living force, effectively reducing the truly active role of the deity, was certainly not Newton's metaphysics!

Then there were purely technical problems with the Newtonian laws of motion. It was only after Newton's day that the second law became $F = ma$ (or in differentials $F = m\, d^2x/dt^2$). Further, Newton's mechanics dealt with point masses, simply a point fixed by two coordinates x and y. But most real bodies are rigid; they have a certain solid shape which is oriented in dimensional space. Therefore a generalized system of coordinates must be described to follow the motion of a rigid body through space. Leonhard Euler, one of the greatest mathematicians of the eighteenth century, derived these general equations only around 1750.

Leonhard Euler! He did everything—mechanics, astronomy, hydrodynamics, pure mathematics, and more! The so-called Newtonian rational mechanics—the movement of bodies treated purely abstractly by algebraic equations—was actually more Euler's creation than Newton's. Euler could calculate effortlessly—as birds fly, the saying went. Born in Basel, Switzerland, Euler did most of his work at the St. Petersburg Academy in Russia and the Berlin Academy in Prussia. At the age of sixty he was totally blind (he had lost the sight of one eye when younger), yet he continued to produce outstanding mathematical papers.

Newton's method of fluxions was built upon problems of velocity and rates of change with a visual, empirical emphasis. Leibniz's differentials were more popular upon the continent, yet they were still based upon the infinitesimal, as in geometry. Gradually the calculus was separated from its purely physical and geometrical roots. Euler was in the vanguard of this development. In strictly formal terms, the calculus expresses the relationship between variables. Euler realized that this relationship, the function, was the central concept, which, by itself, was strictly formal. We need not visualize it in terms of velocity or geometry but simply as an algebraic expression. In short, we *detach* the algebra and its operating rules from all physical intuition and deal with purely abstract symbols. Thus the calculus in Euler's hands becomes analysis, an abstract language separate from intuitive experience.

So why worry about Zeno anymore or that meddlesome metaphysician Berkeley? We simply accept the formal operations without concerning ourselves about the physical impossibilities of instantaneous arrows in flight or "ghosts"—Euler even admitted that dy/dx was actually $0/0$, but $0/0$ could take many values, since the derivative determines its value for the function involved! This was hardly satisfactory even in formal terms, yet Euler's formalistic approach was a step toward the purely abstract concept of the limit as it was finally established in the nineteenth century.

New mechanical problems also created the need for new forms of analysis. The vibrating string, the motion of sound, the flow of fluids—in short, the movement of continuous media—may be treated by the calculus. In such problems we have functions which include several variables. For example, a volume of gas is dependent upon both pressure and temperature; or a continuous wave motion requires the consideration of time, displacement, and position, all of which are continually changing. Both d'Alembert and Euler set up the first equations dealing with such multivariable problems, called partial differential equations. Partial differentials are equations in which the function is dependent upon two or more variables. Their range of physical application is indeed vast.

Another of Euler's contributions was the generalization of the calculus of variations. Suppose we wish to find the shortest distance between two points on the earth's surface (called a geodesic). Since the earth is not perfectly round, the shortest path depends upon just where the points are located. Of course, intuition (and plane geometry) suggest that the shortest path is a straight line. Yet in 1696 Johann Bernoulli had found that in the case of the acceleration of gravity, the quickest time or least time of descent was not a straight line but a curve, a brachistochrone as it is called. This is because the curve allows more acceleration and hence more velocity than the straight line. So the problem of the least time in fall or the shortest distant on the earth's surface is actually a problem of finding the minimum of some variable whose value is dependent upon another function, acceleration or the varying surface of the earth. Involved here is the calculus of variations, which Euler in fact used to give a mathematical expression to another physical principle, Least Action.

It was an ancient belief that nature, like Ockham's razor, does nothing superfluously. Fermat's principle of Least Time for refracted light was a cousin of this economy of nature. In 1744 Pierre L. M. de Maupertius asserted that *all action*—a mechanical product of mass, velocity, and distance—is always the least possible. For Maupertius this was a wonderful scientific proof of God's existence, a sublime metaphysical principle. In Euler's hands it became mathematical, demonstrated by the calculus of variations. Perhaps, in the end, it was just as sublime and metaphysical, but now the principle of Least Action was also rigorous.

In 1788 the visual and geometrical foundations of mechanics were finally stripped away by Joseph-Louis Lagrange. In the preface to his *Analytical Mechanics* he boasted that no figures appeared in the book: "neither geometrical nor mechanical reasoning is necessary." Now we need only algebraic operations subject to formal developments—the formal relationships between variables presented solely in the form of mathematical symbolism. The principle of Least Action was further generalized to cover more dynamical situations—but abstractly. Gone were metaphysical qualities, in fact all qualitative leftovers.

In astronomy, Pierre Simon Laplace revealed the power of analysis.

From 1799 to 1825 his *Celestial Mechanics* appeared in five volumes, a massive work dedicated to complete mathematical treatment of gravitational astronomy. Laplace showed that the seemingly permanent irregularities of the planets were actually periodic fluctuations. For example, the observed increase in the mean speed of Jupiter and decrease in that of Saturn were not continual but actually were due to the influence of these planets upon one another. He also related the minute irregularities of the moon to the orbit of the earth. In fact, all eccentricities in the solar system occurred within certain limits, like a fixed bank account which can never be overdrawn. Hence the implication was that the entire planetary system was *stable* and there was no need for God's intervention. Poor Newton! First the conservation of living force and now this! God was again becoming Aristotelian! There was, however, one tiny burr in the symmetry: the perihelion of Mercury (the point of its orbit closest to the sun) was in motion due to the attractive influence of the other planets, but there was a very slight discrepancy between actual observation and theory.[1] It was nothing to be alarmed about, yet. . . .

Newton had constructed an entire world based upon the action of forces, yet he had moved from one theory to another, searching for some explanation of these forces and taking in general a very ambiguous stance on whether the forces acting between the planets and parts of matter were due to an intervening aether, or action at a distance, or even God's direct ubiquitous power. In the second edition of the *Principia*, edited by Roger Cotes, he had firmly rejected the Cartesian vortices and thus seemed to imply that gravity was action at a distance. Yet in the first decade of the eighteenth century he had become interested in electrostatic phenomena, and he wondered if bodies did not contain within themselves some "electric spirit." In Queries he added to the *Opticks* he once again speculated on the aether medium, a medium exceedingly rare, which might account for electricity and magnetism, the vibratory phenomena of light, radiant heat (the transmission of heat through a vacuum), gravity, and even, in Query 24, animal motion "perform'd by the Vibrations of this Medium, excited in the Brain by the power of the Will . . . ."

[1]To be precise, the anomaly was discovered by Urbain Jean Joseph Le Verrier in the nineteenth century. In 1781 William Herschel discovered the planet Uranus outside the orbit of Saturn. At first he had taken it for a comet. Ceres, the largest of the asteroids, was discovered in 1801 between the orbits of Mars and Jupiter (in the preceeding year the philosopher G. W. F. Hegel had "proved" that there could be only seven planets in the solar system). Calculations based upon observed irregularities in the orbit of Uranus by the same Le Verrier (and independently by John Couch Adams) led Le Verrier to predict the existence of yet another planet. This was Neptune, discovered in 1846, the first planet to be discovered on the basis of theoretical calculations. In 1859 Le Verrier presented his calculations of Mercury's anomaly. Naturally he suggested the possibility of a planet between Mercury and the sun. Yet this planet was never observed.

Thus Newton's Queries were tremendously influential in the development of the concept of imponderable or subtle fluids, which appeared around 1740. Subtle fluids were not like ordinary matter—the flow of heat from a hot to cold body or the transfer of electric effects did not seem to cause any changes in mass. Newton had also thought of these phenomena in terms of forces acting between atoms, yet unlike gravity, they could not be measured. On the other hand, experiments in the eighteenth century made it convenient to think of such things as subtle fluids which carried the physical property, the density of the fluid being proportional to the intensity of the effect. Therefore, Newton's aether theories could be reinterpreted as lending support to the concept of imponderable fluids, and his gravity could be rescued from the troublesome action at a distance.

One problem with Newton's aether was that its density had to be unbelievably low as to allow the planets to pass freely through it (in the *Opticks* he said that the elastic force of the aether in proportion to its density must be above 490 billion times greater than that of air). This meant that action at a distance was still required for the forces *between* the aether particles. However, in this sense, activity could be conceived as intrinsic to the matter of the aether—not gross matter, as Newton had tried to avoid; yet on this scale it could not be measured.

On the other hand, the concept of an imponderable fluid did provide some guide to quantification. The thermometer, for example, measured the concentration of heat fluid, or later the caloric, in a body. Since these fluids had no weight—they were not matter—it could be assumed that they carried the physical properties and as such were the "something" measured in experiments. Imponderable or subtle fluids were in this way long-lost ancestors of the old "sympathies" and "antipathies" of the Hermeticists—as perhaps gravity was for Newton—yet unlike the vagaries of the alchemists they proved amendable to quantification. And further, perhaps most important, the great Newton could be called upon as an authority and the speculations in his *Opticks* redefined into fluids. Electricity was a good example of this process.

The ancient Greeks had known that rubbing amber gives it an attractive virtue (*elektron* was the Greek word for amber). Gilbert had shown that this was true of a wide range of materials. Here, again, was the vexing problem of action at a distance. Boyle had considered electrical attraction due to an "effluvia" liberated from the body. However, electricity exhibited other virtues: The attractive virtue could be conveyed to other bodies, and some electrified bodies repelled others. Electricity consisted of attraction *and* repulsion! In 1732 Stephen Gray published in the *Philosophical Transactions* a report of his experiments, which demonstrated that some materials such as metals would *conduct* the electric virtue, while others like glass and silk did not; rather, they preserved it. And two electrified oaken cubes, one solid and the other hollow, produced exactly the same effect. So, unlike the conduction of heat in a body, electricity seemed to be a surface phenomenon.

In France Charles François Du Fay concluded that electricity was really an affair of two fluids, imponderable and weightless. They can be separated by friction and thus repel or attract according to their kind. Du Fay called his two fluids vitreous and resinous electricity. Euler, who followed Huygens' wave theory of light, thought that electricity was simply a disturbance in the equilibrium of the aether. But how to account for its other properties besides attraction?

In the 1740s experimenters in Leyden found that a corked glass beaker filled with water could be charged with a machine that produced electricity by the friction of rubbed globes. Holding the jar by the bottom and contriving a connection to the top, the experimenter received a shock. Obviously some *force* was being passed from one side of the jar to the other. Some physicists, such as d'Alembert, were quite prepared to accept the quantification of mechanical force and yet, in the manner of Cartesianism, doubted the existence of action at a distance. There was always the temptation to posit some sort of visual model to explain force, like the aether. With electricity, a relative newcomer, the idea of a subtle fluid made sense. But here it was the Newton of the *Opticks* who provided the precedent—the speculative and experimental scientist.

Why have two electrical fluids? Benjamin Franklin believed that there was only one and that all bodies contain it. When a body had its normal quantity of electrical fluid, it is in equilibrium or a null state. An electrified body has either an excess of the fluid, a *positive* state, or a deficiency of the fluid, a *negative* state. All electrical reactions are due simply to transfers of the fluid. Take the Leyden jar, said Franklin; pumping electricity into the top disturbs the equilibrium, for at the moment the top is electrified positively the bottom is electrified negatively, ejecting an *equal* proportion of the fluid. The equilibrium of the jar cannot be restored inwardly, only by communication from without. Hence the shock.

Is there not something familiar about this? The loss of electrical fluid is always accompanied by an equal gain? Yes, answered Franklin, electrical fluid, like *vis viva*, is neither created nor destroyed. All electrical activity obeys the universal law of conservation. Thus we have another conservation law in physics. And, as we all know, Franklin demonstrated with his kite that lightning was an electrical phenomenon. Therefore electrical fluid is part of nature and its conservation must be universal. The only drawback is that we have another imponderable fluid. How many different kinds of matter are there?

By the close of the eighteenth century the first steps were taken toward the quantification of such static electrical phenomena. Franklin had reported to Priestley his inability to detect electrification *within* a charged metal sphere. Newton had shown mathematically that a hollow sphere exerts no gravitational force upon an object inside. Here, then, was a slender clue that perhaps electrical force behaved like gravitational. Cavendish, for his own amusement, actually deduced the inverse-square law for electricity yet did not publish his findings. About 1784 the French military engineer Charles Augustin Coulomb invented a torsion-balance and demonstrated conclusively that the inverse-square law held for static electricity. Coulomb also found that the force between two charges was directly proportional to the product of the charges.

It was also near the end of the eighteenth century that moving charges—electric currents—were opened to investigation. Around 1780 Luigi Galvani found that a dissected frog's leg contracted when the brass hook from which it hung was buffeted against an iron railing by the wind. How curious! Was it animal electricity? Galvani thought so, and he even suggested that soup made from the legs might restore health. Yet it had also been known that two dissimilar metals placed on either side of the tongue with their other ends in contact produced a certain pungent taste. Alessandro Volta saw the connection: moist materials placed between dissimilar metals produced electric currents! His piles of zinc, copper, and moist material between produced a "perpetual motion of electric fluid." And obviously there was some connection between this electrical fluid and chemistry, perhaps even the composition of matter.

Forces, powers, activity—everywhere one looked in the physical world they seemed to call out. Indeed, Priestley himself concluded that matter was "inherently active," and so the human organism requires no immaterial soul to endow it with powers. Mind itself is material; matter and force are one! Not extension, not impenetrability, but force is the essence of matter, and matter is a manifestation of powers. Priestley, recall, was a dissenter, and in 1794 he left England for America. In 1791 a mob had smashed his personal laboratory in Birmingham. Evidently some people in England believed his ideas dangerous, a threat to Church and monarchy. Was he not a materialist? Did he not deny the immortal soul? Was he not akin to those French atheists who were destroying society in the name of reason?

Here we must pause and survey the world beyond the scientific garden. With a sigh of relief we pull ourselves away from the twisting, internal furrows of the garden and glance briefly at the countryside.

How different it looks to us now! The new concepts, new knowledge, language—perhaps we do not understand it all. Maybe we are blind to its complexities, the paradoxes and assumptions of our new science. Nonetheless, it certainly seems *as if* everything has become suddenly clear. Nature operates by unalterable laws, and human reason appears to be admirably equipped to discover them. Natural law, reason, and a third word—perhaps the most important—progress; how easily they flow from our lips in the wake of the great Newton. The veils have fallen and all is light!

Of course we exaggerate, for if we are working scientists we know that we only stand upon the seashore and all is not light. But, oh, the glory of reason—the critical, experimental, empirical kind, not the illusionary labyrinthian reason of the scholastics. *Enlightenment!* we want to shout to the countryside. Let us clear the land of shadows, of superstition, of irrationality. It desperately needs clearing—especially if we live in France.

The French called the eighteenth century *le siècle des lumieres*, "the century of light." The term "enlightenment" was first used by the German philosopher Immanuel Kant in 1785, and so the name has come down to us. Reason, the most important word of the Enlightenment, became nearly synonymous with the laws of nature. God was to be known, not through revelation and traditional authority, but through His creation, Nature. The Scientific Revolution seemed to demonstrate that Nature could be known through the laws of reason—not the formal logic of the past but rather the methods of the new science. Henceforth, if the laws of society could be made to agree with the laws of nature, the operations of government, economics, and society could also be transformed into harmonious instruments for the benefit and moral improvement of all. What was needed was a science of society, what Turgot, the Marquis Condorcet, and their friends called a *social science*.

If natural law was derived from the necessary relations that govern nature, then the laws of nations must derive from the laws which govern human nature. Thus one of the most important books of the Enlightenment was Montesquieu's *Spirit of Laws* (1748), which took the view that the human temperament varied with climate (climate affecting the tension of bodily "fibers," of which temperament is a function), and laws which depended upon temperament should therefore be expected to vary with the physical environment of the society. The principles which governed the study of laws did not derive from some arbitrary or authoritative "ought" but from the objective nature of things. In this way, too, the philosophers of the Enlightenment took the first steps toward a scientific history of culture.

Bacon has spoken of the supreme value of science as a means to improve the material welfare of humanity, a vision ostensibly institutionalized by the Royal Society. But it was on the continent, rather than in England, that the scientific academies received material patronage from the monarchies. Here the absolute monarchies perceived the usefulness of science, or at least the prestige and possible power it afforded. Here, too, the new societies and academies provided scientists with position and status, whereas the universities, not unlike Newton's Cambridge, basically neglected scientific teaching and research. The Royal Society and the Paris Academy both served as models for the others: In 1743 Frederick the Great reorganized the Berlin Academy on the lines of Paris; as far away as Russia Peter the Great founded an academy in St. Petersburg and recruited scientists such as Euler. Smaller academies sprang up in cities in England and across the continent, and societies were created which were dedicated to specific professions like surgery and pharmacy, agriculture, and crafts. Many promoted technology and the practical applications of science—and so promoted (the third word of the Enlightenment triad) progress.

However, the usefulness of pure science, at least immediately, to material progress is problematic. The Newcomen steam engine, invented in the early eighteenth century, owed little to either mechanical or chemical science. James Watt's addition of a separate condensor to the engine, making it more efficient, owed nothing to Black's theory of latent heat; Watt was not a chemist but an instrument maker. The improved methods in agrarian production during the eighteenth century owed little to science per se.

Yet the time was passing when the most important truths were other-worldly, when progress meant spiritual redemption, when the ultimate goal of knowledge was transcendental. A new value system was arising—secular, practical, progressive—a value system grounded in this-wordly prosperity. Its champions were the new professions, growing in economic power, in social awareness, in numbers—the bourgeoisie. Surely the two—science and the altered structure of society— were intertwined.

Had society really changed? In England the monarchy was limited by Parliament, and a seemingly more rational system had evolved along with the new science. Not so in France. Here prevailed the elitism and stratification of the Middle Ages, the ignorance and superstition of the Church, the decadence and financial disasters of an incompetent absolute monarchy. Here was an educational system still mired in the mud of scholasticism. Yet the world view upon which it was all founded—the scholastic Aristotelian mirage—had been overturned by scientific reason. Progress in understanding the world had triumphed over the errors and simple stupidity of the past. So why not society? Why not morality? Let us apply the laws and methods of the successful physical sciences to the political, social, and moral world! With a rational and empirical knowledge of the truth, we may progress towards a state of happiness in the world and, finally, the moral improvement of human beings.

Again, this view is all exaggeration. As we have seen, the changes in science itself were complex and the Middle Ages had hardly been barren. And it is surely questionable that knowledge of natural law (or faith in natural law?) and the application of the new scientific methods (who could say for certain in the eighteenth century what they actually were?) would improve social and moral character. This program, nonetheless, lay at the heart of the intellectual movement called the Enlightenment during the eighteenth century. Surely other factors were involved besides science. Nonetheless, the philosophers of the Enlightenment, the *philosophes* in France, did one noteworthy thing: they popularized science. Indeed it was a form of propaganda; yet through them, science entered the mainstream of Western consciousness, where it has remained ever since.

Voltaire's *Letters on the English*, published in 1734, forcefully portrayed the illusions of the past and how they had been swept away by Bacon, Newton, and John Locke. Science had told us something not only about nature but about the nature of human beings themselves. In popularizing Newton's system, here and in later works, Voltaire brought along another Englishman, one whose influence was most important in psychology and the theory of knowledge. This was John Locke.

Written in 1690, Locke's *Essay Concerning Human Understanding* became one of the gospels of the Enlightenment. In Book II of the *Essay* Locke made the paramount point of the entire work: The human mind is like a blank tablet which receives its simple ideas from the sensations of EXPERIENCE—Locke capitalized the word for Cartesians. Complex ideas are the ordering of sensations by reason. Our knowledge is, therefore, limited to the natural world. We have no innate idea of essence, God, or whatever. Reason must always return to the bench of empiricism. Metaphysical rationalism, reasoning about that which cannot be experienced, is simply another form of illusion. Natural law comes to us by way of this empirical reason (Locke was no mathematician!). Thus we can gain knowledge of God only through His laws; He is the author of nature, its First Cause—the clock-maker God.

Voltaire and many other *philosophes* accepted this clock-maker God, but the other aspects, the revelation which had so confounded the medievals, they jettisoned. God was becoming a footnote to the knowledge of nature. Others would eventually erase even that.

From what sensation, asked the Scottish philosopher David Hume, does the idea of causality itself come? Is not causality the result of simple repetition, the constant conjunction of two events, one following the other, which we label cause and effect? Thus a First Cause which we never experience cannot be assumed at all! Hume desired to build a theory of understanding upon strict Newtonianism, yet his foundation was the empirical and experimental Newton, not the pious believer.

In France the Baron d'Holbach bluntly stated that since we can know nothing about supernatural First Causes, they may be regarded as nothing! Persisting in such fruitless quests, humans blunder unto unhappiness because of their corresponding ignorance of Nature. Into the fire with all metaphysical systems, with revealed religion; virtue and happiness derive from the scientific knowledge of natural law! Most were unwilling to go as far as the good Baron, who became known, not surprisingly, as God's personal enemy.

Nonetheless, the *philosophes* felt a definite sense of mission to bring the new learning to the public. One of the results was the great French *Encyclopédie*, published between 1751–1772, a massive work of seventeen volumes, eleven volumes of illustrations, and four of supplements. Many French scientists contributed articles, and for a time d'Alembert was an editor. But the most important editor of the project was Dennis Diderot. Wisdom, Diderot believed, was the "linking" of objective reality with the human intellect. The method of such linking was experiment—the empirical testing of nature. In less than a hundred years, said Diderot, pure mathematicians would disappear; only a handful would remain. Even pure mathematics was too abstract, too metaphysical for his taste.

The program of the Enlightenment is best stated in Diderot's article, "Encyclopedia," which occurs in Volume 5. The aim of the *Encyclopédie*, he wrote, was to gather together all the scientific knowledge scattered about the world. A unified science representing a unified nature would be passed down to posterity so that "our descendants, being better instructed, may become at the same time more virtuous and happy." Thus the search was on for the natural laws of society, of the human mind, of economics (Adam Smith's *Wealth of Nations*)—in short, a philosophy of social science which adopted proven scientific methods to enhance the human condition.

From France the movement spread—to the German states and faraway East Prussia. There in Köningsberg the philosopher Immanuel Kant began what he called a Copernican Revolution in philosophy. The mind is not a blank tablet, said Kant in his *Critique of Pure Reason* of 1781. Rather, the human mind is endowed with *a priori* categories—categories like time, space, Euclidean geometry—that is, Newtonian categories. Experience of the world is an interaction of the mind and sensation, sensation which activates the mental categories which order experience. But reason cannot go beyond its own constructed world, and God and the transcendental world can no more be disproved than proved by reason. Things-in-themselves, unshaped by the active human mind, are forever beyond human thought—to think is to think of something in some way. Kant's own version of Newtonian physics was an acceptance of repulsive and attractive forces as the essence of matter. Yet these are still phenomenal, linked to the categories of our mind. The primary properties of matter, extension and impenetrability, reduce to forces, to activity. Kant's influence was to be immense—not only in philosophy, where he is considered the father of modern critical philosophy, but in physics itself, where he established the epistemological foundations for fields of force.

A clocklike, law-abiding, rational world—such was the world of the Enlightenment, the watershed of science. In such a world there was little place for the miraculous, the mysterious, the otherworldly. Revealed religion was being separated from physical science. And in the next century life itself, the mysterious complexity of the biological world, would be brought into the confines of natural explanation. Even greater revolutions were in the making.

The nineteenth century had arrived. It was born in the convulsions of the French Revolution and the world turned upside down by Napoleon. The violence and barbarity associated with this drastic change were hardly what the *philosophes* anticipated or even desired. In the new century, however, change would become the order of the day. Change would indeed become the principle of physical life itself.

Human life is capable of "indefinite improvement," the Marquis de Condorcet decided. Condorcet wrote these words while in hiding, under sentence of death by the Jacobins of the Revolutionary Convention in France. Optimism!

Science, so the *philosophes* maintained, had destroyed many illusions. Yet, had it not replaced them with others? Or perhaps not science itself, but rather the popularizers of science, the propagandists, had created new myths. Whatever the case, we now enter into a new story, that of modern science. And we enter it with a burning optimism, a faith in the ability of human reason to grasp the reality of the world and so improve the earthly garden. This belief may be the greatest revolution of all.

SUGGESTIONS FOR FURTHER READING

COHEN, I. BERNARD. *Franklin and Newton.* Philadelphia: The American Philosophical Society, 1956.

GILLESPIE, CHARLES C. *Science and Polity in France at the End of the Old Regime.* Princeton: Princeton University Press, 1980.

HANKINS, THOMAS L. *Jean d'Alembert: Science and the Englightenment.* Oxford: Clarendon Press, 1970.

————. *Science and the Enlightenment.* Cambridge: Cambridge University Press, 1985.

HARMAN, P. M. *Metaphysics and Natural Philosophy: The Problem of Substance in Classical Physics.* New York: Barnes and Noble Books, Harper & Row, 1982.

ROUSSEAU, G. S., and PORTER, ROY, eds. *The Ferment of Kn owledge: Studies in the Historiography of Eighteenth Century Science.* Cambridge: Cambridge University Press, 1980.

WILSON, ARTHUR M. *Diderot.* New York: Oxford University Press, 1972.

19

∞

The Blueprint of Creation

It is said that when St. Francis of Assisi lay on his deathbed, all the animals and birds came to mourn. He had loved them, preached to them, called them brother and sister. Every living creature bespoke the hand of the creator; each in its own way illustrated the divine master plan. Many centuries later Newton would utter a similar thought: The wonderful uniformity he had discovered in the physical world had its analogue in the teeming world of the living. Here, however, the plan became exceedingly intricate, for the organic world stood in sharp contrast to the stable, law-abiding physical machine. Plan was submerged in spontaneity, purpose hidden in change and complexity. The physicist had to evoke the activity of the deity a minimum number of times—to establish the law of gravity, perhaps to tinker with the machine. Laplace in the heat of mathematical abstraction decided that he had no need for "that hypothesis." But how could one explain mathematically why a wing was so perfectly adapted to flight, eyes to sight, gills to water? St. Francis and Newton knew: The hand of God was directly involved in everything. He had designed every living thing and its parts, perfectly molding each to a predetermined function.

What a strange dichotomy! God's role in the physical world had been reduced to that of master clock-maker, or even to none at all. The organic world seemed to require His intervention for every tiny insect. And this was not all. The directional flow of time has no meaning in the law of gravity; there is no qualitative difference between past and future in its equations. Not so in the organic world! Organisms develop, they grow, mature, and die. The face of the earth itself changes over time as floods and earthquakes erase both natural and human monuments. The evidence remains buried in the soil. The earth is historical and a repository for history.

Amid such change the naturalist still sees stability: Cats always give birth to cats, dogs to dogs. And yet—the child is never *exactly* the same as the parents. Then, there are diverse species from different parts of the globe which seem related. Change and stability exist side by side in a vast array of patterns. Oh for the simplicity of physics!

Aristotle had known the difficulties of even trying to logically classify the diversity of life. He had used the terms *genos* (genus or family) and *eidos* (form or species) to construct a kind of downward logical classification. Yet Aristotle, the great observer of nature, realized that such artificial *a priori* divisions were too simplistic, like attempting to paint a living landscape in black and white. Nonetheless, against the atomists he had posited a logical and constant universe. Therefore, his artificial species, and the *scala naturae* they formed, provided at least a semblance of order, although he warned of their fundamental inadequacy and considered many others characteristics of organisms. Still, his belief in the logical constancy of the world ruled out the appearance of new species. He would not admit capricious change into organic stability.

The medieval Christians adopted the *scala naturae* or Great Chain of Being. But where Christian cosmology was served by the idea, medieval cataloguing of living creatures, mixed with the mythical, tended to perceive plants and animals in terms of human usefulness.

According to the rules of Aristotelian logic the final cause, the plan which guided the growth of every individual creature, was inherent in the logical order of nature, in the fixed and eternal species. Genesis with its one grand creative act seemed to give this idea divine sanction. The problem was spontaneous generation, seemingly a fact of nature, which suggested new productions. One could, like St. Augustine, conceive creation in the Platonic sense; God had originally created essences, and new species were simply the activation of these essences in matter, still coming forth according to the Divine plan. Essence or final cause—even among the pioneers of the revolution in the physical sciences the idea of some essential plan held sway. In fact, the regularity of the physical world supported it admirably.

Thus the classification of living things was not only a means of identification, it was also a way to reveal the Divine blueprint. Gather all the facts and the plan will emerge—induction will lead us to essence, just as Bacon said. Alas, how difficult that was! At times the species seemed to blend into one another. What factors made up the essence of any species? What was the meaning of "accidental" individual variations?

The Reverend John Ray taught classics at Cambridge. In 1662, rather than swearing conformity to the Book of Common Prayer (as ordered by Charles II), he retired to his hobby—natural history. He wrote books on European botany and was elected to the Royal Society. Like other naturalists, he struggled with the problem of determining species.

Most naturalists tended to seize upon some single characteristic as reflecting the essence of the species of plant or animal. Yet, asked Ray, can we really decide between what is accidental and essential? Further, why not assume that a number of structures represent this essence, and that species represents a category of clear structural affinities between individuals? A species, then, is a group which reproduce offspring with similar (but not exactly the same) structure. Common descent, said Ray, was the most certain criterion of the species.

Ray was certain that the works created by God were conserved to this day in the very condition they had first been made. How far, then, could an individual vary within the eternal species? And where in so many characteristics does one species end and another begin—without leaps. (Ray accepted the notion that nature does not make leaps.) Ray was also forced to recognize that fossils might well be the remains of extinct species. Nature does nothing in vain, and fossils as "sports of nature" (as many believed) would surely be superfluous. Yet, if no new species were created, then none were lost—there is no place in nature, wrote Ray, for "blot or error." In the end he was left wondering if perhaps the fixity of the species was "fairly constant" and not infallible.

In the same century Leibniz added some philosophical speculation to the already immense problems facing the poor naturalist. God is the most good and infinite Being, creating the best of all possible worlds, as befitting his beneficence. In such a world why should one essence exist and another not? By necessity, Leibniz thought, all essences that could exist *must* exist—and not for the benefit of something else, but simply for its own realization. The idea here is called the principle of plentitude. And Leibniz had another principle (as if plentitude were not enough!). True to his infinitesimal calculus, he saw the species merging into one another, so that it was impossible to show where one began and another ended. This is the principle of continuity. Imagine the poor Great Chain of Being: Plentitude filled it unmercifully, and continuity smashed its links together. The Chain became absolutely full, continuous, and fixed—no gaps, no jumps, no "missing links." The species seemed to melt into each other and fade into nothing like the ghosts of departed quantities! The Chain was threatening to become a blueprint which could not be read!

During the next century, debates among embryologists also posed problems for the species and the Great Chain. In 1759 Caspar F. Wolff decided in favor of the doctrine of epigenesis, the theory that the embryo develops by emergent degrees from homogeneous matter. Harvey had also believed in epigenesis. Wolff, like his great predecessor, was thus forced to call upon some kind of organic force which imposed the blueprint of the species upon chaotic material. Did, then, each species possess its own peculiar force? What a number of forces that would be! Or, assuming a continuum and a single vital force, how may we differentiate between the force which produces an elephant and the one which gives a mouse? How may one force do both? If species were to be denoted by common descent, and embryonic development from some vital force, then how do we fit these ideas together?

Another school of embryologists held to the doctrine of preformation. There is no real generation in nature, they argued, only the expansion of something already existing in the egg or the sperm (and they argued among themselves which it was). Egg or sperm, the plan was there, within the matter of the embryo itself, waiting to be activated through fertilization. Preformationists used the word "evolution" to signify this predetermined process of development. Ironically, the evolution of the preformed embryo was a powerful argument for the fixity of the species.

And what a fixity it was! Some preformationists saw the entire human race existing from the beginning of time inside Eve's ovaries. The entire species, fixed from creation, was encapsulated one inside the other, and evolution meant the unrolling or unfolding of each tiny embryo encased in wombs back to Eve. Whether or not preformationists believed in this radical encasement theory (and many did not), their criticisms against the epigenesists were actually mechanistic: Accounting for development by some vital force was occultism! However, if the species were to be defined by descent, preformed evolution demonstrated their absolute fixity.

In the eighteenth century the Great Chain of Being became a much-discussed biological concept, although "biology" as a science and a profession was unknown until the nineteenth century. The study of the organic world was divided among philosophers, physicians, and a host of amateurs. Thus the Swede Carl Linnaeus gave up theological studies to pursue a medical career in order to further his interest in botany. Linnaeus firmly believed that living nature had been constructed upon a pattern which human reason could discover, just as the physical laws of the universe had been unveiled. The primary and most important task of the naturalist was therefore classification, and Linnaeus had a real talent for it. God had created the universe, so the saying went, and Linnaeus named it.

Plants, it had been discovered, had sexual organs, and as early as

1735 Linnaeus adopted sexual differences in plants as a starting point for his classification. Unlike Ray, he had returned to a logical and highly artificial system. However, classification, he soon realized, was only half the task. The nomenclature of plants and animals in his day was a chaos of descriptive terms and individual inclinations among naturalists. So Linnaeus introduced his system of binomialism, using two names, the genus and species, to identify the organism. It was his chief triumph, although he himself did not consider it so. The task sent Linnaeus ransacking through his Latin, while his students, pillaging nature across the world, kept him well supplied with new specimens to name.

The question of the species, the building block of his system, haunted his efforts. The genus was a general artificial group characterized by affinity, but the species, determined by descent, was the backbone.

In 1751 Linnaeus held that there were as many species as God had created in the beginning; no species ever changed into another and none arose by spontaneous generation. While the principles of plentitude and continuity had blurred the lines between species, Linnaeus, by insisting upon their fixity, sharpened them and resurrected the entire issue. Ironically, as Linnaeus himself grew older he began to wonder about the possibility of modification. Plant hybrids seemed to indicate at times something more than variation or "sports"—perhaps new species. It was possible, he speculated, that even the origin of species might somehow be the work of the environment. But how? What was the "natural" mechanism? As more and more examples of hybridization were brought to him by his disciples, Linnaeus wondered if perhaps not all the species had been created in the beginning; in fact, he speculated that they might be "the work of time."

Experiments with hybridization were carried out by Joseph Koelreuter in the German state of Baden. To his immense relief Koelreuter found that his plant hybrids produced no new species but a curious myriad of forms, some even reverting back to the original parental species. Hybrids, he concluded, did not give rise to new species, nor to infinitesimals to fill gaps in the Great Chain. Yet why—in recrossing his hybrids—did subsequent generations revert to the original forms? And why did some revert but not all? Koelreuter held an important clue in his hands. The problem of inheritance, however, would baffle the greatest minds, even that of Charles Darwin.

Inheritance might be baffling, but it was not without speculation in the eighteenth century (for good or ill no scientific problem is without speculation!). In France Maupertuis gave a kind of Newtonian explanation of inheritance: Particles derived from the parents have an affinity for each other, which makes them pair. In every pair the particles of either the mother or the father dominate and account for various traits.

Some particles can be passed invisibly through generations to suddenly reappear. The sterility of hybrids comes from the fact that these particles, having no affinity, are unable to reproduce in stable pairs. And Maupertuis went on to claim that an excess or deficiency of particles may account for the origin of new species!

The Chain of static species also raised philosophical issues during the eighteenth century. The *philosophes* spoke of progress and improvement—this was not the best of all possible worlds. The reformers saw movement; the Chain implied absolute stability. The *philosophes* sought scientific support for progress; the overcrowded Chain, raised now to scientific respectability, provided none. How could the Chain be made to admit progress?

The philosophical answer to this puzzle was to make the Chain dynamic. Perhaps the Chain was really an unfolding plan, a kind of cosmic ladder upon which organisms climbed upwards toward increasing complexity and greater perfection. Thus in the eighteenth century the Chain was reinterpreted; it became a plentitude of possibilities, not all realized at once, but in time. Arthur Lovejoy labeled this reinterpretation the "temporalizing of the Chain."[1] Progress might maintain the Chain only if the plan was considered historically, established at creation like a seed and sprouting in history like a great cosmic tree. Yet underneath this philosophical dictum, perhaps unseen, lay an interesting biological corollary: Did not the flowing Chain imply a flowing species?

The eighteenth century had discovered history—in society and in nature. But the domain of history had been appropriated long before the *philosophes*.

Christianity is preeminently a philosophy of human history, yet it also had a great deal to say about natural history. Most importantly, it set the time frame for the entire history of the earth. The chronology of Biblical history may be traced backward to the first days of creation. In the 1650s Archbishop James Ussher had done exactly that, carefully computing the years until he reached the first day. Creation, he triumphantly declared, occurred on October 23, 4004 B.C.—begining with the preceding night. The date was added to authorized editions of the English Bible. Imagine it! All history, human and natural, was compressed into a scant six thousand years. If species changed at all, they must transform like lightning, like biological alchemy.

The eighteenth century chafed in this temporal straitjacket.

From 1749 through 1785 the Frenchman the Comte de Buffon published an amazing thirty-six volumes of his *Histoire naturelle*, a vivid description of the wondrous diversity of nature. The reader, like Buffon himself, stands in awe and gasps: "Whatever can be is!" Nature as

[1]Arthur O. Lovejoy, *The Great Chain of Being* (Cambridge: Harvard University Press, 1936). p. 244.

portrayed in his volumes becomes a flowering process of organisms gradually blending by imperceptible changes. The Linnaean effort to sharply define and hence stabilize such variety seemed to Buffon the height of absurdity. Species and genera are illusions, he wrote in one volume. Yet, seemingly unable to make up his mind (although who can be totally consistent over 36 volumes?), he considered that the species may be defined by the test of sterility.

Buffon was certain that organisms had a history. There seemed to be, he observed, a process of variation throughout time, variation in which organisms did depart from their ancestors. Yet, beneath, there was also a primitive design, a number of *molds*. Yet again—the molds were not sterile logical categories; rather they were a system of dynamic processes shaped by climate and other factors. They were historical molds, so to speak, combining stability and change. This natural history of living things required a natural history of the earth—*natural* as opposed to supernatural creation.

In 1755 Kant had suggested that the solar system had "evolved" from matter condensed and separated from the sun, and in 1796 Laplace independently developed this theory, called the nebular hypothesis, in greater detail. The sun's atmosphere had extended beyond the orbits of the planets, and gravity had formed this fiery atmosphere into rings, gradually solidifying and cooling into the planets. Thus in the beginning the earth had been a primitive fireball, requiring time enough to cool. Buffon, who thought that perhaps the original fiery state might have been the result of a comet plunging into the sun, calculated the time it would take for a body the size of the earth to cool enough to support life. To be sure, it was longer than Ussher's time frame! In fact, the time required had to be over 74,000 years, and Buffon thought that a million years was not stretching the time line too greatly.

We must not think of the earth's history or creation in terms of days, said Buffon, but *epochs*. Epochs allowed for development, for gradual change, and for another possibility which troubled Buffon, the apparent extinction of some organic forms. Buffon had wondered about what he called the degeneration of some species from others and the disappearance of earlier forms. Now the continual cooling of the earth's surface seemed to indicate a gradual extinction of life itself. Might some species vanish, leaving nothing but fossils, and thereby breaking the links of the Chain? This possibility, though very "extraordinary" as Buffon termed it, loomed large.

During the brief Biblical history of the earth the primary geological event shaping the earth's surface was obviously Noah's Flood. Thomas Burnet wrote in 1691 that the Flood accounted for the irregular terrain of the earth, as it had declined from its original perfection. It occurred to many that perhaps fossils were remnants of the Universal Deluge. Yet marine fossils found in mountains certainly required an immense, if not unbelievable, deluge of water in a flash of geological time. And—the Flood was a direct supernatural intervention, not what we should expect of the perfect clockmaker!

How *did* fossils form? In 1669 Nicolas Steno, a titular bishop living in Tuscany, was comtemplating fossilized shark teeth and asked himself this question (how solids formed in solids, he termed it). In essence, Steno said that the formation of "solids within solids" was a temporal, historical process. Sedimentary strata had been formed successively, layer by layer, and not all at once. Further, marine fossils in the mountains solidified *before* they were entombed: Bones and other hard parts of the creature left impressions in surrounding sediments, hardened, and were later encased. Soft parts, on the other hand, were entombed first, rotted away, and their molds filled by fossilizing material. Now all of this was surely difficult to compress into one great Deluge!

Nearly a century after Steno, Abraham Werner, a professor of minerology at Freiburg, taught that the stratification of rock formations was the result of a universal sea that once covered the entire globe. Beginning with primitive layers of granite and the like, layers of rock strata had been precipitated out of this primeval soup. The surface of the earth as we now see it, Werner taught his students, had been formed progressively in five stages. His theory has been called the Neptunist hypothesis. Through it the Universal Deluge became a natural event. Like the Great Chain, the Flood too became temporalized.

Nevertheless, a flood—be it Noah's waters or Werner's soup—could not adequately account for volcanic deposits, rock obviously formed from a molten state. Neptunists considered volcanic activity to be part of recent history and localized. Yet some volcanic rock was found nowhere near modern volcanoes; other deposits were found in strata, outburst from recent sediment, which antedated aqueous precipitation. The Neptunists had posited a single event, one great universal catastrophe, in order to account for the earth's history. Some causes, however, still seemed to be in operation, sculpting the earth's surface *now!*

James Hutton of Edinburgh stood upon the seashore, much like Newton, feeling the massive interplay of natural forces. Hutton saw the infinitely slow and steady erosion of the land as its rivers and streams carried silt into the sea. And he perceived, too, the richness of the soil, somehow restored, making it fit for agriculture. Hutton entertained a very Aristotelian idea: The purpose of geological forces, their "final cause," was to make the earth fit for human habitation—fit for agriculture (fertile soil) and later fit for industry (coal).[2] How did the natural machine restore itself?

[2]This idea—that Hutton's theory rested on the argument from the ancient doctrine of final causes—is advanced by Stephen Jay Gould in the essay, "Hutton's Purpose," *Hen's Teeth and Horse's Toes* (New York: W. W. Norton, 1983), pp. 79–93.

By 1795 Hutton had found the answer. The geological condition was not the result of a single catastrophic flood, long past; rather the machine was an ongoing operation, a minute dynamic balance of erosion and volcanic upheaval over immense, perhaps infinite, time. Waters decomposing earth and creating sediment; volcanoes uplifting the sediments to form fresh new land masses—all very ancient, all continual, all inferred from contemporary events—such was Hutton's doctrine of uniformitarianism. It was as if the timeless law of gravity had been applied to geology. The geological history of the earth was not directional but revolved in cycles like the planets. This world machine, Hutton wrote, was subject to constant erosion and repair, designed admirably to support human life, yet without "vestige of beginning, prospect of an end!"

Time! Natural history was expanding time much as Copernicus had inflated the cosmic sphere. As time expanded, so did the human imagination. Gazing at the rocks, the fossils, the species themselves, naturalists could now envision the slow and nearly invisible march of time, operating before their eyes, regular yet dynamic. In 1791 the engineer William Smith surmised that the actual temporal succession of strata could be unraveled by the fossils themselves. But the fossils spoke of discarded links in the Chain, rips in the unfolding blueprint. The gaps were becoming absolutely monstrous, and the leaps from extinct to living species, even on a temporalized chain, could no longer be made so easily. They were separated by chasms of time.

By the close of the eighteenth century naturalists found themselves in possession of fossils which had no obvious relationship to contemporary species. Buffon had written of gigantic mammoth bones, bones six times larger than those of normal elephants. Elephant bones found in Europe were once considered remnants of Hannibal's army. But surely not these! Then mammoths were found in Siberia and in North America. In 1796 Thomas Jefferson found the bones of a giant sloth in Virginia. Formations around Paris revealed, interspersed between marine strata, desposits containing remnants of land animals, some of massive size and unknown species. Many people still clung to the belief that extinction did not occur—surely such a thing was foreign to the beneficent God of plentitude.

Finally, in France, Georges Cuvier threw up his hands to speculation: The evidence, the fossils themselves, must decide! And so, from 1795 onward, he worked in Paris, reconstructing the biological past from its bones.

How does one reconstruct the entire animal from a single bone, or a few? Cuvier, by studying the anatomy of modern forms, came up with the answer. It is called correlation: Each part implies the next, and taken together they give the whole. For example, blunt grinding teeth imply a herbivore, which in turn implies hooves instead of claws (so once when

his friends, dressed as devils with hooves, shouted, "Cuvier, we're going to eat you!" Cuvier replied, "Hooves? You can't!"). Fangs implied
claws. In short (although we have somewhat simplified his approach
here) Cuvier used such techniques to reconstruct the living past. He is
the father of comparative anatomy and paleontology.

Cuvier's four volumes published in 1812 demonstrated conclusively
that extinction had occurred and that the history of organic life was very
ancient. Life seemed to have proceeded through cycles, the destruction
of some species being followed by the emergence of others. The sudden
disappearance of species, as indicated by the fossils and the corresponding violent upheavals in the earth's strata, suggested to Cuvier violent
causes—catastrophies. The sudden appearance of new forms suggested
new creations, although Cuvier actually conceived it as the migration of
fauna into devastated areas. Indeed, the earth's surface does suggest
sudden dislocations, and even Hutton had believed that the forces
which shaped the earth had been more powerful in earlier times. Thus
Cuvier, convinced by the evidence of the fossils, adapted the revolutions of Neptunism to the great time scale of Hutton. There had been
many floods, many earthquakes, many extinctions—many gigantic
catastrophies.

"But the Great Chain of Being . . .?" We can almost hear the protests. "Look at the fossil record itself!" Cuvier might have answered.
"Tell me what you see. No—without theology, without speculation!"
And we look, our minds blank tablets. And we see—no gradual blend-
ings, no smooth transitions, no clear linear ascent; we see Nature making huge jumps.

Cuvier tore down the Great Chain of Being.

His careful anatomical study of present organisms indicated wide
discontinuities in organization. So Cuvier invented four large groups:
Vertebrata, Mollusca, Articulata, and Radiata. They were distinct, each
group a sharp bundle of anatomical organization. Not a chain but a bush
characterized the organic world. The blueprint was actually a huge book
of many pages—several plans separated by geological epochs and anatomical peculiarities. Yet Cuvier, leafing through the book, thought he
found the pages numbered, for the strata seemed to indicate a general
progression from reptiles, to mammals, to humans. There was, it
seemed, an ascent in complexity—the bush had flowers.

In the end the species remained fixed. Cuvier the strict empiricist
could not imagine otherwise. The divisions he had found, both living
and dead, were too sharply drawn. But the temporalized Chain was
gone, and gone with it for the moment were all idle speculations on the
modification or transmutation of the species. Such speculations *were idle*.
Cuvier had brought naturalists back to hard facts, back to the truly objective complexity of life. Now, however, that complexity was set upon

an amazingly long continuum of time. Quietly contemplating the evidence, though, Cuvier might have done some wondering. What could fill the awful gaps and smooth the sharp edges he had discovered? And what really did bind the book of living nature, its past and present, together?

SUGGESTIONS FOR FURTHER READING

EISELEY, LOREN. *Darwin's Century: Evolution and the Men Who Discovered It.* Garden City, N.Y.: Doubleday, 1958.

GILLISPIE, CHARLES C. *Genesis and Geology: A Study in the Relations of Scientific Thought, Natural Theology and Social Opinion in Great Britain 1790–1850.* Cambridge, Mass.: Harvard University Press, 1951.

GLASS, BENTLEY; TEMKIN, OWSEI; and STRAUS, WILLIAM L., JR., eds. *Forerunners of Darwin 1745–1859.* Baltimore: The Johns Hopkins University Press, 1959.

GREENE, JOHN C. *The Death of Adam: Evolution and Its Impact on Western Thought.* Ames, Iowa: Iowa State University Press, 1959.

NORDENSKIÖLD, ERIK. *The History of Biology,* trans. Leonard Bucknall Eyre. New York: Alfred A. Knopf, 1928.

20

~~~~~~

The Law of
Higgledy-piggledy

Today he stares out at us from the paintings and the photographs—the long white beard of the sage, the bushy eyebrows, the somewhat sad but thoughtful eyes. Perhaps we are reminded of the ancient Greeks: Is this how Thales looked? Socrates? We struggle with his ideas, knowing that they have changed our world utterly. From his writings, which lie open before us, have sprung entirely new sciences; in a certain sense they are the origin of modern biology itself. *Origin* is the core word in his most famous work: *The Origin of the Species by Means of Natural Selection or the Preservation of Favored Races in the Struggle for Life* by Charles Darwin, published in 1859. He wrote much more, all interconnected. But this *Origin* is the center, the sun around which the others revolve like the planets.

Most of us are aware that Darwin established, beyond doubt, the fact of evolution in the organic world. Less well known is the fact that Darwin's mechanism of evolution fostered one of the greatest revolutions in Western thought, for he dispensed with the ancient and honored doctrine of teleology, replacing it with continuous variation. His was a philosophical revolution as much as a biological one.

How, we wonder, could one person have accomplished such a thing? Perhaps he had help—"shoulders" to stand on. So we search for precursors. Yet Darwin himself, a modest man, warns us in his autobiography that this is apt to prove fruitless. Many before had speculated on evolution and the transformation of the species. In this sense it might be said that evolution was "in the air." However, its mechanism and wider philosophical implications were not. In these Darwin was truly original. And, like few before him, Darwin was the perfect combination of patient observer and rigorous thinker—and not the least a man blessed with a vivid imagination. The facts of evolution were there; Darwin taught how to think about them. Having read the *Origin*, his friend and mighty champion, Thomas Huxley, exlaimed: "How extremely stupid not to have thought of that!" No one before Darwin *had* thought of it.

What *was* there—"in the air"? It is the historian's task, despite Darwin's caution, to set the stage.

First there was embryology. By the early nineteenth century embrylogists were beginning to favor epigenesis. Carl Ernst von Baer saw a certain similarity of all life in the progressive development from the fertilized egg, to tissue layers, and to organs and parts. Yet each species developed according to its own unique archetype, and von Baer, like Cuvier, rejected any Great Chain of descent. He also rejected any form of evolution. On the other hand, epigenesis did seem to indicate a kind of underlying unity, for in the early stages of embryonic development (ontogeny) a common path of differentiation could be perceived in all organisms. And the new cell theory, proposed in the 1830s (Chapter 21), suggested an even more fundamental relationship between all living things, plant and animal. Such ideas in themselves did not necessarily point to transformation, yet they did tend to erase the sharp lines between living things.

Outside biology the idea of progress also tended to suggest some sort of evolutionary change. But Enlightenment progress was analogous with the machine—a preordained mechanism, built from invariable laws, clanking down a well-marked highway. Both machine and highway ultimately reduced to passive, mechanical matter, to molecules in motion. Most of all, progress implied improvement, direction, and hence teleology. While progress did reject Platonic essentialist thinking, its doctrine of improvement still implied preexisting potential—essentialism in another guise.

In Germany the school of *Naturphilosophen* (Nature philosophers) protested against the coldly mechanistic and materialistic philosophy of the Enlightenment. Nature, they held, was more akin to an organism—alive, growing, a dynamic organic unity developing as the material manifestation of the divine. Although these thinkers were mystical and poetic, many of them, like the poet Goethe, sought a primal archetype beneath the organic chaos. Life itself followed an emerging series yet bespoke a fundamental unity. The essence of nature was, according to Frederich Schelling, universal development in which product and process are one. Lorenz Oken held that species formed a developing series of ideal plans or archetypes, evolving in time and according to limits imposed by the preordained archetype. Likewise, Johann Herder said that the species constituted an evolving historical sequence, the destruction of lower providing materials for the creation of higher. In short, the *Naturphilosophen* viewed Nature as a unity in which form was a direct manifestation of a world-soul (*Weltseele*). In the philosophy of G. W. F. Hegel, development was enshrined in a tremendous rational system, describing the history of the phenominal world as a dialectical expression of spirit (*Geist*). Indeed it may be said that evolution was in the German air, but it was also in the clouds. It did not proceed by natural causes; rather it was mystical and metaphysical, the direct activity of Spirit. And it was essentialist, for nowhere did the species transform into others.

In 1794 a book appeared in England entitled *Zoonomia or the Laws of Organic Life*. The author was Erasmus Darwin, Charles's grandfather. It was a work of mighty speculation—yet, alas, (Charles would say) few facts. In brief, Erasmus held that organisms must adapt to the physical conditions of life or perish. Organisms react to the stimuli of their environments—to sensations, hunger, sex, the need for security. They can acquire "new parts" and become modified, adapting themselves to changing conditions, even consciously, and passing their acquired improvements through inheritance to future generations, "world without end." Success meant survival; nature was a brutal struggle for existence. Erasmus also held that all life came from a single organism and that the antiquity of the earth stretched into millions of ages.

Now this *was* a theory of evolution, except that it completely lacked a basis in biological facts and a clear mechanism (conscious striving?). In France, Jean Baptiste Lamarck presented an evolutionary theory complete with biological data and naturalistic mechanism. Even after Darwin, Lamarck's theory would be considered by some a serious alternative to natural selection.

Lamarck began his scientific career as a botanist. In 1793, however, he was appointed to the chair of zoology at the Museum of Natural History. Baffled by problems in classifying the invertebrates, he came to reject the fixity of the species and collided with Cuvier. Lamarck saw a transformation of the species; fossils did not, strictly speaking, represent totally extinct animals, rather they were the ancestors of living descendants. How did such transformations happen?

Animals are historical, productions of time, and boldly Lamarck included humans in this law. The environment is historical, too, and its alterations produce needs which the animal must meet if it is to prosper and survive. But the organism *is not* conscious of these needs; rather it responds to an "inner feeling" or drive. The environment changes, a need arises, an organ develops to satisfy the need. Continual use of the new organ strengthens it, enlarges it, and this *acquired* change is passed on. Thus over long periods of time new characteristics are developed, and eventually the species are transformed. On the other hand, useless organs disappear.

The environment is a kind of pressure cooker, bringing tensions to bear upon organisms by placing them in danger. Such external pressures set internal, vital fluids in motion which stimulate tissues to form rudimentary organs; these organs are fortified through use, the organisms survive, and their newly acquired improvements are passed on (there was nothing new about the idea of acquired traits being passed on to offspring—the greatest of all was the original sin of Adam and Eve). Lamarck's mechanism of evolution, then, is a strictly physical state of affairs in harmony with the natural machine world-view of the enlightenment. Yet in the background there still remained a divine agency, like some sculptor, using the tools of the physical environment and molding organisms according to some preconceived if invisible plan. Teleology was still very much alive, for Lamarck believed in a general ascent in complexity. Such mechanical evolution, coupled with purposeful direction, would always be a welcome alternative to Darwinian selection.

In his time Lamarck was basically ignored. His ideas could be easily misunderstood. His "needs" could be interpreted as "desires," even as the conscious will—as if with an act of mind over matter an organ could actually be willed into existence! That was surely absurd. To Germans he was just another sour mechanist; to the English, disgusted by the Revolution and the antichrist Napoleon, he was an atheistic materialist. At the time of Darwin's youth, evolution was hardly in the English air.

In the early nineteenth century England had returned to the natural theology of Newton. Fearful of having their science equated with French materialism (obviously the cause of revolution—and all evil!), English naturalists, mostly clergymen, argued that the very complexity and wonderful adaptations of organisms to their environment *proved* the existence of God. So complex a world, said the Archdeacon Paley in his book entitled *Natural Theology*, should convince any thinking person of the presence of intelligence. Probably arguing more against the ancient Epicurean doctrine of chance—the arising of the world from fortuitous collisions of atoms—the natural theologians said that chance was ignorance of nature and knowledge clear evidence of design. Evolution as they conceived it was tinged with materialism. Yet what evolutionist had argued against design? Even materialist progress implied it!

It was in this confused atmosphere that Charles Darwin, born in 1809, took his first steps into the study of nature. His beginnings were inauspicious. The son of a successful physician, Dr. Robert Darwin, and grandson of the famous Erasmus, Charles was sent to Edinburgh University in 1825 to follow in their medical footsteps. But medical studies did not interest him. He was bored easily by lecturing; he liked to read at

his own pace, to observe, to collect things, to tramp through the countryside gathering information himself rather than passively soaking it
up like a dry sponge. Further, the agony of a patient during an operation had sickened him. Not only did medicine fail to interest him, it also
made him ill. Nature interested him (along with shooting), but in those
days Natural History was not yet a profession, and only recently had
Lamarck introduced the term "biology." His father fretted that young
Charles might not amount to much.

Nonetheless he pursued his hobbies at Edinburgh, often collecting
specimens and dissecting under the guidance of the zoologist Robert
Grant. From Grant he first learned of the fantastic theories of the
Frenchman Lamarck, and he was surprised to find that Grant seriously
contemplated the transmutation of the species. He read his grandfather's work and was not impressed; he attended a geology class and was
quickly bored (a boredom for which the professor, Jameson, a
Neptunist, was probably responsible). Finally it was decided that he
should enter Cambridge and study for the clergy.

Again he found the prescribed course of study dull. However, he
formed friendships with J.S. Henslow, a botanist, and Adam Sedgwick,
a geologist, absorbing from these men the fundamentals of their respective sciences. He also absorbed Natural Theology (both were clergymen)
and found Paley's logic as convincing as that of Euclid! From the writings of the astronomer John Herschel he learned of the dignity of
scientific pursuit; from Baron von Humboldt he caught the bug to travel
as well as a vision of the magnificence of nature. If only he could make
some small contribution to science! The opportunity arose, perhaps
sooner than he would have liked, for in 1831 Professor Henslow recommended to Captain Fitzroy of the *Beagle* that Charles Darwin was the
gentleman and naturalist the captain was looking for to accompany a
survey expedition to South America. After some hesitation and opposition (from his father) he was aboard when the *Beagle* sailed.

Before he departed, Henslow alerted him to a newly published book
about geology, Charles Lyell's *The Principles of Geology*. Henslow
thought he should read it—though not necessarily believe it. In
Darwin's mind Lyell would become dynamite, his own observations
providing the fuse.

Slowly over the years the idea of evolutionary progression—the unfolding plan of nature in the mind of God—had crept back into the
thinking of naturalists. Cuvier's catastrophism was well suited to this
idea, and its succession of geological ages had triumphed over Hutton's
uniformitarianism. Cuvier, of course, never suggested biological evolution; there was no physical line of descent from age to age. If there was a
connection, an evolutionary design, it was transcendental. However,
catastrophism used in this manner was a compromise between two dis-

parate levels of thinking—miraculous "special creations" and geological upheavals to explain the discontinuity of the fossil record. As such, it was doomed.

Lyell was the geological avatar of that destruction. Extraordinary agents, he wrote, have no place in science. His message was simple: all changes in the geological record, both inorganic and organic, are governed by laws which *are now in operation*. Geology is a slow and steady balance of destructive and creative forces, operating now before our eyes, the summing of which in almost endless time appears to render immense effects. Lyell's emphasis upon the uniformity of natural law through the ages led him to reject progressionism. There is neither progress nor improvement, and thus, no evolution!

The second and third volumes of *The Principles* reached Darwin in South America. In them Lyell presented Lamarck's theories, some discussions of organic life, and an attempt to account for extinctions. He was ready to admit that "amidst the vicissitudes of the earth's surface, species cannot be immortal." Like most naturalists, Lyell was also prepared to concede the species a range of variability; however, when the changes in conditions reached a point beyond this range, the species perished (except humans, of course). Lamarckian transformation of the species was, nonetheless, banished—it implied progression, and progression implied miraculous causes. In short, Lyell would not accept evolution because it smacked of creationism! Yet, finding no natural explanations for apparently new species, he hedged. He wondered if there might not be centers of creation and perhaps some sort of creative principle at work. Transformation, though, was forbidden.

Lyell taught Darwin how to think in terms of slow, continuous change, how to look at present geological formations and project their history. Meanwhile, as the *Beagle* plied its way down the South American coast and out into the Pacific, Darwin's own keen observations seemed to refute a part of Lyell's theory. He was finding clear evidence of some kind of organic succession. In some areas species had become extinct, as evidenced by the fossil record, yet he found living species which bore remarkable resemblances to extinct ones. And he noted another curious fact; on islands near Africa, like the Cape Verdes, there was an overall similarity between the islands' fauna and the fauna of the continent, and the same was true for islands near South America. However, the environments of all these islands were similar if not identical; nearly identical environments had given rise to different forms, yet forms similar to those of the corresponding mainland.

The Galapagos Islands, recent volcanic formations in the Pacific, presented Darwin with even greater mysteries, for the islands were geologically young, yet it seemed as if he had been transported into an "antediluvian" (his term) past. And each island, nearly identical in conditions, offered its own unique animal adaptations. Natives told Darwin that they could tell by just looking at a turtle what island in the chain it had come from. Even the finches varied from island to island. Each individual island ecology, generally similar, seemed to proffer different vacant ecological niches which had been filled by thousands of special adaptations, all apparently recent (geologically).

The islands were a gigantic laboratory, and Lyell was teaching him to think in terms of creeping change. Different species had filled similar niches in similar environments when these environments were separated by natural barriers. In terms of special creation this would mean that the Creator had formed different species to fill identical roles—and that seemed a bit strange. Darwin could not see progression from simple to complex; he found species which were similar to fossil remains; it seemed as if one small group of birds, his finches, had been modified for different ends; and he wondered . . . .

He returned to England in 1836. His specimens, his delightful *Journal* of the voyage, his theory of coral reef formation—all gained him fame among naturalists and election to scientific societies. The species question, however, would not leave him. He began to ponder—for twenty years!—opening his first notebook on the species in July of 1837. He was determined to find the law which governed the origin and transformation of the species.

The first step was to give up the old essentialist concept of the species. Species were groups of individuals, dynamic, interacting biological individuals—reproducing, having the same competing needs, struggling for existence *within* the group. Species formed dynamic *populations* rather than fixed classes. How did they interact? How did they adapt to the conditions of life?

It is probable that Darwin had some inkling of the answer when in 1838 he happened to read a book by the Reverend Thomas Malthus, *Essay on Population*, which had appeared back in 1798. Malthus had thrown down a dismal economic warning (hence Carlyle called economics "the dismal science"). Unlike populations, said Malthus, nature's bounty in cultivated land can be increased only slowly and has an upper limit. Populations, however, always tend to overproduce beyond the basically stable limits of subsistence. Hence there must be checks on unlimited population expansion—disease, wars, and most of all, poverty. This meant that a high level of mortality would exist among the weaker individuals of the population. It was a bleak and terrible specter, but it held an important clue.

Every naturalist recognized that individuals varied within a given species and that many variations were nearly imperceptible. Now add Malthus. In the extremely delicate balance of life—the relations of individual organisms to one another and the environment—would not the smallest variation, if advantageous, give some individuals an edge in the struggle of existence as painted by Malthus? Would not those individuals, possessing whatever advantage, tend to produce more young than those which lost the struggle? And would not a variation which was not beneficial (or a large mutation) probably doom the organism? Now toss in Lyell. Slow changes in the environment would tend to favor ever-so-slight variations, not directly, but simply through the gradual production of more young with this variation through generations.

Darwin saw an analogy in the example of domestic breeders (his first chapter in the *Origin* would deal with "artificial selection"). From the variations of domesticated animals (variations much greater, he thought, than those in the wild) the breeder *selects* those to be continued and allows only this group of the stock to reproduce. Over successive generations, then, a population arises with the desired characteristic, which has been *consciously selected* by the breeder. The breeder, in effect, selects the variation from an immense number of possibilities thrown up by nature. Within the domesticated breeds there are obviously many "selected" characteristics which can hardly benefit the survival of the animal in the wild (imagine the prospects of some lap dogs!). Therefore it seems that many variations arise by chance and are preserved arbitrarily by the whim of the breeder. Progressive complexity or felt needs have nothing to do with the seemingly spontaneous production of variations in a species! So the question now becomes: What preserves variations in nature?

The answer was provided by Malthus—except that Darwin saw within the "dismal" struggle a creative principle at work. Individuals within a given population produce young with minute variations—an unlimited number of tiny differences. In a slowly changing environment some variations would enhance survival, and the young who inherited this modification would produce similar young with, perhaps, further modifications in their inherited advantage. Thus a gradual *modification away from the original type* would occur. After a long period of time the descendants would no longer breed with the old unmodified form. Hence through descent with modification a new species would arise, similar to yet different from the parent stock. The original type might survive in pockets or become extinct. Darwin called this operation Natural Selection. The principle of Natural Selection simply describes a state of affairs in nature: Organisms change in all directions; some changes provide advantages in survival; these changes are preserved through differential reproduction; they are further modified; a new species is the byproduct. Note the distinction from predetermined design: Variations arise *by chance* (how, Darwin did not know); some *happen* to be beneficial and others not; improvement simply means the survival and reproduction of more young with this *chance* advantage.[1] *There is no higher and lower*—the gills of a fish insure its adaptation and survival just as well as the lungs of a mammal, or the teeth of a shark as well as the human brain! Natural Selection is a blind, mechanistic description of change, survival, descent, and extinction.

Here Darwin threw Western thought itself into the maelstrom. Nature becomes a great laboratory of failed experiments and a few successes. The factors determining survival are many—competition, climate, migration, available ecological niches—a vast array of conditions. No one can tell beforehand which chance variation will enhance survival, and variations themselves may arise from characteristics adapted to quite different tasks. The swim bladder of a fish, for example, seems perfectly designed for buoyancy. Even if we accept design over selection, who could foresee that under different circumstances slight modifications in the swim bladder could eventually give rise to lungs for breathing? An organ which fulfilled one function might by a strictly fortuitous conjunction of modification and circumstance fulfill some quite different function. *That* fish, with *that* modification, crawled into the mud and lived—and over the course of eons gave rise to air-breathing land animals. Think of the failures before that lucky throw of the biological dice! Design and purpose in evolution come after the fact of survival. Natural history was the story of many failures punctuated by a few lucky shots. Nonetheless, the failures left evidence in the very structure of animals—in the form of vestigial organs, once serving a function, useless but not detrimental, still inherited.

The maelstrom? No longer may we argue from design to designer. No longer does biology support the doctrine of final cause—teleology. Darwin could now explain the things he had seen in his travels. Differences were due to random variations and their preservation in specific ecological niches; similarities were due to common descent.

There was something else, something more general. Feet, wings, and fins—all serving dissimilar functions—were yet homologous organs; that is, they corresponded in structure. For the creationist this was mere coincidence. In terms of modification and common descent these organs, now serving different functions, could be traced to a common

---

[1]The words "chance" and "random" can be confusing when applied to natural selection. The theory involves two steps: (1) the random production of variations and (2) the naturalistic causal mechanism of selection, which preserves them. Adaptation is definitely not random. And, of course, Darwin believed there must be some cause that produces variations, although he did often enough refer to chance. Natural selection, however, rejects teleological determinism, and randomness does play an important role. The theory is actually an interplay between randomness and selection. Ernst Mayr writes: "It is neither strictly deterministic nor predictive but probabilistic with a strong stochastic element." See Ernst Mayr, *The Growth of Biological Thought*, (Cambridge, Harvard University Press, Mass.: 1982), pp. 519–530 and 683–684.

ancestor. And the fossil record, though incomplete (Darwin's answer to Cuvier in the *Origin* was: How could we expect the record to show every step, since the record itself was arbitrary and strictly by chance?), indicated in vague paths the lines followed through common descent. Another piece of evidence for Natural Selection revealed itself to him: Preserved adaptations were far from perfect! Even the remarkable human eye does not have the power of an artificial microscope. There is always room for improvement—hence variation, modification, and evolution.

Darwin had found the link between uniformitarianism and evolution, the connection Lyell glimpsed but had not realized. He had shown how evolution could occur by natural causes and without progressionism. Like Copernicus, he was able to relate what had merely been coincidence; like Newton, he described a dynamic "force" without miraculous attributes. His was a revolution on two levels: Interacting and varying populations replaced essentialist species; fortuitous development replaced purposeful design. And here Darwin saw the danger in his theory.

What kind of omnipotent and beneficent Creator would fashion a universe by the biological roll of the dice? Darwin could find no evidence for purposeful design. Indeed, if he did, it would destroy his theory: If one single species existed for the sake of another, Natural Selection would be doomed. In fact—and Darwin never stressed the point publicly—Natural Selection was an argument *against* intelligence behind biological development! Blind chance, purposeless mechanism, tremendous biological waste—this was surely not the universe imagined by Natural Theology all the way back to Newton. It is no wonder, then, that John Herschel, upon reading the *Origin,* contemptuously labeled Natural Selection "the law of higgledy-piggledy."

For twenty years Darwin silently built his case, planning to write a huge multivolumed "species book." In 1844, as he was completing an abstract of his theory, an anonymously authored book suddenly appeared entitled *Vestiges of the Natural History of Creation.* Probably with some consternation, and possibly with amusement, Darwin watched ridicule and pious fire descend upon the book. Written by Robert Chambers, a self-educated popularizer of science, the book set forth a doctrine of evolution, although a theory of progression in that lower forms give rise to higher. Unfortunately he mixed science, folklore, and fabulous analogies. Worse, he committed the unpardonable sin of including human beings in his evolution, even to the point of speculating that human mental capacities were the result of organization. Darwin's old teacher Sedgwick was filled with disgust and threw the book down; Huxley attacked it furiously in print. Nonetheless, the *Vestiges* sold well and served to bring evolution to the forefront of public debate. England now waited for a more rigorous presentation of the subject, and. . . .

Darwin embarked on a long study of barnacles! The study did help to sharpen his theory; he saw up close how each small change had to be functional at every stage in evolution to be preserved. By 1854 he had finished the exhausting work. Then in 1858 the famous letter arrived from Malaya.

It was from Alfred Russell Wallace, another naturalist, and it contained a clear theory of Natural Selection, almost as Darwin had written himself. The letter forced Darwin's hand; he must publish now or lose his priority. A short paper of his own, along with Wallace's, was presented to the Linnean Society. Then he got down to work. In November of 1859 the *Origin of the Species* appeared—and sold out in a day. Darwin's theory was about to become Darwinism.

The *Origin* did not really deal with human evolution, although Darwin said near the close of the book that his theory would throw much light upon the origins of man. Confrontations arose nonetheless over the popular question of "descent from the ape." Such arguments have always occurred on the periphery of the real battle. Many scientists and theologians could easily live with evolution. But Natural Selection?

Not only did the theory exclude design and hence all creationism, but the randomness of variations and the mechanism which preserved them reduced even the noble human brain to an outcome of biological fortune. This was a terrible blow to human pride, to a universe created *for* human beings, to the biological uniqueness of the human race. "No higher or lower"—the human brain was no better an adaptation, an evolutionary event, than the specialized organ of an insect. More than Copernicus ever had, Natural Selection divested human beings of their exalted position in the universe.

Even Wallace had second thoughts. Opposed to many racist theories, Wallace affirmed the mental unity of the human race and drew conclusions from it. Surely primitive societies, possessing the same intellectual endowments as Europeans, had latent mental powers which played no role in survival. Why should a "savage" have the same intellectual capacity as, say, a Newton and yet need only a tiny fraction of such mental powers to survive? (Wallace, apparently, never tried surviving in the wild with his wits alone!) Natural Selection, he concluded, could not have played a significant role. Hence design and special creation slipped back into nature through the human mind.

Darwin's answer to this revived creationism was *The Descent of Man*, published in 1871. The book reaffirmed the close physical kinship between humans and the animal world. It also found mental similarities, the differences being only in degree and not kind. Higher animals do show evidences of rudimentary reasoning abilities, communication, even emotions. Operating upon small variations of instinct and mental capacity, Darwin showed how Natural Selection could easily account for the human mind. Yet Darwin, always aware of the tremendous complexity of life, was prepared to admit other factors operating in evolution. Sexual Selection surely was a factor: the most vigorous males and females bred together and rendered a larger number of offspring; sexual selection could account for animal ornamentation which had no obvious survival value; and, as Darwin wrote in a letter to Wallace, a kind of sexual selection has been "the most powerful means for changing the races of man." Darwin was even prepared to consider Lamarckian possibilities such as use and disuse; however, he did not open to the door to conscious volition, and it remained clear that in the end Natural Selection was to be the primary mechanism of evolution.

As with many revolutionary scientific theories, the temptation was great to expand Darwin's theory beyond its original subject matter—biology. Here we need mention Herbert Spencer, who had conceived of progressive evolution before the *Origin* and who popularized the term "survival of the fittest," not only in terms of biology but also in connection with societies. Evolution, according to Spencer, was a cosmic process, beginning in the inorganic world and mounting the ladder of development to the social organism. The expansion of evolution (or social Darwinism) enabled Spencer and others to find "scientific" validity for all sorts of ethical, political, and economic ideologies all the way from *laissez faire* economics to imperialism. Even Marx and Engels were prepared to hail Darwinism as the biological analogue of scientific socialism (although it may be argued that Marxism, while materialist, is in essence progressive utopianism, closer in this way to the Christian view of history—which makes Darwin even more radical). Some writers linked evolution with quantification—the measurement of brain size, body, body parts, etc.—in order to give scientific respectability to racial theories and such concepts as eugenics and racial foundations of intelligence.

Darwin died in 1882. After his death, while evolution was generally accepted, Darwinism went, using Julian Huxley's phrase, "into eclipse." Why? The theory seemed pure speculation and without proof. Searching for causes, some went back to Lamarckian progressionism. Some searched for innate trends directing nonadaptive linear evolution (orthogenesis). In Darwin's own time the physicist Lord Kelvin "refuted" uniformitarianism with calculations of heat loss of the earth and sun, thereby severely limiting the evolutionary time scale (Chapter 22). Though not entirely opposed to evolution, Kelvin like many others harbored a strong distaste for Natural Selection. Darwin largely ignored the physicists' objections—those were for physicists to ponder (even Kelvin had to admit that there may be other sources of energy in the universe—which, of course, there were). But Darwin the biologist could not ignore the problem of variation itself. How *did* variations arise? How were they maintained? How were they inherited? Here was the major difficulty.

How could small variations, no matter how beneficial, avoid being swamped in the population by repeated breeding into the parent stock? This was a simple yet damaging question, and it was also based upon a very large assumption. Darwin's critics held that in the offspring the characters of both parents blended or fused like a mixing of colors. Thus such blending would tend to swamp a particular variation not possessed by the majority of the species.

Darwin was forced to admit that the laws of variation and inheritance remained a mystery. Nonetheless, he moved in different ways to deal with the problem: isolation, large numbers of variants, use and disuse, perhaps the nonfusion of some characters. Certainly he saw the problem long before critics seized upon it. Also he did speculate on the mechanism of variation, proposing a theory he labeled Pangenesis. Germs or gemmules, he said, are tossed off from all parts of the organism into the bloodstream and end up stored in the germ cells (reproductive cells). Variations result from irregularities in the gemmules, or modification by other factors, and are passed on to offspring. However, his cousin Francis Galton, known for his statistical analysis of heredity, transfused blood from one breed of rabbits to another and found no effects in the offspring. So again Darwin was left wondering.

Neither Darwin nor his critics knew that the mystery had already been solved! A Moravian monk named Gregor Mendel had published a paper in 1866 which, using mathematics, described his experiments with simple garden peas. Mendel meticulously segregated various traits in his peas (color, size, texture), studying the appearance of these traits in subsequent generations. Mendel's unique contribution was to apply mathematical probability to the segregation of traits. Say we have two coins and toss them together many times. The probability of getting heads–heads, heads–tails, and tails–tails in a large number of throws is 1:2:1, respectively. Roughly speaking, Mendel imagined heredity to be the occurrence of a particular trait. Based upon the number of traits considered, he computed the probabilities of their appearances in crossed generations of peas. The great hurdle he overcame was the concept of blending, which seems to exist because of the large numbers of traits involved. Segregating these traits, computing their probabilities, and experimenting, Mendel found that inheritance was really a complex mixture of traits, in which some dominated the physical appearance of offspring and others remained submerged—recessive. But mathematics was for mathematicians, not biologists, and Mendel's work remained in a biological nether world, buried in the *Proceedings* of the natural history Society of Brünn, an obscure journal in the Austrian Empire.

Thus the story does not end with Darwin. Objections remained, and alternatives to Natural Selection were proposed. Yet few (if any) scientists seriously considered the divine intervention of creationism. Indeed, many still hoped for some direction in evolution, purposeful development, perhaps some undiscovered plan to exorcise the demon of randomness. But miraculous causes were unknown causes, and no longer would they suffice as explanations of the natural world—the world of the five senses.

Science and religion both were the offspring of mythology. For ages both had trod the same road through the forest, singing the songs of the wood in harmony. Slowly discordant notes crept into the melody. Now science looked about and, behold, the paths had diverged. Its song had changed. Science found that it could no longer sing the old psalms of design, of teleology, of self-centered purpose—rhymes it had learned long ago as it sat with religion at the feet of the great mother mythology. Religion had added many new verses to the song, but it still chanted the ancient euphonious psalmody.

And so, like Newton, Darwin stood upon the seashore and sorrowfully listened to the fading voices drifting in. At last there was silence. All to be heard was the sound of his own voice describing the beautiful shells and stones on the beach. But no longer was he able to sing the old song of teleology. Now, alone in the physical world, humans would have to discover other paths to the Creator. No one knew this better than Darwin.

SUGGESTIONS FOR FURTHER READING

BARTHÉLEMY-MADAULE, MADELEINE. *Lamarck the Mythical Precursor: A Study of the Relations Between Science and Ideology,* trans. M. H. Shank. Cambridge, Mass.: MIT Press, 1982.

DE BEER, SIR GAVIN. *Charles Darwin: Evolution by Natural Selection.* New York: Thomas Nelson and Sons, 1963.

BRENT, PETER. *Charles Darwin: "A Man of Enlarged Curiosity.* New York: W. W. Norton, 1981.

GHISELIN, MICHAEL T. *The Triumph of the Darwinian Method.* Berkeley: University of California Press, 1969.

GILLESPIE, NEAL C. *Charles Darwin and the Problem of Creation.* Chicago: University of Chicago Press, 1979.

HULL, DAVID L. *Darwin and His Critics: The Reception of Darwin's Theory of Evolution by the Scientific Community.* Cambridge, Mass.: Harvard University Press, 1973.

IRVINE, WILLIAM. *Apes, Angels, and Victorians: Darwin, Huxley, and Evolution.* Cleveland: World Publishing, 1955.

MAYR, ERNST. *The Growth of Biological Thought: Diversity, Evolution, and Inheritance.* Cambridge, Mass.: Harvard University Press, 1982.

21

~~~~~~

Life and Matter

Charles Darwin, ever conscious of his limitations, realized evolution did not account for life's initial creation. Thus he concluded the *Origin of the Species* by speaking of "powers" originally "breathed" by the Creator into the forms (or one form) of life. Perhaps the statement was merely an attempt to appease religious sensitivity. Yet there is a mystery after all. Flesh, blood, and sinew all are formed from inorganic materials—carbon, water, and so forth—and like magic in the organism give rise to self-movement, heat, reproduction, irritability—in short, the phenomenon of life. Perhaps there is a kind of force, some vital principle, breathed into matter. Back in the seventeenth century Francis Glisson concluded that within the muscle fibers existed an inherent vital activity, an intrinsic property of living tissue he labeled "irritability." Whatever it is called, some special principle had to be present, many believed, when brute matter became living organisms.

Of course there was always the world according to Descartes: Organisms are machines, "vitalized" by nervous fluids. Even if we accept this, we may wonder what kind of machines Descartes had in mind. Let us share a little joke with the philosopher Fontenelle. If I have two dog-machines, he said in 1683, and if my two dog-machines are male and female, I can be almost certain that if I wait long enough I will find a little dog-machine. But two clocks will never produce a little clock even if I wait until eternity! The naive machine analogy does seem inadequate. So must we give up all mechanical models?

Perhaps, using Newton as a guide, we might admit the mechanical model yet at the same time conceive of vital processes in terms of some special force, a force like gravity. Its causes may be unknown (as are those of gravity); its effects open to study. Thus in the eighteenth century Albrecht von Haller suggested we need only accept that irritability is a "natural" force, a property neither of the soul nor of matter itself, simply a mechanical force given to matter by God. And since the ultimate source of this vital force was God, von Haller, like Newton, struck a blow against rank materialist atheism.

Ultimate materialism lurked behind every mechanist system. The Cartesian soul in the human machine was, after all, a rather gratuitous addition. This was the conclusion reached by Julien Offroy de la Mettrie in the mid-eighteenth century. Could not the soul, or consciousness, be simply an activity of the brain and as such the result of a higher organization of inherently mobile matter? And this might well be true of all vital activities, argued de la Mettrie in a book entitled *Man a Machine*. Add more gears—some flexible well-oiled microgagets—to the clock, and just possibly a little clock would appear! Adopting this viewpoint, physiology as the study of living functions could simply adopt the methods of physics and chemistry.

Here was yet another issue which divided those who studied the functions of life. Could physiology ultimately be reduced to physicochemical processes? Did the same techniques of research apply?

Continued research into the structure and functions of organisms as well as the chemical complexity of organic material tended to show that life was much more intricate than any eighteenth-century machine. Take respiration, for example. Lavoisier claimed that respiration was a chemical process of combustion—as *he* understood it. In the lungs oxygen combines with carbon and gives up its caloric (heat), which is distributed throughout the body. An obvious question arises: Would not the lungs burn up? At least the lungs ought to be warmer than the rest of the body! But experiments revealed that this was not the case. Experiments also showed that animal heat was dependent not merely upon respiration, but upon a complex of activities: circulation, muscle tone, the nervous system, digestion. Physics and chemistry were simply unequipped to deal with such interconnections. Hence the term "vital principle" could be used as a descriptive phrase (rather than some special force) which set off physiology from these sciences.

Many nineteenth-century chemists recognized this distinction. Justus Liebig saw that it was a mutual chemical reaction between respiration and nutrition, the intake of oxygen and food, which produced animal heat and indeed vital activity. Liebig believed that the oxidation of fats and carbohydrates produced the heat, while protein alone was the source of muscular activity (later in the century this was disapproved).

Yet Liebig also believed that, just as fermentation seemed to be a chemical process giving rise to a chemical force, the organization of matter and the internal chemical activities in the tissues give rise to a certain vital force. Hence the vital principle *emerges* from basically mechanical activities. Vitalism need not be a cause but an effect. Nonetheless, whether cause or effect, it is still the breath of life. Carbon, hydrogen, nitrogen, and oxygen mixed together in some solution like a witch's brew will not produce a living being (Mary Shelley or not!).

There is also the example of Liebig's friend and fellow organic chemist, Friedrich Wöhler. In 1828 Wöhler synthesized urea, an organic compound produced in the kidney. Wöhler did it in his laboratory without a kidney—in essence with a "brew" of elements. Was this the end of vitalism? Not at all!

Most chemists already believed that such artificial productions were possible. Yet what could such compounds, produced either in the animal or in the laboratory, tell the physiologist about the self-regulative processes of life? Wöhler was actually interested in isomers, compounds sharing the same atomic compositions but exhibiting different chemical properties. Wöhler and Liebig both went on to study unchanging groups of atoms—radicals in organic chemistry—which as parent compounds exist as units through various combinations. Later, in 1860, Berthelot demonstrated that total synthesis of all organic compounds might be made possible by further advances in the study of organic chemistry (the theory of types). By that time some chemists did begin to doubt vitalism.

In spite of chemistry, vitalism for the physiologist, for the physician, for the naturalist was an effect, if not a cause, which all the synthesizing of organic materials could not explain. And not only were the functions of the animal machine proving infinitely complex, the gears and wheels themselves were evidently quite different from the simple springs and wheels of clocks. In short, physiology also had to contend with mysterious organic form—that strange entity called the cell.

What *was* that tiny structure, which had been seen under microscopes as far back as the seventeenth century? Was it simply a cavity, a space within the cloth of the tissues? Was it unique to plants? Lorenz Oken, who spurned the use of a microscope, argued in 1805 that plants and animals alike were built from "infusorians," which in the tradition of the Nature-philosophers were the primal elements or universal plan of life. Oken was speculating; his infusorians represented more a vague belief in the fundamental unity of life than a solid principle. Yet, speculations have a way of sometimes becoming facts, once they are refined.

Studying plants in Germany, M. J. Schleiden became convinced that every cell was in fact an independent structure and at the same time an integrated part of the organism. Theodor Schwann, studying the skeletal tissues (cartilage) of animals, saw the nucleus enclosed by the cell membrane. Schleiden had described to him the very same structure in plants. Thus, in 1839 Schwann decided that the cell was the fundamental structure of all plant and animal tissues. Each cell had a life of its own, influenced by neighbors to be sure, yet the life of the entire organism was in effect an aggregate of cellular processes. The implication was clear: The cell was not only the primal form but also the ultimate seat of living functions.

Schleiden taught that the tiny nucleus inside the parent cell became the membrane of a new cell. Schwann held that cells arise by free formation in a kind of living soup, cytoblastema, in a way analogous to inorganic crystallization. The founders of the cell theory initially emphasized the importance of the cell wall, for the faint nucleus—sharply defined only before and after cell division—seemed to be a temporary guest which disappeared once the wall was formed. Schwann's free formation was akin to spontaneous generation.

Gradually the picture became clearer. In 1852 Robert Remak concluded that in normal growth new cells arise from the division of parent cells. Only diseased cells multiply freely from cell material, now called protoplasm. Finally Rudolf Virchow, a German physician fighting for the doctrine of localized diseases, emphatically stated that *all* cells are produced from preexisting cells. Diseased cells are simply an altered condition of normal—only from cells do new cells arise.

Taking a metaphor from politics, Virchow labeled the organism a "social mechanism" like the state; it was the sum of all those processes cells have in common. And yet, Virchow added, life itself is both cause and effect, a distinct force which produces organization and at the same time results from the cumulative activities of cells. So, despite organic chemistry, vitalism found a new home in the cell.

Continued study of the cell slowly unlocked the complicated process of cell division. By the 1870s scientists saw the first step in cell division beginning in the nucleus (mitosis). Using new techniques of staining, they found threadlike tendrils forming on an equatorial plane within the nucleus. They saw that during fertilization the egg cell was penetrated by only a single spermatozoon, the nuclei of the two merging to form the new nucleus of the first cell of the offspring, the zygote. Finally in 1883 the significance of the tiny threads became apparent; each parent contributed a single thread—named chromosome in 1888—to the pairs which formed in the nucleus of the zygote. It was now clear that these chromosomes were the agents of reproduction and of the perpetuation of life, and it was also evident that the cell was the key to *all* vital activities. But what was the key to the cell itself? Could we go beyond, into the protoplasm, to its chemical constitution? If we could descend to this level, might we not lose hold of life, finding ourselves afloat in a sea of essentially "dead" matter?

Such questions are not based upon new discoveries; indeed, they were being asked decades before research had clarified the activities of the cell. They are questions of method, of ontology, even of metaphysics. Was vital force a real subject open to experimental investigation? Emil Du Bois-Reymond argued that it was not. It is true, he said, that we may study forces—chemical, electrical, and so forth—which we no doubt experience in nature. But, taking a page from Kant, we can never know their essence. This is forever beyond the pale of science. Asking for the essence of life, for some vital principle, is like seeking the essence of gravity. If physiology is to be a science, it must confine itself to known physical laws, to chemistry and physics. Still, the nagging fact remains that somehow such reductionism must account for the living, seemingly miraculous activities of the organism!

In Paris the physician Claude Bernard quietly carried out experiments, noting the chemical transformations which occurred within the body—the digestive gastric juices, the protein-splitting pancreatic juice, the complex chemical operations of the liver. Operating by known physical laws, these processes yet presented no exact correlation with similar operations outside the body. Rather, the living body seemed to be a unique, dynamic equilibrium of operations in which each chemical event and each organ was interconnected with the others. Bernard, in fact, found himself in general agreement with the vitalists: The organism *does* exhibit processes unknown in the inorganic world. Naive reductionism did allow the miracle of life to slip through its fingers. On the other hand, Bernard saw that this special thing called life *could be* explained by physicochemical principles and studied experimentally.

The solution had been glimpsed by de la Mettrie long ago, only his analogy with the eighteenth-century machine was a false one. The analogy ought to be with the *environment*, said Bernard, an *internal environment* operating by known chemical laws yet under conditions quite different from the outside ecology. The organs, tissues, and cells performed their operations within a medium of fluids and circulating blood, creating an intricate equilibrium—the internal environment—in which every process both influenced and was influenced by another. Animal heat, for example, was maintained by a series of interrelated activities connected to respiration, nerve stimulation, nutrition, and blood flow. The temperature in some part of the body falls—the nerves are stimulated, the blood flows, the equilibrium is restored. Oxygen is carried by the blood to the tissues; nutrition, arising from organic decomposition, is circulated by the blood as well. Afloat amid the influences of the external world, like a ship upon the ocean, the organism draws its sustenance from the winds and yet must constantly maintain its sails in good repair.

Such was Bernard's great insight. The concept of a dynamic internal environment, operating by known natural causes, was akin to Darwin's evolution. Like Darwin's complex ecology of the species and the conditions of life, Bernard's internal environment took the natural laws of chemistry and used them to create compensating mechanisms which, taken together, give life. At first glance the complex vessel of life does seem to violate inorganic laws. But trying to explain this phenomenon by some vital force, like searching for purposeful design in nature, is fruitless. This, too, was a part of that past when science wore a different identity. By the mid-1860s Bernard had taught physiology a new song.

"Nature has been a chemist for a long time," said Bernard. The actual details of nature's chemistry, however, still remained upon the twentieth-century horizon, awaiting the development of biochemistry. The nineteenth century would wrestle with many of them. What, for example, was fermentation? In the 1830s the Swedish chemist J. J. Berzelius had introduced the idea of the chemical catalyst, a chemical agent which, while not consumed in a reaction, must be present for the reaction to occur—or to occur rapidly. Many biological reactions were deemed dependent upon "ferments," catalysts which mediated chemical transformations. Digestion was produced by gastric juices containing the ferment pepsin (later recognized as an enzyme); the breakdown of glucose (sugar) in body occurs by fermentation. What were the natures of these catalysts? Berzelius and Liebig believed them to be inorganic.

To the contrary! asserted Louis Pasteur, the biological catalyst is a microorganism. Inorganic fermentation in biology implied, Pasteur thought, spontaneous generation. The existence of microorganisms accounting for biochemical processes was a fruitful concept for hygiene and earned Pasteur well-deserved fame. It helped to lay to rest spontaneous generation, yet it also erected another barrier between physiology and chemistry—a barrier which was, in fact, an illusion. For within the living yeast cell with which Pasteur had experimented existed enzymes, chemical catalysts which were responsible for fermentation. Biological catalytic activities were, as Bernard had foreseen, chemical reactions, the critical difference being the conditions under which they occurred, the internal environment of the cell.

Thus life existed on a level of organization, and physiology had found its home among the growing specialized branches of science in the nineteenth century. And now we must descend to another level, the level of chemistry itself: the level of matter. Chemical research, especially into organic chemistry, accelerated with blinding speed in the nineteenth century. It gave birth to a wide range of industries, especially in Germany. From its castle of theory chemistry sallied out into the world to become one of the most important applied sciences. Nonetheless, for all of its industrial and applied glories in the nineteenth century, chemistry's central castle still remained unfinished. Some even wondered if the structure's foundations were solid!

After all, the foundations were merely inferred. From laboratory experiments the chemist could still only infer the existence of Dalton's atoms. Whether matter was *really* atomic remained an open question. As a working hypothesis, however, the atom did seem to prove fruitful, and as a bonus it fit nicely into the physical assumption of matter in motion. Although chemists might well view Dalton's little spheres as purely intellectual constructions (according to the imagination!), it would certainly be a boost to the theory if relative atomic weights were firmly established. Yet, the chemist could work only from compounds and had to infer atomic units, the number of atoms combining in the compound. For example, by relative weight hydrogen and oxygen united in water according to the ratio of $1:7$. Was the atomic weight of oxygen 7 and the formula for water HO? Dalton thought so.

Berzelius devoted himself to the tremendous task of finding the relative weights of all the known elements. (He also introduced the chemical symbolism we are familiar with today, making for a simplified chemical algebra—which Dalton thought was like learning Hebrew!) He dauntlessly analyzed more than two thousand compounds, using various qualitative methods. But these methods caused trouble for Dalton's simple multiple proportions.[1]

Cavendish had noted that hydrogen reacted with oxygen to form water in a ratio of $2:1$—by *volume*. In 1809 Joseph Louis Gay-Lussac, studying gases, concluded that reacting gases formed products according to ratios of whole numbers by volume. Here was remarkable mathematical simplicity, yet it clashed with Dalton's relative weights. For example, one volume of nitrogen united with one volume of oxygen to form *two* volumes of nitric oxide gas. According to Dalton's relative weights water ought to yield a single volume of each of its elements, and the nitric oxide gas should be a single volume as well. Berzelius used the law in analysis, yet he hesitated to accept its obvious suggestions: Equal volumes of gases contained equal numbers of atoms or compound atoms. In France, J. B. A. Dumas thought that particles found in identical numbers in all gases under the same conditions were examples of

[1]In 1819 two Frenchmen, Pierre Louis Dulong and Alexis Thérèse Petit, found that specific heat (the quantity of heat required to raise equal masses of various substances an equal number of degrees) multiplied by atomic weight gave a constant (c) for some metals and sulfur. They concluded that the atoms of all elements have exactly the same heat capacity. For some elements the law applied only after they had *halved* the values of then accepted atomic weights. On the other hand, nineteenth-century experimenters found serious exceptions to this law.

physical atoms, yet he was uncertain that the smallest particles involved in chemical reactions were actually chemical atoms, since reactions might be possible only with aggregates of chemical atoms.

There was always the question of what held atoms together in a compound. In London Humphry Davy and his assistant Michael Faraday passed currents from a voltaic pile through solutions (a process called electrolysis), isolating various elements. Berzelius found that when salts were decomposed by electrolysis, the bases gathered at the negative pole (later named the cathode by Faraday) and acids at the positive pole (the anode). We shall hear much more from Faraday in the next chapter. Berzelius, however, decided that electricity had a very important role to play in chemical affinity; it was the force that held atoms together.

Berzelius speculated that every atom has both a positive and negative charge, one of which is in excess of the other (only oxygen was totally negative). Thus atoms adhere in compounds by their opposite poles, like magnets, sometimes neutralized and other times with the compound carrying a predominate charge. Although somewhat confused (the intensity of the charge was confused with its quantity), this so-called "dualistic" theory of chemical affinity seemed to rule out one important possibility: Two similarly charged atoms could not unite with each other. Hydrogen, for example, could not exist alone as H_2, or oxygen as O_2.

From organic chemistry came a direct challenge. Organic chemists found that certain compounds, called radicals, acted like single substances or as a unit in combinations. Auguste Laurent in France developed a theory in which fundamental radicals (called nuclei) became the building blocks for other "derived" radicals. The theory required that in some reactions Berzelius's electronegative elements replaced electropositive ones without altering the compound. Laurent was attacked vehemently by Berzelius, yet his theory was eventually accepted by organic chemists. Later Dumas called it the theory of types. Importantly, the theory of types required that *atoms of the same element unite!*

Despite the confusion—and the mid-nineteenth century was a period of confusion—chemists dug deeper into the structures of organic compounds. Elements, they soon discovered, had invariable combining powers. No matter what the character of the atoms with which they combined, the combining power was always the same number of affinity units. Then came some dreaming! Friedrich August Kekulé in a "reverie" on a bus saw the large carbon atoms forming a chain with smaller hydrogen and oxygen atoms dangling at its end. In 1858 Kekulé showed that carbon atoms had four affinity units. When two carbon atoms bound together, two units were used and six remained open for other atoms. From this skeleton were built the carbon compounds, so important in organic chemistry. Later, in 1865, dozing before the fire in his study in Ghent, Kekulé saw these carbon chains twisting like snakes until one gripped its own tail. From this dream he derived the structure of the benzene type. Not surprisingly, Kekulé thought that dreams may lead us to the truth!

And while all these researchers were seeking it, the answer to the problem of atomic weights and combining volumes rested precisely there—in a dream. This dream languished in an article in a French journal of physics, a priceless gem of an article published in 1811 by an Italian, Amadeo Avogadro. In 1811 it *was* basically a dream, built upon few experimental facts. Thus it was ignored. It was about inorganic gases—so the organic chemists, whose own "dreams" were giving it substance, also neglected it.

In September of 1860 the First International Congress of Chemists met at Karlsruhe. Among the distinguished delegates was a professor of chemistry at the University of Genoa, Stanislao Cannizzaro. At Karlsruhe he began a chemical crusade for Avogadro's dream.

Let us assume, Avogadro had said, that equal volumes of gases under the same pressure and temperature have the *same number* of particles. From Gay-Lussac we know the ratios of combining volumes, which for water is 2:1. Now let us take the ratio of densities of the two gases, hydrogen and oxygen, at equal volumes, which is .07321 and 1.10359, respectively. Assuming an equal number of particles (the dream), this ratio of densities must be the ratio of masses or weights of oxygen and hydrogen, about 15:1. Therefore, our volume of water formed by the volume ratio 2:1 must be the union of *two* hydrogen atoms and *one* oxygen atom! To see this in terms of atoms we must also assume that gases are not composed of single atoms but of atoms of the *same* element chemically united and which separate in reactions. That is, gases are made up of *molecules*—pairs of identical atoms. Since the formula for water must be H_2O, the true atomic weight of hydrogen is half that of the gas molecule, which means that the relative weight of oxygen is not 7, but 15 (roughly). Not assuming molecules, Dalton's relative weight for water was 8, or 7 + 1 for oxygen and hydrogen, and thus could not be made to agree with Gay-Lussac's volume ratio of 1:2. If, however, in an elemental gas the molecular weights were twice the atomic weights, the problem was solved.

"The scales fell from my eyes," reported the chemist Lothar Meyer. At last the distinction between molecules and atoms could be used to confidently compute relative atomic weights. At last the molecules of the organic chemists began to make inorganic sense. The next logical step was to see if some relationship could be found between the elements themselves. In this task Meyer was to participate.

Isolating elements by purely chemical means required large

quantitites of these elements in compounds. But some were rare, occurring in tiny quantities. It was long known that certain metallic salts gave off different flame colors when burned. Within the spectrum of sunlight could be seen dark lines, and dark and bright lines also appeared in various places in the spectra of elements. Each element had, in fact, a kind of special map of such lines, unaffected by the presence of other elements in the compound. In 1859 Robert Bunsen and G. R. Kirchhoff built a new instrument called a spectroscope to map these lines. New and rare elements announced themselves through such maps, and even in the spectrum of the sun the lines revealed an unguessed element, helium. The very chemistry of the heavens now was open to human scrutiny. By 1869 the number of elements had risen to about sixty-three.

In England John A. R. Newlands arranged the elements by atomic weights (computed according to Avogadro's hypothesis) and discovered that every eighth element, starting from a given one, had similar chemical properties. It was, said Newlands, a kind of "periodic" repetition like octaves in music. And this is what he called it—the Law of Octaves—announced in 1866. But some chemists chuckled: "Why not take the order of the elements by their initial letters?"

There is nothing quite as convincing in science as prophecy. Out of Russia came the voice of Dimitri Ivanovich Mendeleev, predicting the existence of strange new elements, their relative atomic weights, even their chemical properties. Madness? Not quite! Working independently of Newlands, Mendeleev had noted the periodic nature of the elements. He found that the elements formed groups or families, arranged in linear series, in which the properties of the elements seemed to be functions of their atomic weights. Each group also revealed its own characteristics: all atoms in Group I united with oxygen 2:1, in Group II, 1:1; and so forth. Coming to a space on the table which no known element would fill and maintain the periodicity, Mendeleev simply left it blank. Knowing the characteristics of the specific group, however, and noting the position of the missing element in the table, Mendeleev proceeded to make his bold predictions. His table appeared in 1869. Independently Meyer brought out a similar table a year later.

There is nothing like prophecy in science *if it comes true!* In 1874 a new element, gallium, was discovered with the spectroscope. Gallium with uncanny accuracy fit the properties of one of Mendeleev's prophecies, one he had named Eka-aluminum because its empty space lay below the known element aluminum in his table. So it went, like a game of bingo, newly discovered elements filling in the vacant spaces. The table had to be perfected, but the Periodic Law finally brought order to the chaos of the elements. And new families were added; in 1895 the gas argon was isolated, an inert gas which did not combine with other elements. Other members of the new family soon followed—krypton, neon, xenon, and finally radon—and as if by magic helium fell into this group of "noble gases." Mendeleev himself believed that a chemical analysis of the mysterious aether (next chapter) was now possible based upon the Periodic Law (he *was* a prophet!).

How interesting! We began with the almost infinite complexity of life and yet, working our way down the levels of organized matter, we discover order on the atomic level. Science indeed had become specialized! Perhaps—and few saw this at the time—the world itself is a series of levels. At each level our methods and presuppositions, maybe even some basic beliefs, change. But no matter! Despite all the differences, we can still picture in our minds what is going on. Surely such pictures are bringing us closer to reality.

Long ago, in 1815, the English physician William Prout suggested that the atomic weights of many elements were whole multiples of hydrogen taken as unity. Like some mystical alchemist, Prout had a vision: If atomic weights were built from the arithmetical weight of hydrogen, then the physical atoms themselves were somehow constructed by hydrogen atoms. Hydrogen, Prout concluded, must be the primal matter of the ancients, the "catholic" matter of Newton. Daltonian atoms have a common *subatomic* structure.

No chemist believed him. That was alchemy! Surely we are done with that! However, just to make sure, let us ask the physicists. . . .

SUGGESTIONS FOR FURTHER READING

COLEMAN, WILLIAM. *Biology in the Nineteenth Century: Problems of Form, Function and Transformation.* Cambridge: Cambridge University Press, 1977.

GOODFIELD, G. J. *The Growth of Scientific Physiology.* London: Hutchinson, 1960.

HALL, THOMAS S. *Ideas of Life and Matter: Studies in the History of General Physiology 600 B.C. to 1900 A.D.*, 2 vols. Chicago: University of Chicago Press, 1969.

HUGHES, ARTHUR. *A History of Cytology.* London: Abelard-Schuman, 1959.

OLMSTED, J. M. D., and OLMSTED, E. HARRIS. *Claude Bernard and the Experimental Method in Medicine.* New York: Henry Schuman, 1952.

PARTINGTON, J. R. *A History of Chemistry*, Vol. 4. London: Macmillan, 1964.

ROE, SHIRLEY A. *Matter, Life and Generation: Eighteenth Century Embryology and the Haller-Wolff Debate.* Cambridge University Press, 1981.

22

Of Aether
and Many Other Things

Classical Physics

"We must reduce everything to mechanical models," some nineteenth-century physicists might have declared. "Particles of matter in motion, governed by forces, strictly determined, and expressed in mathematical formalism—this must be our goal."

A few optimists were ready to proclaim victory: "We are near a complete understanding of physical phenomena. Our task is nearly finished!"

Many more frowned: "Who can say if our mechanical models—and there are so many!—are not simply hypothetical representations of physical reality? Maybe these fine pictures of wheels and springs, or particles in motion, are only visual aids and should not be taken as actually existing things. There is a great gap, you know, a chasm between theory and a reality we may never be able to cross."

Others sighed: "Maybe such models are ultimately impossible to draw. Perhaps we ought to stay with the mathematical formalism alone."

Still others: "The models must come from our aether and our fields of force. Perhaps even atoms are vortices in the aether."

And finally, near the close of the century, some asserted: "Atoms do not exist! Everything is a manifestation of energy."

Such was nineteenth-century physics, called by many historians the "classical synthesis," or the century of mechanism, or even the calm before the storm of the twentieth century. Perhaps it may be described by the first two phrases—surely not the third. Calm it was not. For new discoveries in that century raised questions and prompted reconsiderations. Few believed that the picture was complete. In fact, the century began the long and painful task of rethinking the Newtonian tradition, rebuilding it, changing it. Slowly physics departed the Newtonian legacy, and, behold, in the twentieth century that departure became a revolution.

Although they did not like it, physicists in the first decades of the nineteenth century had learned to live with action at a distance. Laplace's *Celestial Mechanics* contained, in effect, a program for the universal application of molecular motions governed by forces. Phenomena including heat, light, electricity, and magnetism could be reduced, like gravity, to mathematical mechanical models based upon forces and matter—and imponderable fluids. Action at a distance was the one burr in the symmetry.

Take heat, for example. Despite Rumford (who had never quantified his experiments and had committed the sin of extrapolating the generation of heat in about two hours to eternity), heat was a fluid, caloric, surrounding particles of matter. When a gas was compressed, caloric was squeezed out. The attractive forces of particles and the repulsive forces and quantity of caloric determined the properties of gases. Heat, like matter, was neither created nor destroyed (so much for Rumford!), it was conserved and transferred. Imponderable fluids in general were characterized by tiny repulsive particles and subject to sophisticated methematical treatment.

And light? Here Newton had only suggested that light rays may be particles, since shadow boundaries appeared to have sharp edges, and, as we all know, waves "turn corners"—that is, they propagate behind objects. Of course light particles were another form of matter. Further, sunlight carries heat, and electric currents make wires glow and become hot—perhaps all were made of the same stuff. Newton's "hypothesis" became authority. In 1800 astronomer William Herschel, using colored filters to reduce the glare of the solar spectrum in his telescope, discovered that the heat increased as he moved from the violet to the red—and continued to rise *beyond* the red! The spectrum, Herschel concluded, contained "caloric rays" which obeyed Snel's law. Thus this "radiant" heat, like light, could be reduced to molecules of different masses governed by forces.

However, it was already known in the seventeenth century that at the edge of shadows existed weak patterns of alternating dark and light fringes. Newton's rings had shown similar patterns. This phenomenon—diffraction—did seem to suggest that light could bend, which in turn suggested waves. Huygens had thought so, and at the beginning of the nineteenth century the English physician (and later Egyptologist)

Thomas Young decided to settle the issue with experiment. His experiments showed beyond doubt the phenomenon of interference of light which could only be explained by waves. In 1819 Augustin Jean Fresnel presented a paper to the Paris Academy on the undulatory or wave theory of light, again confirmed by experiment. Some physicists were still not convinced, for here indeed was a rebellion against Laplacian mechanism. Nonetheless, by mid-century the mechanics of waves had triumphed over the particle theory of light. Why?

Imagine a wave spreading over a calm surface of water. It is a period ic phenomenon of crests and troughs, and it is also a state of motion through the medium. Mathematically we may call the height of the wave its amplitude, the length between crests its wavelength (denoted by λ, the Greek letter lambda), and the number of crests passing a fixed point in a given time its frequency (f). The velocity of the wave is determined by multiplying the frequency times the wavelength. Suppose we add two waves of the same amplitude together. If they are in phase—crest to crest and trough to trough—their amplitude is doubled. If they are out of phase, they cancel (see Figure 22.1).

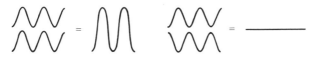

Figure 22.1

Young's experiment simply took advantage of these facts about waves. He made two narrow slits in a board and projected a beam of light through them to the wall behind. When the waves arrive at the wall in step, they combine to produce a wave of greater amplitude, hence a band of light. When they arrive out of phase they cancel, producing a band of darkness (imagine, light added to light produces darkness!). The pattern of light and dark bands is called interference. It can be explained only by waves, Young and Fresnel believed.

Now if waves are simply states of motion through some medium, what is the medium of light waves? Well, we must "imagine" one for light. Let us call it the *luminiferous aether*. Thus a question comes to mind: What are the properties of the aether?

Young and Fresnel initially thought that light waves were longitudinal, meaning that the disturbance was in the direction of propagation, as it is in sound waves. Therefore the aether must be a kind of imponderable fluid of tiny particles like gas or liquid. Yet it was known in Newton's time that a certain type of crystal (the Iceland spar) splits a light ray in two (called double refraction). The two rays are different from each other and from ordinary light, for when they enter a second crystal they do not recombine into an equal ray. Rotating the second crystal might even cancel one of the rays. This is called *polarization* (Newton accounted for it by saying light had "sides"). To explain polarization by waves Fresnel in 1821 concluded that unlike sound, light waves were transverse; their disturbance was in a plane perpendicular to the direction of propagation of the ray. Such waves occur in all directions at right angles to the ray, and some are blocked by the polarizer. Unfortunately, the old fluid medium or aether was not rigid enough for the necessary transverse propagation of waves *through* it. The aether had to be an elastic solid of great rigidity to transmit at the enormous speed of light waves. In fact, it had to be a continuum of "infinitely close" particles carrying the wave by contact. Action at a distance did not fit the optical aether. Yet, using the methods of the calculus, physicists found that they could live mathematically with the rigid aether.

Could astronomy live with such a thing? The aether must fill the universe, for it exists wherever light exists. Now if the aether is an elastic solid, how do the planets move through it without some friction and hence loss of velocity? Many models were proposed. For example, in the 1840s George Stokes suggested that the aether might act as a solid with respect to light and a fluid with respect to matter, a kind of waterlike jelly. Such rather odd explanations might cause scientists to doubt the wave theory. The problem, however, was that by mid-century Leon Foucault had shown experimentally that light travels slower in water than in air—a contradiction of the particle theory, which held the opposite because of attraction. Astronomy had still other objections.

Back in 1676 Olaf Römer had calculated the speed of light from observations of eclipses of one of Jupiter's moons. As the earth in its orbit approached Jupiter, the eclipse came early; as the earth receded, the eclipse came later. Römer reasoned that the light took time to cross the earth's orbit. Knowing the diameter of the orbit (as estimated at that time), he was able to calculate the speed of light. Although that speed is enormous (today it is given as 186,000 miles per second—Römer's figure was slightly less), we have already seen the wave theory and the aether account for it. However, they had to account for another observation. In the 1720s James Bradley was searching for the old stellar parallax and at first believed that he had found it. The tiny displacements he found in the stars, however, were not in the expected direction, nor at right angles to the direction of the earth's motion. Instead, Bradley determined, because of the motion of the earth the light coming down from the star was slightly displaced, and the star would seem a little displaced in the direction of the earth's movement. Calculation of this displacement depended upon the ratio of the orbital speed of the earth to the speed of light. The phenomenon, an optical effect, is called aberration, and Bradley was able to determine the speed of light. Bradley held to the particle theory of light and had no problem explaining aberration—like the earth running into a shower of light hailstones!

Could aberration be explained by the wave theory? Yes—but on the condition that the earth did not drag along the aether. For if the aether was carried along with the earth, light would be also, and there would be no aberration. So the aether must be stationary and, in fact, freely pass through matter. Wait! An absolutely stationary aether ought to remind us of something else. Does it not sound like Newton's absolute space? Is it not a privileged frame of reference, this aether, and may we not by optical experiments determine the relative motion of the earth with respect to it? Following this line of reasoning, Francois Arago as early as 1818 attempted what is called a "first order" experiment—that is, simply determining the earth's velocity (v) from the refractive index of a lens, which depends upon the speed of light (c). If, say, the motion of the earth is toward a star, c should be greater by that velocity in the lens (relative to the substance of the glass) and less by the same amount six months later as we move away. But Arago found no such difference!

He then told Fresnel. Fresnel found an ingenious solution. The aether did flow freely through matter, he agreed, yet more aether is trapped within the glass, the amount depending upon the refractive index of the glass. Using this idea of entrapped aether moving through the stationary aether, Fresnel calculated a formula for the speed of light in moving glass which seemed to answer Arago's negative experiment. Later, in 1851, H. Fizeau confirmed the formula by measuring the speed of light in moving water. The only trouble was that Fresnel's theory required a different amount of entrapped aether for every color, since the refractive index depends also upon the frequency of light.

Notice that the wave theory, however well supported by experiment, has embroiled us with the throny problem of mechanical movement. This would have profound implications.

At the same time, another rebellion was brewing in England, this one against the imponderable fluids of electricity and magnetism. In Humphry Davy's laboratory worked Michael Faraday, Davy's assistant since 1812, studying the effects of an electric current upon solutions. Faraday, the son of a blacksmith and one-time apprentice to a bookbinder, was a sure-handed experimenter. Largely self-educated and having little mathematical grounding, he nonetheless possessed a vivid physical intuition, a mind which thought in pictures rather than abstract formulas. He preferred to call himself a natural philosopher—and so he was.

Faraday the natural philosopher, experimenting with electrolysis, became convinced that far from being an imponderable fluid electricity was somehow an inherent property of matter as a force. In Germany the other Nature Philosophers viewed the world in terms of a unified force, in contrast to the French view of the primacy of matter. Kant taught that since we know substance only through forces, not matter but force was the phenomenal ground of science. The English poet Coleridge, who was interested in science, absorbed this philosophy in Germany. Coleridge returned to England and visited Davy, a scientist who wrote poetry. And there was Faraday.

At the University of Copenhagen Hans Christian Ørsted also thought about forces. Take electricity and magnetism—two separate fluids. Perhaps, thought Ørsted, inspired by Kant, they are really manifestations of a single force (people knew that lightning magnetized iron). During the winter of 1819–20 Ørsted attempted to demonstrate their unity. At first he failed. He had placed a magnetized needle perpendicular to an electric current and found . . . nothing! No attraction. Then, perhaps in frustration, he positioned the magnet parallel to the wire. The current was turned on—the magnet spun—to the perpendicular! Amazing! Unlike Newtonian attractions, which acted on a line, electromagnetic effects were cirucular.

In France André-Marie Ampère thought that perhaps magnetism was in fact electricity, the poles being two electric fluids revolving in tiny circles concentric to the axis of the magnet. Thus he brought two closed circular currents of electricity near each other. If the currents were in the same direction, attractive force resulted; if opposite, repulsive. The current–current force could be formalized by the inverse-square law. Here was action at a distance, communicated perhaps through the aether, between the moving electrical fluids. So Ampère believed.

Faraday was not convinced. His electrochemical experiments seemed to contradict fluids. So he experimented—and discovered electromagnetic induction.

He found that he could induce the flow of a current in a wire by magnetism (as Ampère had shown, electric currents gave rise to magnetic effects) but *only if* the magnet was moving with respect to the wire, or if the wire was moving relative to the magnetic field. Also, an electric current could be induced in a wire by another current only when the first current starts and stops—a varying current. Again we are dealing with motion. Almost by accident Faraday had discovered the principles behind electric motors and dynamos. How to explain them?

By pictures, not mathematical abstraction—this is how Faraday proceeded. He imagined invisible *lines of force* spreading out into space, magnetic lines of force sprouting from the magnet's poles and electric lines of force from electric charges. At first he thought of them as the alignment of polarized particles of some medium—like the pattern of iron fillings sprinkled about a magnet. Yet he had already rejected one set of imponderables—fluids—so why introduce another? And vacua must exist between particles—action at a distance. He had begun to doubt that. And his magnetic lines of force seemed to curve back into the magnet like a net. And the lines could be imagined as so numerous that as in the calculus they merge together forming a smooth *field of force*. And if force was the primary phenomenon of human experience . . . .

By the 1840s Faraday was prepared to consider his lines of force as forming a real physical field in space in a state of tension. When a wire is moved across magnetic lines, or the magnetic field is moved across a wire, the number of lines threading the wire changes, and we have induction. Roughly speaking, all electromagnetic effects can be explained as the interaction of fields, their states of tension excited by movement. Why, then, continue the dualism of force and matter? Might not matter be a collection of forces and all space be filled with lines of forces like a fish net? The propagation of forces may well be due to vibrations, say waves, along these lines. For Faraday the void slowly disappeared. Atomism followed. Only fields of force remained.

Professor William Thomson of Glasgow (who in 1892 became Lord Kelvin) pondered Faraday's lines of force. Kelvin was a man of extraordinarily wide interests and a mathematician. He exchanged letters with Faraday and applied mathematics to the latter's lines of force. He noticed interesting analogies between his equations and those of Joseph Fourier, who in 1822 had mathematically studied the flow of heat in solids. Kelvin was cautious—mathematical analogies only suggested physical relationships. Yet there was something else. Faraday had observed a slight rotation in the plane of polarized light passing through a block of glass in a magnetic field. Here, in Kelvin's mind, was the suggestion of some vast unity between heat, light, and electromagnetism. Unlike Faraday, Kelvin turned to the aether. In 1858 he proposed a complicated theory of a dynamic aether analogous to the mechanics of fluid motion, based upon an idea of vortex atoms or eddies in the fluid. The spider web had again become the aether.

Kelvin encouraged another Scotsman to study Faraday's work. This was James Clerk Maxwell, who had been born in 1831 into a wealthy family in Edinburgh. Like Kelvin, Maxwell was a man of wide interests. He pursued an academic career, although he was wealthy enough that he really did not need to earn a living by science. In 1871 he took the first Chair of Experimental Physics at Cambridge, planning the development of the famous Cavendish Laboratory. He was possibly the greatest physicist of the nineteenth century (if not the equal of Newton). His physics, however, was of connections, of relationships. As a student at Edinburgh he had heard Sir William Hamilton (not to be confused with the Irish mathematician), a professor of metaphysics, teach a kind of Kantian relativity of knowledge. We can know only the relations between things, Hamilton had said, never the things in themselves, the essence. Maxwell, in short, approached physics like Faraday—only Maxwell was a superb mathematician.

Kelvin had attempted to reduce Faraday's lines of force to the dynamics of the aether. In the 1850s Maxwell started on a similar road, elaborating a geometrical model of these lines and imagining them like tubes filled with an incompressible fluid (to explain the intensity of the force). Significantly, Maxwell cautioned that his fluid *was not* even a "hypothetical fluid," simply a mathematical analogy with fluid dynamics. Soon other analogies came to him, strange but fruitful.

In 1861 he imagined a honeycomb or cellular picture of the electromagnetic medium. Now the field (and it *was* a field!) was a fluid aether filled with individual cells or vortices whose geometrical arrangement gave Faraday's lines. The cells spun like tiny pinwheels, their angular velocity corresponding to the intensity of the field. Between layers of these cells Maxwell pictured "idle wheels" like ball bearings. The idle wheels serve to keep the cells rotating in the same direction, as in some common machines. Was this science? Whatever it was, the model contained surprises.

The rotation of the cells caused centrifugal forces, making the cells contract and exert radial pressures—tension. These pressures constituted magnetism. The idle wheels were identified with electricity, free to move in conductors, fixed in insulators. The cells were elastic and so were subject to strains in the field. A steady current in the idle wheels should not distort them, but a sudden change, a starting and stopping of the current, would communicate an impulse by elastic distortion of the cells. So conduction is a flow, induction a strain—the communication of a strain through the cells. The whole thing seemed hopelessly grotesque, and Maxwell knew it. He admitted that it was "awkward." But there was something very interesting about the model. The elastic distortion of the cells caused by the shifting idle wheels meant that in effect the elastic force of the current was no longer confined to fluids moving in a pipe. It was as if a momentary impulse could spread freely into the surrounding field of cells.

The "awkward" model was merely a first step, and so Maxwell began to develop equations to describe the elastic field as well as incorporate all the known electromagnetic laws. Slowly the cells and idle wheels faded, replaced by pure mathematical formalism. One thing did not fade—that momentary impulse of elastic force. Maxwell discovered a beautiful symmetry between his equations for electricity and magnetism only if he retained the momentary flow of electric charges. It had to count as an electric current, only one "displaced" from any conductor. And that is what Maxwell called it, a "displacement current," not a current as in a wire, "but the commencement of a current." Because of the displacement his equations for electricity and magnetism proved nearly identical.

And there was more. Maxwell saw that his equations for the field predicted that there should be electromagnetic waves! Further, these electromagnetic waves were transverse—like optical waves. Thus they must propagate with some velocity. Now in Germany physicists had sought to find how many electrostatic units of force (the units of electric force between two charges at rest) were in one electromagnetic unit of electricity (the magnetic force between two moving currents). The ratio they computed between these two units was 3.1×10^{10} cm/sec. Foucault's velocity of light was 2.98×10^{10} cm/sec. The two agreed "pretty nearly." The velocity of Maxwell's electromagnetic waves also agreed "pretty nearly." And Faraday had shown that light was subject to the influence of a magnetic field. Thus Maxwell made one of the greatest connections in the history of science: Light consists of transverse waves in the same medium as electromagnetism. Light was an electromagnetic phenomenon! Optics and electromagnetism were now united, indeed radiation in general. The wavelengths and the frequencies of the waves determined their character. From such "awkward" beginnings have sprung radio, television, and a host of modern technological marvels.

In his *Treatise on Electricity and Magnetism,* published in 1873, Maxwell presented his series of equations (later reduced by other workers to a basic four) describing the properties of the electromagnetic field and the propagation of electromagnetic waves. The mechanical pictures had faded into the background, replaced by generalized functions and incorporating the laws of Coulomb and Ampère into Faraday's field of force. His equations gave quantitative statements for electrostatic and magnetic fields, the circular electric field set up by a changing magnetic field, and the circular magnetic field set up around a steady current or changing electric field. Thus a changing electric field gives rise to a magnetic field, and since this magnetic field also varies in time it gives rise to an electric field, and so forth—electromagnetic waves.

For many physicists, including Lord Kelvin, Maxwell in the *Treatise* had surrendered physical understanding for ghostly abstractions. And Maxwell himself admitted that many mechanical models of his equations were possible. Yet they did demonstrate a unique harmony—they did give relations between effects. Maxwell certainly believed that there must be an aether to carry the waves and hold the energy of the field. His theory was not simply his equations, as Heinrich Hertz would say (Hertz also was making the distinction between the formalism and mechanical models, and believed the aether necessary). Unfortunately Maxwell did not live to see his theory vindicated. In 1888 Hertz succeeded in producing the waves and found them to behave just as Maxwell had predicted—they interfered and could be reflected. It was also Hertz who simplified the equations, avoiding all references to mechanical or "concrete" models. Yet Hertz, too, recognized the necessity for the aether. And here again—postulating the aether lands us back into the problems of accounting for the mechanical motion of bodies (like the earth) through it. Evidently more refined experiments were necessary, and physicists, ever ingenious, concocted them.

Later we shall see what they found! For nineteenth-century physicists had another great synthesis to achieve and another imponderable to tame—the concept of energy and the caloric, heat. Fourier, studying the conduction of heat through solids, thought that its effects and its essential nature could not be subsumed under mechanical force. Yet heat obviously played an important role in the universe and was responsible for the mechanical operation of the steam engine. The difficulty was not really the caloric—Fourier refused to say what the nature of heat was—rather, the problem was the theory that heat was conserved and hence could not be transmuted into something else.

The steam engine produced _work,_ and engineers defined work as the product of force times distance. How, then, does heat generate work (also called duty or power) in the operation of the engine?

In 1824 a French army officer, Sadi Carnot, published a small book (118 pages) entitled _Reflections on the Motive Power of Fire._ At this time Carnot accepted the caloric theory and its conservation. He saw in Watt's engine that the essential condition was temperature difference: Caloric supplied to the boiler produced an expansion of steam in the cylinder, and the same quantity of caloric was absorbed when the steam passed to the cold condensor. The process, Carnot perceived, was analogous to a waterwheel—the fall of temperature or the passage of caloric from a hot body to cold produces mechanical work just as the fall of water turns the wheel. And, like water, caloric is not consumed by mechanical work.

Consider the Carnot cycle. Work is generated by the fall of caloric—a one-way process. We may, however, conceive a reversible process in an idealized situation—the passage of heat from cold to warm. Yet this would take more work than the engine itself produces, since it is striving, so to speak, against the natural gravity of heat flow. Therefore, two engines at the same temperature levels must have an upper limit of efficiency, since a greater efficiency would drive the lesser backward, leaving the thermal conditions unchanged yet producing a net excess of work. Perpetual motion would be the result, and that Carnot thought absurd. The directional flow of heat and its conservation seemed to be the foundations of his cycle. However, before he died (of cholera) in 1832 he had begun to wonder: Perhaps heat was not conserved.

James Prescott Joule, the son of an English brewer, pondered the same question in the 1840s. Let us take another engine, one that produces an electrical current. Work produces the current, and the current produces heat. Assuming (like Rumford) that mechanical work did generate heat, Joule, using his machine, was able to assign a mechanical value to heat. An exact equivalent of heat is always obtained, said Joule, from a certain quantity of work. The two are converted. Joule also performed a brilliant experiment with gases. Compressing a gas required work, and the result was a rise in the sensible temperature of the gas. Was this sensible heat really the result of latent caloric squeezed out? Expanding a gas required heat to be pumped in, hence a sensible heat loss. Basically what Joule found is that when air expands without doing work (say against a piston), no heat is lost! But expanding a gas *against* some external force requires more heat, hence work, hence a conversion of heat to work. Heat had a mechanical value, and by implication an absolute temperature scale could be established based upon the mechanical value of heat. Heat was not conserved; it was transmuted into work.

Does it all seem very confusing? It did to William Thomson, our future Lord Kelvin. He accepted the Carnot cycle as it had been revived and mathematized by Emile Clapeyron in 1834 (he also coined the term thermodynamics). He saw how Carnot's conversion of caloric contradicted Joule's transmutation. And he noticed something else. Had not Fourier shown that when heat is conducted through a body, from hot to cold parts, it flows into equilibrium? True . . . but what becomes of the work (Carnot) this *fall* of heat should produce? What *was* conserved?

In Germany physiologists had begun to suspect that heat generated by organisms was proportional to work expended (Robert Mayer). Hermann von Helmholtz argued that the so-called vital forces must be simply modifications of natural forces, for clearly, an independent vital force could make animals into perpetual motion machines. All forces—animal heat, mechanical work, and so forth—were convertible yet conserved, subject to a universal conservation principle (his book, published in 1847, was *On the Conservation of Force*). Let us take an example from mechanics. In the collisions of inelastic bodies, some of our old "living force" is consumed—like Joule's heat. Ah, says Helmholtz, but it is simply converted into "tensional force" as in the winding of a watch. However—and here was a significant thought—some of the living force generates heat! And heat is another form of force. Hence the total conservation of force must be the sum of living force and tensional force. Like Kant, Helmholtz believed that the essence of nature was force, not matter.

The use of the word "force" for so many things was becoming imprecise and cumbersome. A new term was required, one which would represent the interconvertibility of all forces and the higher level of their conservation. And so by mid-century one was "created," *energy*. The energy of the universe is constant, neither created nor destroyed—the First Law of Thermodynamics. Tensional force was thus *potential energy* (the potential to do work) and the energy of movement was *kinetic energy*, measured by work as force times distance (take Newton's $F = ma$, Galileo's distance formula $S = \frac{1}{2}at^2$, multiply, substituting $v = at$, and the formula for kinetic energy is $\frac{1}{2}mv^2$).

What then is the significance of the Carnot cycle? Is it mistaken? No, answered Rudolf Clausius in Germany, the principle of heat flow is sound, and so is Joule's principle that heat is consumed in work. What must be given up is the conservation of heat. Energy, the higher abstraction, is conserved. Adding together Carnot and Joule, Clausius simply showed that the two processes may happen at once—some heat is converted into work, some passes to a lower state. The Carnot cycle actually represents a second law, the irreversibility of heat flow. Clausius, too, denied the possibility of perpetual motion.

The answer also dawned upon Lord Kelvin. The Carnot cycle as modified by Clausius actually represents the irreversibility of heat flow in natural processes as well as Fourier's phenomenon. Some heat must dissipate into the environment, forever lost but not destroyed. Through friction or radiation it is lost to use and becomes unrecoverable. Now, heat is a form of energy, the total energy in the universe is constant, all physical processes required energy, energy must be oganized to be useful, yet some is always lost through heat dissipation as heat seeks a constant equilibrium in the universe. Hence available energy is forever being lost. This tendency, Kelvin declared, is universal, and only Divine intervention could restore dissipated energy. All history must therefore be directional. As we saw, Kelvin calculated solar heat loss to the dismay of evolution—that is, Darwin's evolution!

In 1865 Clausius, having developed a mathematical equivalence value of transformation for irreversible processes, coined a word for this universal tendency. He called it *entropy*. The tendency of the total energy of the universe to achieve equilibrium is designated by a positive value of entropy. Thus we have the famous Second Law: The entropy of the universe tends toward maximum. At some future time all the energy available to perform work will reach a state of equilibrium in the universe, all natural processes will cease, and, as Helmholtz said, the universe will be in an eternal state of rest from that time on. The world, it now appeared, did have an end—"heat death."

We now know, too, that we cannot have a material caloric, for heat may be converted into other forms of energy, and only alchemists believe in material transmutations. Long ago Newton and others had speculated that heat was actually the physical sensation of molecules in motion. Molecules colliding in a gas and rebounding from the sides of a container give temperature and pressure; molecules vibrating in a solid or liquid conduct radiant heat. During the 1850s and 1860s Clausius and Maxwell gave this process a mathematical description. The molecules in a hotter body are moving faster—have greater kinetic energy—and when their kinetic energy is lost to slower molecules, we have a transfer of energy which manifests itself as heat.

Consider the molecules of a gas. The entire kinetic energy of the gas is actually the sum of the kinetic energies of its molecules. Since kinetic energy is the energy of motion, the molecules of the gas are pictured as rigid bodies colliding and spinning in the gas—at random! Theoretically we could, by Newtonian dynamics, follow the complex path of a single molecule. Of course this is impossible. Thus we must extrapolate from the gross properties of the gas. This is how Maxwell developed his kinetic theory of gases.

The individual molecules, moving at random, must gain or lose kinetic energy *at an instant*. Thus it follows that some will have more and others less at an instant. Imagine a gas whose total kinetic energy is constant. At any given instant its individual molecules all have different energies and are changing to the next instant. How may the total kinetic energy be calculated from this chaos? Simple, said Maxwell: We take averages. In short, taking the average value over long periods of time, we may *assume* that the total kinetic energy of the gas is "equipartitioned" among the various degrees of freedom (translation, rotation, vibration) of the individual molecules. Therefore we are dealing with what the American physicist Josiah Willard Gibbs called "statistical mechanics." Maxwell's work was continued and expanded by Gibbs and Ludwig Boltzmann of Vienna. For practical calculations, statistical mechanics becomes exceedingly abstract.

Statistics posed problems. The equipartition theory, for example, did not seem to work well when applied to the specific heat of gases. The spectroscope, Maxwell conceded in 1875, revealed considerable complexities within molecules, which required the addition of more degrees of freedom. When physicists attempted to take into account these additional variables, the averages given by the equipartition theory became much too high for experimental determinations of specific heat of gases which the theory ought to predict. It was also known that the Dulong-Petit constant encountered curious exceptions and even decreased at lower temperatures. In short, it had to be assumed that there were restrictions on degrees of freedom, that somehow a number were lost. Boltzmann thought that in solids these losses might be due to atoms' "sticking together" at neighboring lattice points and at low temperatures. How could they stick together in a gas? Perhaps we must consider the aether, said Boltzmann in 1895. The aether-gas cannot come to thermal equilibrium and thus the theory does not apply to a system of combined aether and gas.

Worse was entropy. Imagine a partitioned chamber. In one portion we have a hot gas and in the other a cold. Since kinetic theory deals only with averages, we must concede that some molecules in the hot gas will be moving slower than others and some in the cold gas faster. Now, says Maxwell, imagine a being of very sharp faculties standing at the barrier between the gases, opening and closing a frictionless door. This being—called a demon by others—only allows fast molecules of the cold gas to enter the hot and slow molecules of the hot gas to enter the cold. Surely entropy has no love for this demon, for like the sleeping maiden in medieval times entropy is violated by Maxwell's incubus! Kinetic energy with the help of the demon has actually passed from a low state to a higher one, thus reversing entropy. Maxwell imagined his "being" to prove a point: Entropy, too, is statistical, depending (as he said) upon the nonexistence of the demon. The Second Law is not absolute.

Even greater paradoxes "haunted" entropy. In time, entropy says, all mechanical systems pass from high organization—high energy and low entropy—to disordered states of high entropy. Thus there is a basic distinction between past and future states of mechanical systems. Yet in Newtonian mechanics we may substitute negative values for time, which means that any given system of molecules may run equally as well backward. There is no preferred previous possible state of microscopic systems in time, only changes in spatial coordinates. On the macroscopic level this implies that a decline in entropy is equally possible—which is strictly forbidden by the Second Law. A strict interpretation of entropy does not agree with the mechanics of molecules in motion because of this *reversibility*.

Very true, said Ludwig Boltzmann in 1877, only we have not exhausted the mathematical possibilities of dealing with the puzzle. If we take the macroscopic state of a gas as a collection of nearly infinite conceivable molecular states, we may say that entropy is based upon *probability*. Roughly speaking, entropy is like tossing dice. The probability of getting two sixes in a throw of two dice is $1/36$ ($1/6 \times 1/6$)—very low. It increases for other combinations while it remains low for *ordered* pairs in a given throw. Increase the number of faces of the dice—to nearly infinity as in our molecular states. Entropy is simply a high probability that a given macroscopic state will be disordered. (Since entropy is the addition of microscopic states and probability is multiplication, Boltzmann's formula as written by Planck is $S = K \log W$, where S is entropy and W is the number of microscopic states corresponding to an observed gross state). Entropy, then, is not absolute, and there even exists a tiny possibility that it may be violated (as in enough throws of the dice).

Boltzmann even speculated that perhaps the sense of time in our world is determined by the direction of entropy. Thus it is equally conceivable that in some other part of the universe entropy—hence time—may run in reverse!

Some scientists had had enough. Physics had become full of "spooks!" First statistics, now probabilities—what was the kinetic theory coming to? Undoubtedly a large part of the problem was caused by hard material atoms and their molecules. They were the basis of the kinetic theory (matter in motion), and the kinetic theory could not live with strict interpretation of the energy laws. Yet, these atoms and molecules were still only hypothetical—spooks! Away with spooks! shouted Ernst Mach and F. W. Ostwald. Away with atoms! Let us accept the rigor of the energy laws without them. Energy, not atoms (or matter?) is the essence of existence. Mach, in fact, was very critical: spooks such as absolute space had been haunting physics for centuries.

Rebellion was leading to revolution in the empire of classical physics. Perhaps the revolution had already begun. Surely physics with its fields, energy, and probability was neither Newtonian nor calm. Rather it was laying the foundations for a new empire, one still to be completed. And while it had lost one of its master builders, Maxwell, who died in 1879, it would gain another in the same year. For on March 14, 1879, in Ulm, Germany, Albert Einstein was born.

SUGGESTIONS FOR FURTHER READING

ARIS, RUTHERFORD; DAVIS, H. TED; STUEWER, ROGER H., eds. *Springs of Scientific Creativity: Essays on Founders of Modern Science.* Minneapolis: University of Minnesota Press, 1983.

BELLONE, ENRICO. *A World on Paper: Studies in the Second Scientific Revolution,* trans. Mirella and Riccardo Giacconi. Cambridge, Mass.: MIT Press, 1980.

BRODA, ENGELBERT. *Ludwig Boltzmann: Man—Physicist—Philosopher,* trans. Broda and Larry Gray. Woodbridge, Ct.: Ox Bow Press, 1983.

BRUSH, STEPHEN G. *The Kind of Motion We Call Heat: A History of the Kinetic Theory of Gases in the Nineteenth Century,* 2 vols. New York and Amsterdam: North Holland Publishing Co., 1976.

————. *The Temperature of History: Phases of Science and Culture in the Nineteenth Century.* New York: Burt Franklin and Co., 1978.

CANTOR, G. N., and HODGE, M. J. S. *Conceptions of Ether: Studies in the History of Ether Theories.* Cambridge: Cambridge University Press, 1981.

GOLDMAN, M. *The Demon in the Aether: The Story of James Clerk Maxwell, the Father of Modern Science.* Edinburgh: Adam Hilger, 1983.

HARMAN, P. M. *Energy, Force, and Matter: The Conceptual Development of Nineteenth-Century Physics.* Cambridge: Cambridge University Press, 1982.

HESSE, MARY B. *Forces and Fields: The Concept of Action at a Distance in the History of Physics.* London: Thomas Nelson and Sons, 1961.

MERZ, JOHN THEODORE. *A History of European Scientific Thought in the Nineteenth Century,* 4 vols. (1904–1912). Gloucester, Mass.: Peter Smith, 1976.

WHITTAKER, EDMUND T. *A History of the Theories of Aether and Electricity,* 2 vols. New York: Philosophical Library, 1951–1953.

WILLIAMS, L. PEARCE. *Michael Faraday: A Biography.* New York: Basic Books, 1965.

23

Strange New World

Relativity

Despite the problems faced by physicists, the mechanistic and material-ist basis of physical science seemed quite obvious and certain by the time of Einstein's birth, and most of the action and thinking of the era took it very much for granted. Scientists in general could look to the fu-ture with confidence, trusting not only in apparent internal progress, but also in a rising scientific culture which afforded them a more perma-nent and important position in Western society.

In France, following the revolutionary Reign of Terror, a centralized system of educational establishments and government offices was set up, providing careers for scientists. Some of these actually had their ori-gins during the old regime, when specialized schools for the professions had arisen. During Napoleon's Empire scientists emerged as a class, be-coming part of the official elite, and even being given posts in the gov-ernment. The combination of the *philosophes'* campaign for science as rel-evant to the social, political, and technological needs of the country; the new schools, which in some cases were granted research facilities; the emergence of an influential class supporting the scientistic movement—all meant that after the Restoration of 1815 science in France had become permanently institutionalized. Science now became recognized as an in-trinsically valuable pursuit, to be supported for the material and social welfare of the country; and in France, roughly from 1800–1830, science achieved a level unsurpassed in the rest of Europe.

Yet, as the nineteenth century progressed, the government monopoly of higher education, the extreme centralization of the French system, made it rigid and thus tended to thwart organizational flexibility, in turn hindering science by inhibiting the kind of cooperation and technical and industrial innovations which increasingly characterized the patterns of scientific research.

In England the Royal Society, though more of a club consisting of amateur philosophers and naturalists, spawned other metropolitan societies—such as in Edinburgh in 1783 and in Manchester in 1831, and one-subject societies such as the Linnean Society in 1788, the Geological Society in 1807—which helped to increase the social status of those who "cultivated" science. Around 1830 there was an outcry and debate over a so-called "decline of science" in England and thus a need for recognition and public support of science. In 1831 a new British Association for the Advancement of Science met, originally dedicated to giving a more "systematic direction to scientific inquiry" and to drawing national attention to science. The latter goal seemed to be its most important function for scientists themselves, since it provided a public forum for scientific debate and discussion as well as contact between different disciplines. It was at the Association's annual meeting in Oxford in June of 1860 that Huxley met Bishop Wilberforce for their famous debate over Darwin's "hypothesis."

However, a call for government support for fundamental research— research that had its own value besides applied science—led to various proposals in England between 1850 and 1868 for the reform of British education, especially at Oxford and Cambridge. Later in the nineteenth century and on into the first decade of the twentieth, supporters of this movement joined forces with other influential groups and called for government and private endowments of research in which science would be cultivated as a national resource. While substantial industrial and governmental aid did not come until after the First World War, when research, both fundamental and industrial, became a national priority, the sense of community and professionalism had grown among English scientists.

It was in Germany most of all that science approached a professional status in the nineteenth century and research became a necessary qualification for a university career. Until 1871 Germany was a melange of small states, kingdoms, and free cities in which, as in Prussia, the autocratic rulers cultivated science and yet the intellectuals could have no pretensions to political leadership. Therefore learning and knowledge became ends in themselves, concerned with the inner or "spiritual" (*geistig*) values of life, rather than with utilitarian and political reforms as in the Western countries. Added to this, the Napoleonic invasions caused a certain reaction to French ideas, especialy in education, for the feeling was that German strength lay in the realm of its *Kultur*, its culture of national literature and especially philosophy. Thus the new universities—Berlin in 1809 served as a model—were first organized around the philosophical faculty, who, steeped in idealism and *Naturphilosophie*, tended to inhibit empirical science.

The faculty of universities formed an elite class dedicated to preserving and cultivating *Kultur*. Existing for no other reason, they became an autonomous aristocracy of tremendous prestige in German society. While the state assumed financial reponsibility for the universities, with professors ranking as civil servants, academic affairs remained solely in the hands of this elite, a kind of guildlike corporation. The oligarchic tendencies which resulted from such a structure were, however, counteracted by competition among the universities in the recruitment of high-quality appointments, the requirement for appointment being the *Habilitation*, an original contribution to knowledge based upon some original research.

Throughout the nineteenth century the physical sciences grew in importance, and while the older professoriate sometimes resisted them, the competitive system among the universities made it possible for individual scientists to initiate innovations. Thus an informal scientific community was established, based upon a network of communication among individuals or small groups working in various universities. In this situation there was a regular market for scientific research, since creative work was vital for entrance into the academic market. The scientist could also use such bargaining powers to obtain laboratories to train would-be researchers. The laboratories in various German universities gained world-wide recognition in the nineteenth century, and some became the virtual centers of research in particular scientific fields.

Scientific research in the German universities did not have to prove itself for any practical purposes; the goal, as in the older philosophic system, was to create valid new knowledge for its own sake. But in the second half of the nineteenth century, with the establishment of the German Empire and the rise of industry, government and industrial research institutes were also founded. The Imperial Institute of Physics and Technology was established in Berlin in 1887 and the Kaiser Wilhelm Society in 1911, as well as industrial laboratories for chemistry and physics. This professionalization of scientific research in Germany was carried over to the United States, where it was transformed into the concept of the graduate school and the belief that training for a profession such as medicine, and others, should be based upon research and scientific theory.

Slowly a process had begun, especially in Germany, which in the twentieth century would eventually make the scientific community an interest group, competing with other groups for resources. Science was becoming recognized as an important national resource, playing an important role in the improvement of the quality of life as well as the quality of the human mind. Western culture was on the road to becoming scientific culture. Perhaps August Comte had foreseen its approach in the early nineteenth century, when he declared that the age of science and positive fact would replace the ages of religion and metaphysics. Humboldt had written that a serious pursuit of science would prevent political decline. The scientific horizon appeared bright, the sun of progress casting its friendly rays upon both the internal and external landscapes of scientific endeavor. Secure in their growing professionalization, certain of their foundations in the materialist, mechanistic "world picture," and with some utopians even contemplating the abolishment of war through the increasing rationalization of European culture, it is no wonder that many nineteenth-century scientists looked to the future with a sense of confidence and pride in their accomplishments and their social worth.

Yet thoughtful physicists who pondered the mechanism realized that some facts were troublesome. Maxwell's field seemed to answer the old dilemma of action at a distance; nonetheless, the greatest of all action at a distance, gravity, stood outside untamed. Maxwell himself had written to Faraday of the need to bring gravity into the folds of the field and weave a web across the universe. Moreover, the field was a continuum, and matter, if atomic, was discrete. How do continuous fields react with discrete matter? And the most pressing problem of all: How can electromagnetic motion be made to agree with Newtonian mechanical motion? Those who harbored a confident belief in mechanism had overlooked the inherently theory-laden nature of their facts, much as some utopians had forgotten the irrationality which existed like a silent but deadly volcano beneath the sunny European rationalism. What followed would seem to both groups a revolution and a sobering shaking of confidence.

Albert Michelson sought to answer the burning question of motion with an ingenious experiment. If the earth was moving through the stagnant electromagnetic aether, Michelson reasoned that an aether wind due to the motion of the earth should be apparent. Yet, as we have seen already, no first-order experiment could detect the wind. Michelson devised an instrument called an "interferometer" to overcome the difficulties of measurement presented by the enormous speed of light. Imagine a light source at rest in the earth's frame of reference, hence moving through the aether. Send a ray of light in one direction to a mirror so that the reflected light returns. At the same time send a ray the same distance, yet perpendicular to the first. Upon return the two rays ought to yield an interference pattern. Now, the ray sent against the aether current should take a bit longer to complete its journey than the one traveling transversely. Allowing for the transverse aether drift, and using the Pythagorean theorem, Michelson computed that the time for light to travel perpendicularly to the aether wind should be shorter by a factor of $\sqrt{1 - (v^2/c^2)}$ than the light traveling parallel. The difference, a second-order one, should cause a displacement in the interference pattern which would yield the factor.

The Michelson-Morley experiment is diagramed in Figure 23.1. The time for a light ray to go from O to A and O to B when there is no aether

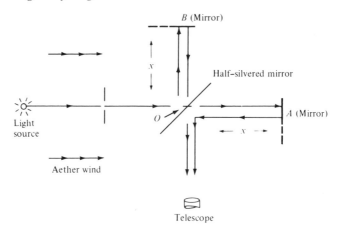

Figure 23.1. The Michelson-Morley experiment.

wind is measured by $t = x/c$ (where c = speed of light), and the round trip is $t = 2x/c$ for both, which arrive in phase.

Taking into account the aether wind with velocity v, the time from O to A is $x/(c + v)$ and from A to O is $x/(c - v)$, and the total time is the sum, which works out to

$$t = \frac{2xc}{c^2 - v^2}$$

For the perpendicular light ray in the aether wind we must use vectors and the Pythagorean theorem:

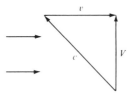

$$c^2 = v^2 + V^2$$

and, solving for V:

$$V = \sqrt{c^2 - v^2}$$

Thus the time for a light ray to travel in the aether wind from O to B and back is

$$t' = \frac{2x}{\sqrt{c^2 - v^2}}$$

The interference of the two rays depends upon the time difference which is expressed by perpendicular time (t') = parallel time (t) $\sqrt{1 - (v^2/c^2)}$, or the perpendicular time is shorter than the parallel time by the factor $\sqrt{1 - (v^2/c^2)}$, and the two rays will be out of phase.

Michelson attempted the experiment in Berlin in 1881 and a second time in the United States in 1887, assisted by Edward Morley. No such differences were found! What was worse, Fresnel's successful formula could not account for second-order effects of the size $(v/c)^2$. An aether dragged along with the earth would not account for aberration.

There is a great deal at stake here. In an inertial frame of reference, one moving uniformly like Galileo's ideal ship, no physical experiment performed in that frame can decide if the frame is at rest or moving uniformly. As stated by Newton, this is the principle of Newtonian relativity. Yet Newton had also introduced absolute space as the seat of inertial forces and absolute time flowing on without reference to any other object. Theoretically, from the idea of absolute space, rest and uniform motion *are not* relative; but in practice they are. The physical laws which hold in an inertial system must also hold in one at rest, and there are rules for transforming the mechanical equations from one to another without altering their fundamental character. These transformations are called Galilean, and they simply allow us to link two such systems or frames of reference moving relative to each other. The transformations assume absolute time and the background of absolute space, and while we may transform our equations from one inertial system to another in relative motion, we cannot measure either system's relative motion to absolute space without some visible absolute reference point, some buoy.

Let us take two reference systems moving with a uniform motion relative to each other, and let us simply label them by coordinates x, y, z, and x', y', z'. The Galilean transformation may assume x, y, z at rest and x', y', z' moving relative to the umprimed observer with velocity v. The primed observer converts the unprimed coordinates into primed by the relation $x' = x - vt$, $y' = y$, $z' = z$, and times are equal, $t' = t$, since time flows at the same rate in each system. To convert primed into unprimed coordinates the relation becomes: $x = x' + vt'$, $y = y'$, $z = z'$, and $t = t'$. Now if we take the unprimed reference system (x, y, z) to be at rest in absolute space and the primed system (x', y', z') to be moving uniformly, say it is a laboratory on the earth, the Galilean transformations show that both must be exactly the same, because we cannot measure v, which is the relative motion of the earth to absolute space, in our laboratory. Hence in all mechanical experiments performed upon the earth—a reference system moving relative to absolute space—the earth's movement, v, does not enter into our transformation equations, since there is no experimental way to detect it.

In electromagnetic experiments, however, there is a way. The earth is moving through the stagnant aether, which must constitute a buoy in absolute space. Therefore the velocity of the earth *does enter* into Maxwell's equations when we transform them from rest relative to the aether to our laboratory on the moving earth. We should, in other words, be able to detect our uniform motion on the earth from experiments *inside* our laboratory and transform Maxwell's equations according to Newtonian laws. The Michelson-Morley experiment demonstrated we could do neither, for this time Fresnel could not come to the rescue. The fundamental laws of mechanics, which many if not all physicists viewed as nearly tantamount to science itself, are in trouble when applied to electromagnetism.

In Ireland G. F. FitzGerald found a solution, and independently, the Dutch physicist Hendrick Anton Lorentz arrived at the same idea in 1892. FitzGerald suggested that objects might be *contracted* in the direction of the movement through the aether. Lorentz vigorously pursued the mathematics of the contraction and demonstrated that it had to be by that tiny missing factor $\sqrt{1 - (v^2/c^2)}$ in order to effect transformation. In 1899 Lorentz found that another assumption had to be made: For a uniformly moving frame of reference time had to slow by a tiny factor as well to achieve transformation. Lorentz called this time "local time" as opposed to Newton's universal time. Thus by 1904 the mathematical problems of electromagnetic mechanics had been solved.

Did material objects actually contract as they moved through the aether? This was Lorentz's physical interpretation. On the other hand, the French physicist and mathematician Henri Poincaré had begun to wonder. An awful patchwork of assumptions had been made to save what seemed to be a seriously listing ship. We really have "no direct intuition" of universal time, thought Poincaré, and without additional assumptions no observer can distinguish absolute rest from uniform motion. He spoke of the "principle of relativity" and speculated that someday the aether might be tossed off as useless. In 1905 he presented the Lorentz transformations and concluded that perhaps *all* mechanical laws needed modification.

Thus by the turn of the century important scientists such as Poincaré were beginning to doubt the absolute validity of classical mechanics. Lord Kelvin referred to the Michelson-Morely experiment as a "cloud" over the dynamics of light. Yet such doubts did not immediately lead to rejection; the experiment itself *could be explained* in the context of classical mechanics, and even Michelson kept the aether. The patchwork required did disturb some thoughtful physicists; nevertheless, classical concepts such as absolute space, time, and the aether were too strongly embedded in the nineteenth century's parcel of facts. Poincaré sensed that all concepts by which we comprehend the world were at heart things of convenience, yet even he found it difficult to surrender such well-established ones.

Here was a situation similar to the one Copernicus faced: Nature appeared to be a disharmonious jumble of discordant notes, not the beautiful simplicity and esthetic concert of the Pythagoreans. Copernicus, had he been alive, would have probably found his cosmic religious sensibilities outraged. Science is often motivated by such deep-seated drives—as was the science of Albert Einstein.

Einstein was not unlike some of the other scientific innovators we have met; he preferred to follow his own intellectual inclinations and disliked formal instruction. He graduated from the Zurich Polytechnic Institute but failed to find an academic position. In 1902 he was employed as a technical expert in the Swiss patent office in Bern. Strictly speaking he was an outsider to professional physics, not unlike the scientific amateurs of earlier times, not unlike the isolated Newton of 1666. He studied mostly on his own, and in the course of his readings he discovered another (self-professed) outsider, Ernst Mach.

Mach needed no optical experiments to be convinced of the defects in classical mechanics. How, he wondered, could inertia arise from absolute space as Newton held? For according to Newton's own Third Law there must be an opposite action of a body upon absolute space. This seemed to Mach absurd: How can absolute space be affected by anything? Centrifugal forces arise from absolute space (Newton's bucket)—but to the contrary, asserted Mach, they arise relative to other masses like the stars. Who is to say that the stars themselves are not actually rotating, giving rise to the same phenomenon?

Mach believed that the historical development of mechanics had carried along with it a great deal of metaphysical baggage. All we may say with certainty is that every body stands in some definite relationship to some other body in the universe. All else, like absolute space and motion, is physically meaningless. As a student, Einstein read Mach's *The Science of Mechanics* (1883), and his faith was shaken. Later Einstein would write that concepts are introduced because they are useful, and yet when such concepts assume a great historical authority we tend to forget their origin and accept them as facts.

So it was that in 1905 Einstein published five articles in the German *Annalen der Physik*. The fourth paper was entitled "On the Electrodynamics of Moving Bodies." Later it would be known as the restricted or Special Theory of Relativity. At heart the paper was about concepts, about how we think of space, time, and motion. The argument proceeds with clarity, simplicity, and yet profundity. We might well paraphrase Huxley: "How silly not to have thought of that!"

Notice the peculiar "asymmetries" in Maxwell's theory of induction, Einstein begins. Induction depends only on the relative motion of the magnet and wire loop, yet Maxwell treats each case differently depending upon which is a rest. Both, however, are still in relative motion only, and rest is assumed. Here is Mach. Next, Einstein continues, consider the unsuccessful attempts to discover the earth's motion relative to the aether. All this seems to suggest that in mechanics as a whole there is no such thing as absolute rest. Thus in his first two paragraphs Einstein strikes to the heart of the matter: Newtonian relativity holds universally for *all* rest or uniform motion. Since physics has no way of determining absolute rest, we must simply accept its nonexistence. Einstein's second postulate seems just as obvious: In all reference frames the speed of light in vacuum is always a constant. Such innocent postulates led to startling results.

But where is the Michelson-Morley experiment? Did Einstein know of it? Did it influence him? Yes.[1] The harder question deals with its significance. Einstein was, in general, aware of all such nineteenth-century failures to detect the motion of the earth relative to the aether (as he said in his paper). He was also aware that the failure of such experiments *could be explained* in classical mechanics without dropping classical concepts. That was the rub. The mechanical carrier of electromagnetism (the aether), absolute space and time—all created difficulties which could be overcome only by some rather arbitrary (and ugly) assumptions. It is tempting to compare Einstein's motivations to those of Copernicus: Not any single thing but the entire situation seemed hopelessly entangled and in need of some fundamental and radical reform. He might have asked himself, "If I were God, how would I have created the universe more beautifully and simply?" Newton had gone far in his day, now Einstein perceived a more wonderful uniformity, a more aesthetically satisfying harmony in nature. Experiments alone do not a revolution make!

We must look to the consequences of Einstein's two postulates for the "revolution" (or completion, perhaps). Let us take two inertial refer-

---

[1]See Abraham Pais, *'Subtle Is the Lord' . . .: The Science and the Life of Albert Einstein* (New York: Oxford University Press, 1982), p. 116. (Most physics textbooks—like the one I used as a student—imply that the failure of the experiment led directly to Special Relativity!)

ence systems (say ships, trains, spaceships, or whatever) moving uniformly relative to each other. For both the speed of light, say from a distant star, is the same whether they are rest, moving away, or moving toward the star. Yet the old transformations tell us that *c* bought to greater or less by the velocity of the system (like the earth). Now, however, it appears according to Einstein's principles that we may no longer use these transformation laws. How then do we transform the measurements of time and space made by the two observers moving relative to each other? Simple, said Einstein, the Lorentz contractions and "local times" apply universally to all physics (he deduced the mathematics himself). Physically this means that if one system (*A*) is moving uniformly relative to another (*B*), say near the speed of light, each will see the other's measuring rods contract and clocks tick slower. Physicist A in system *A* will notice nothing unusual in *A*, yet glancing over to *B* will find *B*'s rods contracted and clocks slower. Physicist B in system *B*, will notice nothing unusual in *B* until he or she glances at *A*. Both must use the Lorentz transformations for their measurements in relation to the other. "Your lengths are shorter and your clocks slower," says A. "No, yours are," replies B. Who is correct? Both, says Einstein.

And that is the surprise. All measurements are relative to inertial frames of reference, but this relativity is not due to some physical compression; rather it is inherent in the very nature of space and time. Space is determined by measurement, and yet there is no absolute measurement—absolute space—to serve as a universal yardstick. There is no privileged frame of reference; the aether, Einstein wrote, will prove superfluous. No matter how fast a system is moving toward or away from a light source, light waves from the source will go by with the same speed, *c*, and the speed of light is the "speed limit" for any object of intrinsic mass.

We must also give up absolute simultaneity. Two events occurring in the same reference system may be simultaneous for an observer in that system; the same two events viewed by an observer in motion relative to that system are not. Say we have two spaceships, *A* and *B*, and a physicist in *A* flashes a lamp exactly in the middle of his or her cabin, sending light rays fore and aft. The light waves travel at the same speed in both directions, no matter what the velocity of the spaceship (like our moving laboratory earlier in the Michelson-Morley experiment) and strike fore and aft simultaneously. The physicist in spaceship *B* does exactly the same and finds the same simultaneity. However, the physicist in *B* glances over to *A* and discovers *A* moving uniformly relative to his or her spaceship; thus to physicist B it appears that the back wall of the cabin in *A* is moving toward the light waves and the front wall away from them. Physicist B, then, sees the light waves reach the back wall *before* the forward waves reach the front; and hence the two events—

light striking fore and aft in *A*—*do not* occur simultaneously when viewed from *B*. Physicist A, on the other hand, will see *B* moving backward and light waves reaching the front *before* the back, and again the two events will not be simultaneous for physicist A. The shock is that, according to Einstein, we must say that the two events in each spaceship moving uniformly relative to each other are simultaneous for observers in those reference frames, but they are not simultaneous when viewed from the other spaceship—and yet both observers are correct in their observations! Simultaneity is relative.

In terms of Newtonian absolute time we could decide which events are simultaneous and which are not by simply seeing if they occurred at the same absolute time. But if local times apply universally to all reference systems moving uniformly relative to each other, then each physicist in his or her system will see the other's clocks running slower by the same factor $\sqrt{1 - (v^2/c^2)}$. If both were at rest relative to each other, v would be zero and there would be no slowing (the factor having the value unity). If v equaled c—moving at the speed of light—the factor would be zero, and it would appear to each physicist that the other's clocks had stopped. Yet each clock runs at its normal rate for each observer in his or her system, and if all systems must be treated equal (no privileged system or buoy of absolute time and space), then both are correct in their own measurements and the simultaneity of events for their systems. This means that there is no master clock with which to synchronize all other clocks universally, and t does not equal t' for observers moving uniformly relative to each other. Thus there is no absolute "now" across the universe, no time flowing universally, since all reference systems find the speed of light the same in each and hence are moving or at rest relative to another system, which is also relative.

In 1908 Hermann Minkowski, Einstein's former mathematics professor in Zurich, illustrated how we must think of the reformed concepts. Separate time and space are doomed, said Minkowski; there are separate times *and* spaces. For every x, y, z in Euclidean three-dimensional space there is a t. But the ensemble of all such "world points" (x, y, z, t) do not exist in isolation (all find the speed of light a constant); rather they constitute the "world." Hence this "world" must be *four-dimensional*, a world of *spacetime*. Spacetime can be handled mathematically, yet we cannot visualize it. It is the end of visual mechanical models.

Spacetime, in a sense, changes with motion as the coordinate points on a graph change with the position of a line. Only the speed of light is fixed. Each point is called an *event* in spacetime, and the history of an object is a collection of such events, which form a *world-line* in spacetime. World-lines themselves do not move; their spacetime coordinates change. Past and future expand motionless before us as time meshes with space.

Look at Figure 23.2. Holding space to two dimensions, we have a three-dimensional picture of spacetime. Point *E* is an event. The light rays traveling from this event form a light cone with a future and a past. The world-line of any body is forbidden to cross its own light cone, since this would constitute traveling faster than the speed of light. Thus any event, such as E_2, can be reached from *E*, and all observers would agree that it occurred later than event *E*. However, event E_3 cannot be reached by anything within the light cone, for to do so would require a speed greater than light. Therefore, event E_3 cannot influence or be influenced by anything in the cone of *E*, and for some observers *E* will be later than E_3 while for others it will be earlier. Likewise, anything in the past of the light cone of *E* can influence what takes place at *E*. It can be seen that the light cone divides events causally.

Such are the strange deductions from Einstein's two innocent postulates—and there were others.

Showing how Maxwell's equations conform to his principle of relativity, Einstein presented formulas for the relativistic increase of electron masses as their speeds increase relative to the observer. Characteristically he then pointed out that such increases hold for all "ponderable material points." Now if velocity reaches *c*, the relativistic mass becomes infinite. Since inertial mass is measured by force and mass becomes infinite at *c*, it would take infinite force to give a body the speed of light (since resistance to force, inertia, becomes infinite). Such a thing (infinite force), is impossible, hence *c* is the speed limit for all bodies. And further, mass too is a relative measure.

Force conveys energy to a body as it increases the body's velocity, and a body moving relative to an observer also increases in mass. Einstein's final paper during 1905, a mere three pages, asked if the iner-

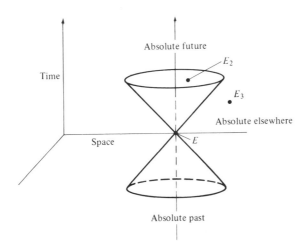

Figure 23.2

tia of a body depends upon its energy content. He then calculated that if a body gives off radiation (say energy in the form of light), its mass is diminished by E/c^2—a tiny amount, since c is so large. All forms of energy are interconvertible, hence Einstein concluded that the law applies universally: Mass is actually a measure of the body's energy content. *Mass and energy are equivalent!* Here was yet another profound simplification and harmony. Mass equals E/c^2, and in 1907 he wrote the corresponding equation, $E = mc^2$. In a way Einstein settled the debates between atomists and energeticists such as Ostwald—both were talking about the same thing.

The equation showed that the energy content of any rest mass of even the most tiny amount must be enormous. Ideas pass into achievement, artistic talent into concrete forms, scientific theory into technical applications. The application of $E = mc^2$, though difficult and based upon a greater understanding of the atom, has for good or ill changed human history.

A few years after his burst of creative genius in 1905 Einstein finally received recognition and academic positions. Recognition was one thing, but acceptance was another. Special Relativity played havoc with some very fundamental physical concepts. It was no repair; it was a major redesign. Poincaré, who had foreseen much of it, never accepted the theory. Lorentz, whom Einstein respected, remained quietly ill at ease and clung to the aether and dynamical contraction. Michelson once expressed regret that his work had created this "monster."

The monster was still incomplete. Relativity dealt with only uniform motion. The theory *was* restricted, leaving outside a very important form of motion—acceleration. Should not all motion—uniform and accelerated—be relative in spacetime? Einstein's thoughts must have returned to Newton as he pondered the problem of generalizing relativity. Let us imagine him in his study or at the office in Bern around the year 1907. He is contemplating the problem, uncertain where to begin. The smoke from his pipe forms a blue cloud about him; it seems to congeal and, behold, the ghost of Newton appears.

"How much you are like Leibniz and the others," whispers the spirit. "They too criticized my absolutes. Yet is not acceleration absolute? Do you not feel it *within* your accelerating frame of reference?"

The living man sighs. The ghost relentlessly continues: "And you also chose to relativize inertial mass. Fine—but there is gravity, pulling you down at this moment, giving you weight. What of that? What of gravitational mass? We always found the two equivalent, and refined experiments in your day have confirmed their equality beyond doubt."

Suddenly Einstein blinks: "I feel acceleration, I feel gravity. What if I should fall from the roof of this house?"

"You would accelerate uniformly, as Galileo proved."

"Well, then, let us proceed like Galileo and neglect resistance and all other surroundings. If something falls with me, say a ball, would not the ball seem to be at rest relative to me? And as I fell, would I feel my weight? No, gravitational effects would be canceled by acceleration! And say this house was in far space, accelerating uniformly at the same rate that gravity produces here on earth. All mechanical experiments performed in the house in space would yield exactly the same results as on earth. But if I could not look outside, how would I be able to tell whether the floor was rushing up to meet a suspended ball or the ball was falling in a gravitational field? Inside my house I could not say that acceleration was absolute. How could I tell that I was not really in a gravitational field? Newton, *the two are equivalent!*"

"A happy thought," replies the ghost.

"Yes," answers Einstein, *"the happiest thought of my life"* (he actually said this). "Now we know why inertial mass and gravitational mass are equal; acceleration produces the former and gravity the latter, and gravity and acceleration are equivalent. It was no coincidence, it was actually a clue to the beautiful subtlety of nature. So now that the two are equivalent and hence relative, there are good reasons to expect their relativistic effects on time as well. And another thing—if I sent a beam of light across the house in space, from one side to the other, would not the beam seem to bend as if I had thrown some object horizontally? By the principle of equivalence, then, light should also bend in a gravitational field."

The ghost laughs softly: "Did you read my *Opticks?* I said the same thing in my first Query; that light, composed of particles, would bend when near a heavy body, being attracted as all particles are to one another in the universe. But now you say it bends because of your principle of equivalence which dissolves my gravity into . . . what? I don't think you know—and you sound like the scholastics with all these imaginary things. Remember, my Queries were simply speculations, not rigorous experimental science!"

Einstein frowns. "Well, I think some may be tested. The relativistic effect of a gravitational field on time should be to slow its rate. Take an atom radiating light at regular oscillations: In a strong gravitational field, like that of the sun, the lines of its spectrum should shift toward the red, since the gravitational field of the sun lowers its frequency as compared to the same atom here on earth. The atom is a clock, its rate being the frequency of the light it emits. In a strong gravitational field the atom-clock should run slower and we may measure this by its gravitational red shift in its spectrum. As for the bending of light, I believe that it's more than speculation and may be tested . . . but I'll have to think . . . ."

The spirit roars: "While you do, think of this: Your cherished velocity of light is no longer a universal constant in gravitational fields! And if gravity somehow 'warps' time, since you have welded time to space it must 'warp' space too. *Imagine* the result of that. Your coordinate systems would bend and stretch and twist in various fields, your rods twisting and bending and stretching with them . . . you'll lose all contact with all physical measurement!"

The happiest thought of Einstein's life, the principle of equivalence, was also a first step into a maze of problems which he would later say made the Special Theory seem like "child's play." Actually it was not until 1912 while in Prague that he realized space must be "warped" by gravity. In a large gravitational field, like that of the earth, lines of fall would converge toward the center: Two balls dropped in his accelerating house would seem to fall parallel, yet the same two falling in a large gravitational field (the earth's) would converge. Thus equivalence holds only locally. However, the bending of light, the distortion of bodies (a solid body is distorted—flattens and elongates—as it falls in the gravitational field), the slowing of clocks—all represent relativistic effects, not on the things themselves, but upon the spacetime coordinate system. Varying fields would mean varying coordinate systems, all relative, no single one a privileged system—even the speed of light is no longer fixed. Spacetime would thus seem to be shattered into an infinity of separate physical systems, each having its own form of physical laws. This posed a serious mathematical difficulty, for the laws of physics must be expressed in a form that is *the same for all systems of coordinates in spacetime.* Einstein called this the principle of general covariance. But how does one express the form of physical laws mathematically so that they remain unaffected by changes in coordinate systems? It must have seemed hopeless.

Einstein returned to Zurich from Prague in 1912 and sought help from his friend Marcell Grossmann. It was a fateful encounter, for with Grossmann's assistance he found the mathematical tool required to solve his problem. It was the tensor calculus, and its history takes us back to the very roots of Euclidean geometry.

For centuries mathematicians had tried to reform Euclid's parallel postulate with its assertion of what happens at infinity. After countless failures, the great German mathematician Karl Friedrich Gauss saw that the postulate was in fact independent, and the consistency of Euclidean geometry has to include its infinity assertion. Yet Gauss also found that if he changed the parallel postulate, say to one that contradicted Euclid, he was able to construct a self-consistent *non-Euclidean* geometry. Gauss did just that in the early nineteenth century. He called his geometry "astral," since it could hardly be the geometry of physical space (nonetheless he experimented to make sure), and he kept it in his desk drawer.

The first two non-Euclidean geometries were published in the 1820s by Nicholas Lobatschewsky in Russia and Janos Bolyai in Austria, and twenty years later Bernhard Riemann constructed a geometry in which there were no parallel lines. Riemann's geometry may be imagined as a geometry which belongs to the surface of a sphere. For example, in Riemann's geometry the sum of the angles of a triangle is greater than 180° and increases as the area of the triangle increases. The smaller the area, the closer the sum gets to 180°. The shortest distance between two points is always (and obviously) a curved line, and there are no such things as straight lines on the two-dimensional surface of the sphere. All measurement and all geometry are intrinsic to the surface. Riemann envisioned curved surfaces of many dimensions and geometries in which curvature itself varied from point to point.

Consider a Cartesian coordinate system in two dimensions. It is a grid of uniform squares in Euclidean space. Say we have two points, each labeled by x and y coordinates. A line connects them. Imagine this line to be an arrow having direction and magnitude. It is called a *vector*. The differences between the coordinates of its two points are called the *components* of the vector. In whatever way we wish to rotate the entire system, the coordinates change, the components change, but the vector does not. It is actually independent of the coordinates. However, if we distort our mesh system, say to curved lines so that the squares are no longer uniform, the coordinates of the points no longer have any significance as to distances and we lose our independent vector (Figure 23.3).

In 1827 Gauss developed a method for expressing the distance between two points on such a two-dimensional surface. In essence he generalized Cartesian graphs to curved lines by the application of differen-

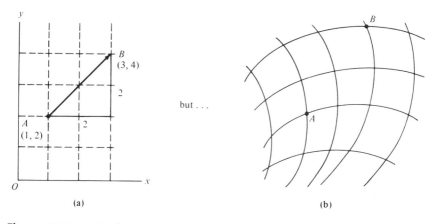

Figure 23.3. In (a) the components of vector AB are (2, 2) and represent actual lengths. In (b) this is not the case.

tial methods to geometry. In brief, the method requires more components than those of a vector in order to find distances on curved surfaces. In the case of two dimensions we require three independent components in any coordinate system. Riemann extended the method to manifolds of many dimensions, and other mathematicians perfected a calculus which enabled them to convert coordinate differences into actual distances in any coordinate system whatsoever. This is called the *metric tensor*.

The metric tensor in general frees us from dependence upon coordinate systems, but in order to do so it requires more components than a vector, collectively denoted by the symbol g_{uv}. A vector in three-dimensional space has three components; a metric tensor has six. In four dimensions, like spacetime, the metric tensor has ten separate components. Thus a gravitational field in spacetime, when expressed in the metric tensor, required ten gravitational potentials to describe the force which would act upon a body in that field (Newton's physics required only one). Therefore, Einstein had to find ten tensor field equations which conformed to his principle of general covariance. It was a tremendous undertaking, and when he left for Berlin in 1914 he still did not have a relativistic theory of gravity. Yet he had not lost his faith in the beautiful harmony he sought, even with these herculean tasks before him and as the world about him plunged into war.

In late 1915 the pieces finally fell into place. Einstein found that the metric tensor contained an operation that Gauss had described for two dimensions back in 1827 and that Riemann and Elwin Christoffel had extended to many dimensions: from the metric tensor one could build a curvature tensor representing the curvature of the surface at any point. This meant that representing gravity by ten tensor field equations actually revealed the intrinsic structure of spacetime itself. From this amazing mathematical result Einstein reached a startling physical conclusion: gravity was not a separate force after all but the intrinsic structure of spacetime! Spacetime is like a nonrigid expanse of rubber deformed by the presence of stars and planets and their accompanying gravitational fields. The planets do not orbit the sun because of some force, some action at a distance acting instantaneously; rather they follow the shortest paths (called geodesics) in curved or warped spacetime around the massive sun. Thus, according to Einstein, *gravity and geometry are the same thing*. Maxwell would have been pleased, for his web had become space itself. Newton's question—what is gravity?—had finally been answered. The gravitational attractions between bodies result from an intermeshing of spacetime curvatures caused by mass.

Gravity, of course, does not enter into Special Relativity, and so spacetime is flat; on the other hand, gravity is present in General Relativity and therefore spacetime becomes curved. In General Relativity it must be concluded that here are no "straight" world-lines, only less curved world-lines which represent the shortest distances between two points, and these are the geodesics. World-lines of bodies are given by geodesics, which means physically that the planets (or any other bodies) do not follow their courses because of some force acting upon them; rather, the geometry of spacetime is curved by the presence of these bodies, and their motions are along the shortest paths in spacetime, the geodesics. Hence light does not bend by some mysterious force of attraction (Newton's Query 1), but light too moves along the shortest paths in curved spacetime. In fact, gravity as a force in the Newtonian sense, like the aether, becomes superfluous in Einstein's General Relativity.

Here was a magnificent unification of concepts. We conclude that non-Euclidean geometry is the geometry of the world! In 1915, however, as Newton might have protested, it was still indeed an abstract "world on paper." How may we know that it actually applies to the phenomena?

The equations had to represent the planetary movements. This they did, agreeing closely with Newton. However, they contained a bonus. Recall that the perhelion of Mercury (the point of its orbit nearest the sun) was known to advance, and after all Newtonian corrections a mere 43 seconds of an arc per century could not be accounted for. Einstein's equations yielded this exact amount and in the proper direction. It was due to a relativistic effect, the curving of spacetime close to the sun. No arbitrary assumptions had to be made. It was an immediate triumph for the General Theory of Relativity. If Einstein had any doubts (which is doubtful) they were dispelled. But the rest of the world took little notice, for in 1915 the rest of the world was fighting the First World War.

The war had disrupted the European scientific community, and Einstein watched in horror as nationalistic hatreds seized hold of many of his fellow scientists. His own views were closer to those of the cosmopolitan *philosophes* of the eighteenth century. Ironically the war, coming when it did, proved to be an advantage for his theory.

Back in 1911 Einstein had conceived of a way to test the bending of light in a gravitational field. A ray of light coming from a distant star and passing near the sun should be deflected by the curvature of spacetime. On earth, specifically during an eclipse, the deflection should be detected as the difference between the apparent position of the star and its actual position. In 1911, without his field equations, Einstein calculated the deflection as 0.83 seconds of an arc. The war prevented German astronomers from testing his prediction (in 1914 an eclipse occurred in Russia—Germany's enemy).

And a good thing it did! With his new field equations Einstein discovered that the value ought to be 1.7 seconds. Once the war had ended, the English astronomer and physicist Arthur Eddington led an expedition to the island of Principe off the coast of West Africa to photograph an expected eclipse in May, 1919. His measurements confirmed the new value given by the field equations. In November of 1919 the results were announced to the world, and Einstein became an instant celebrity. The *London Times* declared a revolution in science and the overthrow of Newtonian ideas.

And Einstein himself? Lorentz had sent him a telegram in September informing him of the expedition's results. Surely he had expected them and naturally he was elated. Yet years later a student, Ilse Rosenthal-Schneider, remembered asking him at the time what if his theory had not been confirmed. His answer, as she reported it, is revealing: "Then I would have been sorry for the dear Lord—the theory is correct."

At the price of higher mathematical abstraction Einstein revealed a new and profound unity and simplicity. Relationships are invariant—the same for all coordinate systems—but measurement, space, and time themselves flow and ebb about these relationships, depending upon the convenience of the observer. Convenience, nonetheless, is not absolute. Given the restricted range of human experience, it was natural to assume that Euclidean geometry represented the structure of the world, which it does approximately in our tiny region of spacetime. Yet concepts derived from gross experience and theory-laden facts cemented in our minds by historical tradition and authority tend to become absolute tyrants. Science, Einstein wrote, is not merely a collection of facts and laws; it is a free creation of the human mind, forming a "link" between our sense impressions and the world. In this sense Ptolemy and Copernicus were both "right"—the argument was over convenience. Copernicus established a more harmonious link, a bridge to nature from which the "facts" flowed more economically, more naturally, and, we might add, more esthetically.

In this way, Einstein, like Copernicus, was close to the ancient Pythagoreans. His years of struggling with General Relativity taught him a soaring respect for the power of mathematics. Guided by mathematical thinking, human beings could uncover the sublime structure of the world which was hidden from immediate experience. In this view he departed from Mach: Pure thought could, he came to believe, reach beyond mere economical association of sensations. Beneath those sensations existed a world constructed by a mathematical plan open to the human mind, a mind which Kepler had believed thought in harmony with the thoughts of the Creator. The strength to uncover this plan, Einstein wrote, comes from a "cosmic religious feeling." Such was the feeling of the Pythagoreans.

Yet Einstein was not finished, and that strength would be sorely tested. Riemann and the English mathematician William Clifford had suggested that matter itself might well be the curvature of space. Such became Einstein's view: A Unified Field links spacetime to matter and incorporates the electromagnetic field such that matter becomes something like a hill in the geometry of spacetime. He was to struggle with the Unified Field Theory for the rest of his life. To some, relativity might have appeared to be a revolution ushering in a strange new world. Einstein, however, found there the wonderful uniformity of Newton—a deeper harmony, to be sure, nonetheless a determined universe operating by strict causal laws as Newton himself had believed. But lurking in the world of the atom, a world Einstein himself had helped to create, was a demon. The demon would challenge his cosmic religious feeling, and he would never stop thinking about it. The demon's lair was the strange world of the quantum.

SUGGESTIONS FOR FURTHER READING

BEN-DAVID, JOSEPH. *The Scientist's Role in Society: A Comparative Study.* Chicago: University of Chicago Press, 1984.

BONOLA, ROBERTO. *Non-Euclidean Geometry: A Critical and Historical Study of Its Development.* New York: Dover, 1955.

BORN, MAX. *Einstein's Theory of Relativity,* rev. ed. New York: Dover, 1962.

CANNON, SUSAN FAYE. *Science in Culture: The Early Victorian Period.* New York: Dawson and Science History Publications, 1978.

CAPEK, MILIČ. *The Philosophical Impact of Contemporary Physics.* Princeton, N.J.: D. Van Nostrand Co., 1961.

EINSTEIN, ALBERT, and INFELD, LEOPOLD. *The Evolution of Physics: From Early Concepts to Relativity and Quanta.* New York: Simon and Schuster, 1938.

GRAVES, JOHN C. *The Conceptual Foundations of Contemporary Relativity Theory.* Cambridge, Mass.: MIT Press, 1971.

HOFFMANN, BANESH. *Albert Einstein: Creator and Rebel.* New York: The New American Library, 1972.

————— . *Relativity and Its Roots,* New York: Scientific American Books, 1983.

HOLTON, GERALD. *Thematic Origins of Scientific Thought: Kepler to Einstein.* Cambridge, Mass.: Harvard University Press, 1973.

PAIS, ABRAHAM. *'Subtle Is the Lord . . .': The Science and the Life of Albert Einstein.* New York: Oxford University Press, 1982.

SWENSON, LOYD S., JR. *Genesis of Relativity: Einstein in Context.* New York: Burt Franklin and Co., 1979.

WOOLF, HARRY, ed. *Some Strangeness in the Proportion: A Centennial Symposium to Celebrate the Achievements of Albert Einstein.* Reading, Mass.: Addison-Wesley, 1980.

24

~~~~~~

The Story of *h*

Strange as it may seem, physics up to the 1920s held firmly to a doctrine which can be traced back ultimately to Aristotle. This doctrine may be summarized by the assertion: "Chance is ignorance." However, where Aristotle had concluded that chance could not be a part of science, scientists had learned to include the beast by taming it with mathematical probability and statistics. This accommodation did not contradict a statement made by Laplace: With the laws of physics and a complete knowledge of every body's position and velocity at an instant, an omnipotent mind could deduce the shape of everything past and future, for the universe is strictly determined with every effect pointing to an antecedent cause. We use statistics and probability to deal with the kinetic theory, for example, because we cannot *practically* make the necessary observations—in short, because we are ignorant. Theoretically nothing is left to chance, and everything is determined even in relativity. The *quantum theory* challenged this doctrine and began a long debate, which continues today.

Its beginnings were inauspicious. Back in 1859 (surely an important year in the history of science) Gustav Kirchhoff conceived of a perfect radiating body, a "black body" which absorbs and emits all radiation with perfect efficiency (meaning that it is completely efficient at turning the energy of heat into electromagnetic radiation and so is not really "black," for it may be red, white-hot, or whatever). Then he asked a question: Heating the black body to any given temperature, what would its spectrum look like? How would it glow? Consider Maxwell's waves as they fall upon the body; they impart energy continuously, so at the proper temperatures the body's glow begins in the red, moves to the white, then still higher to the blue. The distribution of colors forms a radiation curve, and the problem was simply to calculate it, using classical concepts, so that it agreed with experiment.

The problem was similar to that of the specific heat of gases; it was assumed that the energy absorbed by the black body was equipartitioned among its atoms and the atoms were vibrating at all possible frequencies. Thus some would always be emitting energy more effectively than others, and their frequencies would be higher and wavelengths shorter. However, formulas based upon these and other assumptions failed. One, computed by Lord Rayleigh, implied that however high the frequency of absorbed radiation, an infinitely greater radiation should be given off, eventually reaching into the ultraviolet range and beyond, like a catastrophic burst of energy. The result was called the "ultraviolet catastrophe," yet the formula worked for low frequencies. Another formula, computed by Wilhelm Wien, worked for the high range but failed in the low. Experiments showed that most energies radiated in the middle band of frequencies (the energy radiated is directly proportional to frequency, and frequency is the inverse of wavelength; thus the higher energies are radiated at high frequencies and short wavelengths). While the peak of the spectrum shifted toward the shorter wavelengths as the body grew hotter, there was always a limit, and measurements did not go off the scale at the short-wavelength high-frequency end.

Contemplating the problem in Berlin, Max Planck was inspired by Boltzmann's probability treatment of energy distribution in gases. Boltzmann had calculated the most probable ways discrete molecules with fixed total energy were distributed in abstract mathematical space (called phase space cells). Planck, on the other hand, was dealing with what he called atomic oscilllators absorbing and emitting continuous energy—discrete oscillators and continuous streams of energy absorbed in quantities of any amount. But what if, taking a cue from Boltzmann, Planck assumed that the total energy of his oscillators was made up of discrete "energy elements"? In the autumn of 1900 he did something wild, referring to it as an act of desperation. He gave his energy elements physical significance, which required him to assume that his oscillators absorbed and emitted energy *in bundles!*

It was a kind of mathematical trick, yet it worked. The higher-frequency oscillators absorb only larger bundles, as in a quota system; hence without sufficiently large packets of energy they would not be activated and not radiate. The lower-frequency oscillators, activated by smaller bundles, radiated lower frequencies with longer wavelengths. To get the correct formula Planck had to fix his energy bundles by multiplying the frequency (v) times a constant which is called the *quanta* (energy bundles) of action (h). Planck found the value of h to be very small—6.55 × 10^{-27} erg/sec (today it is 6.63). His desperation nonetheless agreed with observation!

The tiny size of h belies its very large consequences. It was totally at odds with all classical mechanics, with common sense itself. For some inexplicable reason, h said that when continuous radiation interacted with matter it did so in spurts, in jumps, in certain jumps but not others. We cannot see the quantum of action because it is so small and the jumps are blurred. Still, theoretically it is there. This seemed a bit silly, and Planck himself tried different ways of getting around it. For roughly five years after its introduction, h seemed to most physicists to be a kind of effective joke. Then in 1905 came Einstein's paper on the photoelectric effect.

In 1881 Helmholtz had suggested that electricity might be atomic, and even earlier a unit of electricity, the "electrine," had been proposed to account for chemical affinity bonds or valency. Working in the Cavendish Laboratories, J. J. Thomson subjected cathode rays—a current seen issuing from the cathode pole in evacuated glass tubes—to careful experiment. In 1897 Thomson announced that these corpuscles of electricity were tiny particles of matter, much smaller than hydrogen atoms, carrying a negative charge. Atoms, he concluded, were not uncuttable but made up of something more primordial: They were *electrons* embedded in a positive universal stuff like plums in a pudding. Prout had foreseen a great truth, declared Thomson. Experiments conducted by C. T. R. Wilson and Robert Millikan established the electron as a fact.

Another curious little fact had been noted by Hertz in 1887. In the course of producing Maxwell's electromagnetic waves he discovered that ultraviolet light falling upon the apparatus caused the sparks of electricity to come slightly more freely. If the sparks were electrons, ultraviolet light somehow knocked them off metal surfaces. This was the photoelectric effect.

The curious thing about it was that no matter how much light shone on the metal, if the light did not reach a certain frequency, nothing happened. How to explain that? Einstein saw that the answer might be found in Planck's quanta and by an analogy with statistical mechanics. Planck, using his quanta, had related frequency to energy by $E = hv$—the higher the frequency of a "bundle," the greater its energy. Yet outside matter his radiation with its energy still came in waves. Einstein reasoned in 1905 that radiation outside matter might also come in bundles like a gas. He named them "light-quanta." Later they were renamed *photons*. Increasing the frequency of a light-quanta meant, according to Planck's rule, increasing its energy, which in the photoelectric effect meant hitting the electrons with more powerful projectiles. So off they came. Increasing the amount of light simply meant firing more light-quanta, but if their frequency was low they would not have the energy required to dislodge the electrons. If they did, more light-quanta knocked off more electrons, and the greater the frequency the faster the electrons came flying off. Here was a second triumph for *h*, and now it could no longer be taken as an effective joke.

Yet—Einstein's solution was intolerable! It explicitly stated that light was corpuscular, as Newton once held. What about Young and Fresnel and all the wave experiments? And in the same year Einstein himself had used Maxwell's wave equations in his own Relativity paper. Worse, the quantum theory began to invade other areas of physics. In 1906 Einstein used it to explain the anomalies in the theory of specific heats. The quantum theory was becoming too successful and even Einstein was forced to admit that with each success the quantum looked "sillier." Then it found the atom.

In 1895 a German scientist, W. K. Roentgen, was experimenting with cathode rays when he discovered a new form of radiation, which to his surprise penetrated through solid matter. He named these rays X-rays. The new X-rays aroused the interest of H. A. Becquerel in France. Experimenting with certain uranium salts, Becquerel supposed that the penetrating rays were given off by exposure to light. Then by accident he stumbled upon a very curious effect: The rays were being emitted even without exposure to light—spontaneously pouring forth *without* an outside source of energy! How could that be? And in Paris, Marie and Pierre Curie discovered two new elements which spontaneously emitted radiation: polonium (after Marie's native Poland) and radium. Radium's output of energy was fantastic, and the element always maintained itself a little warmer than its surrounding environment. There could be no doubt; certain elements emitted enormous amounts of energy without visible outside support. The Curies called the phenomenon radioactivity.

Working at McGill University in Canada, Ernest Rutherford and Frederick Soddy found that radioactive elements were actually *transmuting* themselves into other elements—proceeding, in fact, through a sequence of elements until finally they reached the stable state of lead. The radiation consisted of beta-rays, alpha-rays, and gamma-rays. They found that the beta-rays were electrons, and Soddy showed that the alpha-rays were actually helium ions carrying a double positive charge (gamma-rays turned out to be the more penetrating X-rays). Here was some modern-day alchemy! But what did it mean for Thomson's picture of the atom?

The alpha particles were very fast and heavy, like massive projectiles of the microworld. Now Rutherford had an idea: He would probe
the interior of the atom with a barrage of alpha particles. In 1909 the experiment showed that while alpha particles mostly passed through thin
foils of metal, a few were deflected, and some even came flying back. No
plum-pudding atom here! Obviously the alpha particles occasionally
collided with another massive positively charged particle in the atom. In
1911 Rutherford announced that the atom was similar to a tiny solar system: a positive nucleus surrounded by orbiting electrons whose combined negative charge balanced the positive charge of the nucleus. Two
years later chemists saw that the charge of the nucleus could be used to
place elements on the periodic table. Thus Soddy believed that while
certain elements could vary in atomic weights—he called them *isotopes*—
the number of the nuclear charge, their *atomic number,* was a sure
classification of the element.

Here was a pretty classical picture of the atom. But there was a problem. How could it be stable? Would not the orbiting electron emit its
energy and plummet into the nucleus? Further, we know that all atoms
when excited emit light at specific frequencies which appear as lines
when seen through the spectroscope, each atom having its own identity
card. And back in the 1880s J. J. Balmer had found that the four visible
lines of the hydrogen spectrum formed a neat mathematical series, like
the rungs of a ladder. Yet we should expect Rutherford's atom to emit
light at all frequencies as the electron spins into the nucleus—which it
does not do! Here was yet another situation similar to the black-body
problem, and so . . . .

From Denmark came Niels Bohr to study with Rutherford, who was
now in England. As we might have guessed, in 1913 Bohr reformed
Rutherford's atom with the quantum. He declared that electrons moved
about the nucleus in *discrete* orbits, certain specific orbits and no others.
Energy is related to frequency by h, therefore the lines of the spectrum
correspond to photons (light-quanta) of specific energies emitted from
the atom. The emission of photons may be seen as electron "jumps" between stable orbits; therefore the orbits themselves may be quantized as
stationary energy levels. Therefore—photons of a given frequency are
emitted when the electron "jumps" from a higher energy level (further
from the nucleus) to a lower energy level (closer), and a photon of the
proper frequency absorbed by the atom pushes the electron into a
higher permitted orbit. Bohr then went on to calculate the values for
these energy-level orbits: The formula he found showed that the angular
momentum of the electron is equal to $h/2\pi$ times a number n ($n =$
1,2,3, . . .). The number n is a quantum number, and according to its
value we get the specific quantized orbits. Bohr's calcuations of emission
values based upon the permitted orbits worked well for the "observed"

hydrogen spectrum. Yet, as we might expect by now, there were problems.

Bohr's atom was really a potpourri of classical and quantum ideas,
not unlike the amalgamation of different body parts that made up the
Ptolemaic hybrid. Angular momentum was the speed with which the
electron moved around its orbit, the picture being that of a planet orbiting the sun. Then, suddenly, the "planet" disappears and instantaneously reappears in another orbit, not just any orbit but one "permitted,"
emitting into space a pulse, not continuous but discrete, of radiated energy. Only certain energy levels or "shells" are allowed, and other
zones are forbidden. And, for the sake of argument, if the electron's
jump from one permitted state to another is not instanteneous, then it
must pass through the forbidden zones in time. How does it know
where to stop? A more urgent problem was to account for the intensities
of the observed spectrum lines. To calculate them (as well as polarization) Bohr was forced back to classical methods. Thus he argued that
there was some connection, a necessary one, between classical physics
and the new quantum physics; for example, at higher energy levels the
transitions become so tiny that they are actually smooth. Bohr called this
the *correspondence principle.*

There were still other problems to consider. In a magnetic field the
spectrum lines split into groups of three (called the Zeeman effect) and
splitting also occurred in an electrostatic field (the Stark effect). Refined
analysis showed that even when undisturbed the spectrum lines were
actually composed of bundles of lines, finer lines yet. So two more quantum numbers were added to account for these properties: In visual
terms, one number described the shape of the orbit and another number
gave the orbit's orientation. Then, in late 1925 and early 1926, a fourth
number was suggested to account for the splitting of spectral lines. It
was called "spin," and yet any visual or classical picture of an electron's
spin would be misleading. The electron had an intrinsic spin in two
values, either $+\frac{1}{2}$ or $-\frac{1}{2}$, and thus its angular momentum or spin vector
could be thought of as pointing either "up" or "down." Yet to picture it
like a tiny spinning planet is difficult, because the electron must spin
around *twice* to get back to where it began. For the hydrogen spectrum
these additions seemed to work, yet a finer splitting of the Zeeman effect (the anomalous Zeeman effect) ruined everything, and for the spectra of normal atoms with more electrons the theory broke down.

Still another problem with the Bohr atom had troubled young
Wolfgang Pauli, who at the age of twenty (he was born in Vienna in
1900) had published a major article on General Relativity. Bohr had the-
orized that electrons filled shells at various energy levels, each shell becoming "full" followed by a "filling" of the next shell and so on, building up the periodic table of elements. But why, Pauli wondered, were

not all the electrons found in the ground shell of an unexcited atom, and why did the first shell contain *only two*, the next eight, and so forth? In 1925 he realized that this could be explained because if Bohr's shells corresponded to a set of quantum numbers, then a full shell exactly corresponded to the number of *different sets* of quantum numbers belonging to that shell. This meant that no two electrons could have the same *four* quantum numbers—they were "excluded" from having identical quantum numbers. Pauli's new rule is called the *Pauli exclusion principle*. But *four* quantum numbers? Pauli had inserted a fourth quantum number before that number was labeled "spin"!

The exclusion principle applies to all particles of half-integer spin, called fermions after Enrico Fermi who worked out their rules with Paul A. M. Dirac in 1925. It does not, however, apply to full-integer spin or zero-spin particles like the photon, called bosons after Satyendra Bose who worked out their rules with Einstein. The exclusion principle is of vast importance in explaining the structure of atoms and thus the periodic table; only—why do electrons obey it? There was no classical reason.

Experiments, assumptions—more experiments, more assumptions; the pace is furious and our once-solid picture of the solar-system atom is becoming vague as the quantum reveals more and more puzzles. Perhaps it is time to play Plato and turn our backs upon pictures, upon the shadows in the cave, for the pure light in our minds—the mathematical light.

Werner Heisenberg decided on this course. Like Pauli, Heisenberg belonged to the new generation that came of age in scientific research after Planck's "desperation." This generation also grew up during the First World War (1914–1918) in which many scientists from the preceeding generation had left for the trenches, many never to return. In general we may speculate that the new generation to which Heisenberg, Pauli, Dirac, and others belonged was less tied to the principles of classical physics, principles which their elders held nearly tantamount to the foundations of science, and they had also witnessed the disillusioning collapse of some of the moral and utopian progress which the nineteenth century assumed to be the result of rational European civilization. The war had caused science to lose some of its cosmopolitan internationalism; scientists themselves were not immune to nationalistic hatreds. On an even more general level, the war created a gulf between the nineteenth and twentieth centuries; it created a new Europe devoid of the three great monarchies—Austrian, Russian, and German. In Heisenberg's own Germany the new Weimar Republic was a great experiment, at odds with the monarchical and authoritarian German past;

it was, as historian Peter Gay has said, "an idea seeking to become reality." The world seemed less determined, European civilization less certain of itself; the extraordinary and perhaps even the inconceivable were becoming palpable.

In philosophy a new positivism, or logical empiricism, had arisen (Mach playing a role), a philosophy which disdained metaphysics and speculation, insisting upon statements that were logically correct and empirically verifiable. In the 1920s its motto could well be expressed by the Austrian philosopher Ludwig Wittgenstein's famous phrase: "Whereof one cannot speak, thereof one must remain silent." Perhaps the time had come to remain silent on all classical pictures, to dispense with all mechanistic assumptions, to simply accept the logic of mathematics and the empiricism of experiment.

After spending a year with Bohr in Copenhagen, Heisenberg went to the University of Göttingen, a place famous for its mathematics and physics and where young physicists gathered about theoretical physicist Max Born (Pauli also had gone there). In the summer of 1925 Heisenberg dropped all attempts to picture the atom. Instead he decided simply to accept the numbers given by experiment and see if he could make some sense of them. He set the numbers in arrays, columns, and rows like square tables. The square tables were constructed with the rungs of the energy ladders, and all the information on jumps, their frequencies, intensities, and so forth was packed into them. Each square table was to represent an aspect of a particle; one its position, q, and another its momentum, p. Now the question was to determine mathematically the values for any given position and momentum of an electron.

Imagine a wave repeating itself with a single frequency ⌒⌒ ; it is called a sine wave. Now imagine a wave of different frequencies ⌒⌒⌒⌒ . A hundred years earlier, Joseph Fourier had demonstrated that the latter could be analyzed into constituent sine waves of various frequencies, like writing out a list. This was the method Bohr had used to define his orbits. In short, according to Fourier analysis Heisenberg's p's and q's could be analyzed into lists of their constituent parts and handled mathematically, like unscrambling and recombining sine waves. But Heisenberg's p's and q's were not lists; they were square tables called matrices, and the rules of Fourier analysis did not work with them. When Heisenberg showed them to Born, the latter recalled a lecture from his student days dealing with the rules governing these matrices. Heisenberg had discovered them himself, and now he, Born, and Pascual Jordan began working out the details.

The results were a shock. While Fourier analysis obeyed all the standard rules of algebra—for example $p \times q = q \times p$, the commutative law—the matrices did not obey: *qp does not equal pq* in matrix calculus! In fact, Born and Jordan published a paper which proved that the difference between them was proportional to Planck's constant: $qp - pq = h/2\pi \sqrt{-1}$. The mathematics worked, giving the proper experimental values. But what did it mean physically that pq did not equal qp?

Not only have we surrendered all pictures, we seem to have lost all common sense. The story of h is becoming a weird fantasy. Yet in France, Prince Louis de Broglie found another way of telling it. He got the idea from relativity: Particles have mass and mass is energy; energy is related to frequency and frequency is wavelike; light may be particles or waves—so, particles of mass may display a wave character too! Electrons, according to de Broglie, should behave like waves, and in 1925 at the Bell Laboratories in New York experimenters discovered that electrons did behave like waves, waves of tremendously small wavelengths like X-rays. (From this discovery came the creation of electron microscopes—streams of electrons providing far greater detail than longer-wavelength visible light.) Matter was wavelike! De Broglie formulated a theory for free particles moving in space. In 1926 Erwin Schrödinger, a Viennese physicist in Zurich, applied the wavelike electron to the atom.

Schrödinger's matter-waves could not be free to travel like de Broglie's, rather they had to be *standing waves.* A standing wave may be thought of as waves traveling up and down a fixed string rather than spreading out into space. Because the ends of the string are fixed, say like a violin string, its vibrations form a fixed pattern and only those patterns are possible in which the ends of the string are fastened. For example, the frequencies can be one, ⌒﹍﹍⌄ or two ⟨﹍⟩⟨﹍⟩, but never half ⟨﹍⟩⌄. Standing waves give discrete patterns corresponding to certain frequencies, and thus Schrödinger's waves were limited to definite patterns, though not boundaries in the physical sense, since they vibrated in abstract space.

Applied to the atom, the standing waves, vibrating at designated frequencies, stored energy and hence corresponded to definite energy states. A change of pattern could take place only within a limited sequence of possible patterns; thus Bohr's energy levels and restricted transitions were no longer arbitrary, since only specific quantities of energy could alter a pattern to another. The exclusion principle meant that no two identical patterns could exist, and the possible number of patterns at a level accounted for the number of electrons in that shell: two for the first, eight for the second, and so forth. (Quantum mechanics was used by Born, Linus Pauling, and others to explain chemical bonding and valence. In terms of the wave function, referred to as an orbital, molecules form by the overlapping of specific patterns into more stable patterns.) Schrödinger's wave equation therefore accounted for

the facts as well as Heisenberg's matrices, and Schrödinger was prepared to give up particle electrons; electrons were matter waves,
wavicules. Matter waves are real and the physical world is formed by
them.

Yet the "real" matter waves existed in multidimensional mathematical space! Also, experiments clearly indicated that electrons behaved
like particles, and other evidence (the Compton effect) seemed to demonstrate the particle nature of light—photons. Such wave mechanics
were long familiar to physicists and hence preferable to the cumbersome
matrices; yet in the winter of 1925 Paul A. M. Dirac, working independently in England, had refined the matrices mathematically into an elegant theory in which classical mechanics naturally passed into quantum
mechanics. Thus two different theories adequately represented the
facts! Could they be united?

In June of 1926 Max Born suggested that Schrödinger's waves in
their mathematical space were not waves in the physical sense at all.
Rather, if we interpret the formula in terms of statistics, and keep
particle electrons, Schrödinger's function (specifically, the amplitude of
the wave squared) actually represents the statistical probability of
finding an electron in a given place. What are Schrödinger's waves? *They
are waves of probability*! And during the same year Schrödinger, Dirac,
Heisenberg, and others worked out a mathematical unification of the
wave function and matrices. From such onslaughts a strange new picture of the atom emerged.

Consider again a wave. Mathematically it can be shown that a large
number of waves, differing successively in wavelengths by tiny
amounts, may be combined together as a single wave. But there will be
only a small region in which they are in phase. In this region the amplitude of the wave is great, while in others it is not. The greater the range
of successive wavelengths, the smaller the region (or wave packet) and
the greater the *probability* of determining the position of the particle
⁀⌣⌣⌣ . Yet here is the catch: the wave is traveling and the particle has
momentum; momentum is related to wavelength; a precise value for
wavelength gives a precise value for momentum. So if our wave is made
up of sharply defined wavelengths (unlike those above), we have a good
chance of determining its velocity and hence the momentum (mv) of the
particle. On the other hand, this sort of wave has nearly a constant amplitude (nearly equal wave packets), and the position of the particle may
be anywhere! In short, nearly equal amplitudes ⌒⌒⌒ mean nearly
equal probabilities of finding the electron; a precise amplitude means
that the composite wave is constructed of a nearly infinite number of
wavelengths and hence the electron's momentum takes on nearly
infinite values. *Only probabilities can be determined;* in fact, only probabilities in the atomic world *are* determined. Chance is inherent in the very

structure of matter! And—it is not our ignorance—it is nature! We have found the demon in his lair.

Confusing? Yes—and exasperating too. Worse is yet to come. The mathematics forces us to conclude that a precise knowledge of the electron's position means a loss of knowledge of its momentum and vice versa. And now, in 1927, Heisenberg saw the significance of the strange matrix relations between his p's and q's. Since matrices were mathematically equivalent to Schrödinger's probability waves, the relationship $pq - qp = h/2\pi i$ represents the *uncertainty* of being able to fix precisely and simultaneously both the position *and* momentum of an electron. This means that there is an inherent indeterminacy in the atomic world which even Laplace's omnipotent mind cannot overcome. This is Heisenberg's famous *principle of uncertainty* or *indeterminacy*. It says we cannot know specific events (our p's and q's) with exact certainty. We know only certain probabilities, which change from instant to instant and are based upon which ones we choose to measure. Our predictions are statistical predictions; the future of the system is a collection of probabilities. The uncertainty exists because Planck's h exists, for if h were zero there would be no uncertainty. Yet for all this, the theory worked. It solved the problems Bohr's atom was unable to solve. It was successful for atoms other than hydrogen. Dalton's neat little spheres became strange, purely abstract mathematical entities ruled by an equally strange thing called *quantum mechanics*.

The uncertainty principle implied a rather shocking conclusion for physics. Somehow the very act of observation alters the thing observed. In deciding to measure p we automatically affect the probability of establishing q or vice versa, and thus the observer (human or instrument) must be included in any complete description of the object. Not only does the strictly causal and determined Newtonian world collapse in the microworld or level of probabilities, but the old wall separating the observer from the object observed also crumbles. And this is not due to the clumsiness of our instruments (although it may be pictured as such); it is inherent in the mathematics.

We may also view the uncertainty principle in relation to the wave-particle duality of the electron. Position is more of a particle property, fixing the electron in a given location at a given instant; momentum is more of a wave property, the waves havng no precise position but a definite momentum. So the more we know about waves, momentum, the less we know about particles, position, and vice versa. When we experiment to find particle properties of the electron, why, we get particle answers; when waves, wave answers. There is no "reality" in the classical sense on the quantum level. Strictly speaking, there is an aggregate of probabilities like the "spooks" of which nineteenth-century physicists complained. When an experiment is performed to find the electron, the array of probability waves vanish except for one wave-packet which describes the electron. This is called the "collapse of the wave function," and in a sense it means that observing the system forces it into "reality." We look for an electron in one energy state, we look again and find it has "jumped" to another; when we are not looking or experimenting it could be anywhere—it *is* anywhere—and from one observation to the next we cannot even say that it is the *same* electron!

Perhaps Niels Bohr of all the physicists was best prepared to interpret the meaning of the quantum theory. As a student he had been introduced to the thought of another Dane, a nineteenth-century philosopher named Søren Kierkegaard. An opponent of all-encompassing systems (like Hegel's), Kierkegaard believed that to explain something was also in some way to participate with or influence the thing. The division between objectivity and subjectivity, taught Kierkegaard, is inherently arbitrary; it is, in fact, a human decision.

So what does it mean, Bohr wondered, that light can be wave or particle? Simply this: Photons or electrons may without contradiction be described as both because the two concepts are necessary to account for atomic processes as given by experiment—*they are complementary*. Wave and particle are contradictory concepts developed from the gross human experience of the macroscopic level of being; on the microscopic level they are complementary aspects of the same thing. Objectivity and local causality are concepts of the classical macroworld and work well there; but in the quantum world, probability and uncertainty are the necessary concepts, and describing "reality" on this level requires both. We are in a sense free to select within the range of material possibilities the conceptual nature of reality.

We enter the quantum world, so to speak, with our parcel of classical concepts and ask our questions based upon them. The answers we get arise not from "objective reality" but from our questions and are colored by our concepts—ideas which, in essence, do not apply to the atomic world. Moreover, we must interfere with the quantum world to observe it. When we are not making our observations, the atomic world simply dissolves into a ghostly realm of probabilities; in a sense, it does not "exist" until we look at it, and it is meaningless to ask what specifically is going on when we are not looking. The philosophical implications of all this are tremendous, for we must not forget that Bohr was talking about the "stuff" that we, all of us and the world itself, are made from. "Anyone who is not shocked by quantum theory has not understood it," said Bohr. Physicist Richard Feynman stated simply that no one understands quantum mechanics (in the 1940s Feynman adopted the familiar relativity spacetime diagrams to particle interactions, showing that it was quite possible to describe elementary particles going *backward* in time).

Bohr's *principle of the complementarity* became the basis for what is called the Copenhagen Interpretation of the quantum theory. Other interpretations have been proposed. In the Many Universes Interpretation, first proposed by American Hugh Everett in the 1950s, all probabilities are realized as an infinite splitting of myriad alternative worlds. These constantly splitting realities, each unaware of the other, make up a kind of "superspace." Nonetheless, the Copenhagen Interpretation of the nature of the quantum theory seems to be the most generally accepted one. At last the inherent difficulties of translating abstract symbols into concrete pictures had been realized. Bohr made it plain that the symbols and concepts of physics are what humans *say* about nature; they are a human language which translates nature's whispered clues about the world. But they are not nature itself. To ask physics for more is to ask too much.

And here Einstein parted company with his contemporaries. He could (and did) appreciate the tremendous successes of quantum mechanics. He did not believe, however, that it was the final word. His most famous and oft-quoted statement on the theory was: "God does not play dice!" He still believed in the possibility of finding a model of reality which was causal and determined, not a thing of probabilities influenced by human experiment and observation on the microlevel. Beneath quantum mechanics he hoped to find an "objective reality" within a deeper theoretical construction—perhaps even a "hidden variable" which would reintroduce causal determinism to that level. He conceived of all sorts of paradoxes, not to refute quantum mechanics but only to show that it was incomplete. Yet Bohr and others showed that the paradoxes were natural outgrowths of a consistent quantum theory. The quantum haunted Einstein's attempts to formulate a Unified Field Theory. When he died in 1955 in Princeton, having come to the United States as an exile from Nazi Germany, his quest lay unfulfilled.

Quantum mechanics had been constructed within the domain of electrons and photons. Meanwhile Rutherford continued his assault upon the mysterious atomic nucleus. Finally, in 1919, firing a powerful beam of alpha particles into nitrogen, he accomplished the first *artificial* transmutation in history (excepting whatever the alchemists did or did not do). The nitrogen nucleus swallowed an alpha particle, became unstable, and ejected a positively charged hydrogen nucleus, leaving an oxygen isotope (O_{17} with charge 8). Rutherford already suspected that all atomic nuclei contained a positive hydrogen nucleus, and in 1920 he named this positive subatomic particle the *proton*. The proton accounted for most of the atom's mass and had one unit of positive charge. Matter was built from electrons and protons. . . .

Until 1932! What accounted for isotopes—the varying masses or atomic weights of elements having the identical charge? Were there more protons in the nuclei whose excess charge was somehow canceled by "nuclear electrons"? The helium nucleus, for example, had a charge of two yet a mass of four. Rutherford suspected that the helium nucleus contained two protons only and no electrons. Thus there had to be some other particle, uncharged, having nearly the same mass as a proton. In 1932 Rutherford's student, James Chadwick, found the missing particle by bombarding beryllium with alpha rays. It was christened the *neutron*, a chargeless particle nearly as heavy as the proton.

The next problem was to explain natural radioactivity. The nucleus obviously emitted particles *spontaneously*, and no one could predict when the nucleus of a single atom, say a single radium nucleus, would decay. So again physicists called upon statistics. Within a given period of time and for any given quantity of material, physicists found that exactly half its atomic nuclei will have disintegrated, emitting radioactivity. This is called the half-life of a radioactive substance. For uranium it is about 4.5 billion years, nearly the age of the earth. For an individual nucleus it is random; the half-life is a statistical prediction for large numbers. But *how* does the nucleus decay?

It was becoming obvious that there are energy levels within the nucleus itself—two repulsive positively charged protons were somehow bound together in the tiny nucleus by a tremendously "strong force" which overcame their mutual repulsion, and the nature of radioactivity seemed to indicate jumps going on inside the nucleus. Yet experiment seemed to indicate that the nuclear particles did not have the required energy to escape the nucleus. How did they get out? In the late 1920s physicist George Gamow found the answer in the newly created quantum mechanics. Let us consider the nuclear particles in terms of waves. Imagine the nucleus itself to be a kind of deep well and the matter waves within the well to be vibrating at some definite energy level. Now this wave vibration may penetrate the walls of the well, analogous to sound waves pentrating the walls of a room. But the waves are probability waves, and thus their penetration is in terms of probability. The greater the amplitude of the probability wave *outside* the wall, the greater is the possibility of finding a nuclear particle outside the nucleus. The phenomenon, called quantum tunneling, explains the statistical nature of natural radioactivity. It is another example of quantum randomness. Yet electrons also came flying out, and electrons were not only too large (roughly) for the nucleus, they also upset the idea of the nuclear energy ladder.

In 1928 Paul Dirac returned to the electron. He was looking for a mathematical equation which would bring its wave properties into line with relativity. When Dirac calculated the energy levels using his relativistic quantum mechanics, he discovered two sets of solutions, one positive and the other negative, and this seemed to indicate that there were negative energy levels or states. But if there are negative energy states, then we should expect electrons to fall into them and hence disappear. Such obviously did not happen, and since Dirac chose not to dismiss the negative states, he had to explain them.

Dirac reasoned that electrons were fermions and thus obeyed the exclusion principle. Therefore, the negative energy states were actually full of negative-energy electrons! If that sounded a bit strange, Dirac's next hypothesis was almost unbelievable (but not for quantum mechanics): given enough energy, one of these "negative" electrons should jump to a higher level, high enough to become an ordinary electron— pump enough energy into the invisible negative sea, and out comes an electron of ordinary mass and charge. Now real electrons carry a negative charge; a newly promoted electron should leave a "hole" in the negative-energy sea; and since this "hole" is really the absence of a negatively charged particle in a negative-energy sea, well, the "hole" should behave like a *positively charged electron*. It could not be a proton, since it must have the same mass as an electron.

At first it all seemed a kind of mathematical chicanery, these so-called anti-electrons, until in 1932 in California Carl D. Anderson discovered the anti-electron while investigating cosmic rays, energetic particles of enormous energy bombarding the earth from space. In them Anderson found particles with the same mass as the electron but with a positive charge, *anti-electrons* or, as they were later named, *positrons*. This was the first discovery of the phenomenon of *antimatter*, a kind of mirror image of normal particles only with opposite signs. In principle, any elementary particle may be created from energy accompanied by its antiparticle, and when matter and antimatter collide, there is a mutual annihilation of particles in a burst of energy. Elementary particles were no longer immutable, for they could be "created" (more than two hundred have been created in high-energy accelerators since) and they could be unstable, decaying into radiation and other particles.

Thus a new vision of elementary particles emerged from Dirac's theories. Elementary particles themselves transmuted; energy converted to mass or vice versa, as Einstein had shown; particles interacted, creating new particles and energy; yet charge and energy remained balanced. So some particles could decay into others, which was the answer to the electron problem of radiation: A free neutron having tunneled out of the nucleus decayed into an electron, a proton, and . . . to balance the energy books Pauli postulated yet another particle, which in 1933 became the *neutrino*. A neutron decayed into a proton ($+1$) by shedding an electron (-1) and a neutrino. Particle decay thus entered the lists of elementary particle description, and only the electron, proton (so far), neutrino, and photon were observed to be stable. Also, a new force was held responsible for nucleur disintegration, the "weak force" which governs natural radioactivity.

In the 1940s another strange phenomenon sprouted from Dirac's quantum field: Physicists were forced to postulate something called a "virtual particle." Imagine a normal elementary particle surrounded by an invisible cloud of "unreal" ghost particles; suddenly, one of these "unreal" particles flashes in and out of existence for a brief instant. It is created *ex nihilo*, out of nothing, out of the vacuum. Where does the energy come from in order to promote these "spooks" into reality? The answer was Heisenberg's uncertainty principle, which also applies to energy and time: For brief time intervals, again when we are not looking, there is a great enough uncertainty in energy to accomplish the creation. Hence we have virtual particles existing for brief instants according to the uncertainty of energy and time. For example, an electron may be thought of as accompanied by a ghostly cloud of virtual photons; in a brief enough period of time there exists an uncertainty of energy, and for this instant energy conservation may be violated, and one of the ghost photons promoted into a "real" photon. Almost as quickly as it is created it must be absorbed to balance the energy books. Thus we have an electron surrounded by a seething cloud of virtual photons popping in and out of existence. In fact, an elementary particle can suddenly erupt into a whole batch of virtual particles, interacting and recombining in a brief instant of energy uncertainty, and then all of them disappearing. And, theoretically, at least, virtual particles may suddenly spring into being out of empty space! But why have them?

If we imagine an electron surrounded by a cloud of virtual photons, and the electron drops from a higher to a lower energy state in an atom, we may visualize the excess energy given off to a virtual photon, which then flies free. In the same way an electron may absorb a free photon and so energy. Thus a force based upon this virtual-particle exchange is established, the virtual photon creation and annihilation setting up a field around the electron. This force is none other than the electromagnetic force which "glues" the atom together, and in the quantum sense it may be conceived in terms of photon exchange. The quantized field is called *quantum electrodynamics (QED)* and it describes the electromagnetic interaction in terms of quantum particles—photons (see Figure 24.1). Roughly speaking, we imagine two electrons repelling by the exchange of photons, emitting and absorbing photons. The creation and annihilation of virtual quanta accounted for slight discrepancies in the energy of an "orbiting" electron (the original reason for them). In general, all forces between particles, both attractive and repulsive, may be

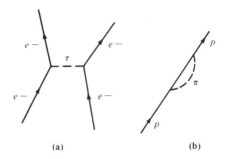

Figure 24.1   (a) A Feynman diagram of electron interaction by photon exchange. (b) A proton creating and absorbing a virtual pion.

thought of as virtual-particle exchange. Dirac's unification of relativity and quantum gave birth to a whole new physics, relativistic quantum field theory.

One problem with QED was that, in setting up equations corresponding to the electron plus its cloud of virtual particles, the solution of the equations gave "infinite" answers: mass, charge, and energy. Physicists were able to ignore these infinities by a kind of mathematical trick called "renormalization" in which the infinities canceled. Some, however, like Dirac, did not approve of this seemingly arbitrary procedure. As yet there seems to be no solution that everyone can agree upon.

Ironically, the new physics strangely altered the old vacuum idea. The vacuum now became a kind of Cartesian plenum, but not vortices of lifeless matter, rather a seething ocean of creation and destruction governed by quantum randomness. It is a vacuum and not a vacuum, the being and not-being of virtual particles all mixed in one, almost as if physics had finally found the answer to Parmenides.

Back in 1935 the Japanese physicist H. Yukawa had postulated an exchange particle for the strong force in the nucleus which binds together the repelling protons. His reasoning, however, was based upon the formation of molecules by exchange of electrons. He called his particle a *meson*, and in 1946 it was found in cosmic rays and was named the *pi-meson* or *pion*. And so another force was found to have an associated quantum particle; in the case of the strong force it is the exchange of virtual pions betwee protons and neutrons. The world of elementary particles was becoming crowded, and worse was yet to come.

In 1932 E. O. Lawrence designed the first machine, called the cyclotron, to probe the far reaches of matter. Using magnetic fields to deflect particles, it could accelerate them in a circle, giving them tremendous energy, and send them smashing into nuclei. In the 1950s and 1960s new and more powerful accelerators spewed forth all sorts of new particles whose lifetimes were but an instant. Here was the beginning of another new physics, high-energy particle physics. New quantum states of matter were discovered at high energy levels. Created from the proton beams, these new forms were collectively called *hadrons*—referring to all elementary particles participating in the strong interaction—and there seemed to be no end to them. The once uncuttable atom had disintegrated into realms of transmuting particles beyond the wildest dreams of any alchemist.

And in its transmutation the atom itself yielded powers beyond that which any amount of alchemical gold could purchase. The story has been told many times before: It incorporates the Second World War, the Nazi threat (imagined or real) of gaining nuclear energy, the exiles from fascism who came to the United States, men such as Einstein and the Italian physicist Enrico Fermi, Leo Szilard, and others. Beginning in 1943, J. Robert Oppenheimer organized in Los Alamos, New Mexico, one of the greatest "weddings" of theoretical and practical or applied physics.

Bombarding a uranium nucleus with neutrons broke it apart, liberating enormous amounts of energy as mass transmuted. In a discussion with Bohr, Fermi had mentioned the possibility that this could be made into a chain reaction in which additional neutrons are emitted, causing other uranium nuclei to transmute. The process is called *fission,* and it can be pictured as pumping neutrons into a "water-drop" nucleus until it bursts. The technical difficulties were stupendous, for the chain reaction had to be initiated with slow neutrons and the fissionable material was a rare uranium isotope, U-235. Yet these and many more difficulties were overcome, and one of the primordial powers of nature fell into human hands.

Lighter nuclei such as hydrogen will unite at sufficiently high temperatures in a much greater burst of energy. This is called *fusion,* the source of stellar energy and the answer to Lord Kelvin. It, too, was obtained here on earth, using the temperatures generated by fission of uranium. Both processes realized Einstein's $E = mc^2$; both resulted in atomic and hydrogen bombs.

Gold was once the road to power. It was an indirect road, and that power was ephemeral. The modern alchemists have sought and found the primal essence of power. Whether modern humans will also seek the spiritual wisdom of the ancient alchemists remains to be seen.

SUGGESTIONS FOR FURTHER READING

GAMOW, GEORGE. *Thirty Years That Shook Physics: The Story of the Quantum Theory.* New York: Dover, 1985.

GUILLEMIN, VICTOR. *The Story of Quantum Mechanics.* New York: Charles Scribner's Sons, 1968.

HEISENBERG, WERNER. *The Physical Principles of the Quantum Theory,* trans. Carl Eckart and Frank C. Hoyt. New York: Dover, 1949.

————— . *Physics and Philosophy.* New York: Harper & Row, 1958.

HOFFMAN, BANESH. *The Strange Story of the Quantum*, 2nd ed. New York: Dover, 1959.

JAMMER, MAX. *The Conceptual Development of Quantum Mechanics*. New York: McGraw-Hill, 1966.

MEHRA, JAGDISH, and RECHENBERG, HELMUT. *The Historical Development of Quantum Theory*, 4 vols. New York: Springer, 1982.

MOORE, RUTH. *Niels Bohr: The Man, His Science, and the World They Changed.* Cambridge, Mass.: MIT Press, 1985.

PAGELS, HEINZ R. *The Cosmic Code: Quantum Physics as the Language of Nature.* New York: Simon and Schuster, 1982.

25

Things Past, Things to Come

Cosmology

Once they were simply philosophers, walking the dusty paths of the market in Athens or sitting in the cool evening shade of the Academy. Today they are specialized: astrophysicists, astronomers, particle physicists, and so on. Today they inhabit the great research centers, the universities, the company laboratories, the military. They command immense resources (and never enough) in order to construct their massive accelerators, giant telescopes, and satellites. In Plato's day there were the gods and the giants; now, roughly speaking, there are the pencil-and-paper theoreticians and the experimentalists. And yet all, past and present, still ask the same questions: How did the universe begin? What is it like? How will it end? Or, is it eternal?

Pascal shuddered before the prospect of an infinite universe. Newton and others welcomed it: an infinity of stars spread more or less uniformly through infinite Euclidean space. Such a universe befitted the majesty of the Creator. However, infinity is a paradoxical concept. How may universal Newtonian gravity be combined with a universe of infinite mass? Infinite stars must shine with infinite radiation. With the discovery of the gas laws in the nineteenth century, scientists could easily imagine stars like particles of a gas, running off into infinity in a kind of cosmological entropy.

While physicists grappled with the dynamics of infinite space, astronomers studied its structure. In 1750 Thomas Wright of Durham speculated vaguely on the structure of the Milky Way, but it was at the turn of the century that the great William Herschel, guided by painstaking observations, formulated a precise picture of the Milky Way as a disclike island of stars with the earth in the center.

Beyond this island galaxy, our home, observation revealed dim and wispy patches of light, not unlike the Milky Way before Galileo's telescope had resolved its luminous fluid into stars. Kant had proposed that these patches, called nebulae, were galaxies like the Milky Way. Herschel agreed, although the fact that he could not resolve some of the nebulae led to him to suppose that they were a kind of unknown "shining fluid." Other objections were soon raised, and throughout the nineteenth century the arguments shifted back and forth over the existence of these "island universes." One of the greatest problems was estimating distances. Nebulae could just as well be within our galaxy.

An important breakthrough came in 1912. Henrietta Leavitt and Harlow Shapley of Harvard discovered that there were certain stars, called Cepheid variables, whose uniform periods of varying luminosity could be used to calculate distances. The Cepheids were, in short, cosmic mileage signs. In 1923, on Mount Wilson above Los Angeles, Edwin Hubble fixed a giant one-hundred-inch telescope upon Andromeda, the closest nebula. He was able to resolve Andromeda into its stars, and among them he found the Cepheids. Hubble calculated the distance of Andromeda to be nearly a million light years away (the distance light travels in a year is a light year). Since the Milky Way was estimated to be about 100,000 light years wide, Hubble was convinced that Andromeda and the other nebulae were island universes. Shapely placed the sun near the edge of the Milky Way, and recalculation has today doubled Hubble's value for Andromeda. Nonetheless, it was agreed after some debate that the nebulae were island universes or galaxies. Today there are thousands of millions of galaxies; the Milky Way with its roughly billion stars is a rather typical one. Pascal indeed had cause to shudder.

Yet Einstein had claimed in 1917 that the universe was finite. All along he had intended General Relativity to apply to the universe as a whole, and in 1917 he took the first steps towards a relativistic cosmology. Yet infinite space caused relativistic trouble and clashed with the Machian principle of inertia. How could inertia, relative to all other masses in the universe, not become infinite? Thus Einstein proposed that the universe was finite as to its spatial coordinates, closed but without boundaries. The problem was to fit his equations to the model.

Banishing infinity created a strange situation; the universe became unstable as if the galaxies were rushing off. Like most cosmologists in 1917, Einstein believed that the universe had to be static. But to achieve a static closed universe he was forced to insert a small quantity into his equations, a cosmological constant he denoted by the Greek letter lambda.[1]

Nonetheless, there proved to be other solutions and hence cosmological models. In Holland, Wilhelm de Sitter found a static-universe solution, but one which implied the universe was empty of matter! Worse, de Sitter's universe began to expand with the addition of mass. In the early 1920s the Russian mathematician Alexander Friedman found a whole range of solutions which gave static, expanding, contracting, and expanding/contracting universes. And in 1927 Georges Lemaitre of Belgium constructed a model incorporating all three types—static, expanding, de Sitter—one following the other. Thus a variety of possible models existed, depending upon the value of the cosmological constant.

Back in 1914 Vesto Melvin Slipher had reported a peculiar discovery to the American Astronomical Society meeting in Evanston, Illinois. Slipher had detected the spectroscopic Doppler effect of light wavelengths for some dozen galaxies and found them shifted toward the red. What could it mean? Were the galaxies actually rushing away? In the audience sat Edwin Hubble. Later he heard of de Sitter's strange results, and in 1929, pursuing the clues, Hubble found that every galaxy whose distance he could measure showed the red shift. Further, he calculated a proportionality between their distances and rates of velocities—and found that the greater the distance the greater the velocity of recession! This ratio, the same for all galaxies, is called Hubble's constant. If the universe is the same everywhere, homogeneous, then Hubble's constant must hold for every galaxy as seen by every other. In short, observation seemed to indicate an expanding universe.

The red shift alone did not *prove* that the universe was expanding, yet Einstein, who initially found the concept repugnant, eventually dropped his cosmological constant. In relativistic terms we must imagine the expansion as an inflation of space itself, like expanding a balloon. The galaxies may be seen as dots uniformly distributed on the surface of the balloon, rushing away from each other because the rubber surface itself expands.

So if the universe is expanding, then it must have expanded from something. Simply put, it is logical to suppose that the expanding universe had a beginning in time! Here was a strange situation indeed for

---

[1]Physicist John Wheeler has emphasized the possible philosophical motivations behind the cosmological constant. Einstein was a student and admirer of the seventeenth–century philosopher, Benedict de Spinoza, who taught the perfect eternity of the universe and denied the Biblical story of creation. See John Archibald Wheeler, "Beyond the Black Hole," in *Some Strangeness in the Proportion: A Centennial Symposium to Celebrate the Achievements of Albert Einstein* (Reading, Mass.: Addison-Wesley: 1980), p. 354.

modern cosmologists, for such a beginning further implied a distant point in time when all matter was compacted together. And the Friedman models also implied a number of possible scenarios for the future. If the composite value of matter in the universe is above a certain value, the expansion of space must halt at some point in time and the universe then begin to contract—cosmologists call this the "closed future." If matter is below the critical density, the universe will continue expanding to infinity—the "open future." But computing the mass density of the entire cosmos is no easy task! There may exist unknown masses, and since no signal travels faster than light, we can only know events within the parameters of light reaching us from the beginning of the universe. Anything else is beyond our *horizon*. The problems are by no means solved; answers exist mainly in the form of opinion polls among cosmologists.

The idea of creation was even more interesting—and controversial. Since the galaxies are receding from us with speeds proportional to distances, in principle we should be able to calculate backward and find a figure which would give us a rough approximation of the age of the entire universe. Hubble's initial value was small, from one to two billion years. Again geology howled: how could the earth be older than the universe? Other cosmologists, none too happy with the prospect of creation (rather theological!) and concerned with the age problem, suggested static model alternatives which would incorporate the red shift. Such a model was proposed in 1948 by three Cambridge scientists, Herman Bondi, Thomas Gold, and Fred Hoyle. Matter, they suggested, was being continually created at random throughout the universe at a given rate, and although the mass is infinitesimal, it still fills up the gaps left by receding galaxies. Hence at any given time in a large area the number of galaxies remains constant—no beginning, no end. Here was cosmological uniformitarianism.

Improved calculations stretched the Hubble age (today it is about 15 billion years, although this is more of an average best guess). Nonetheless, until the 1960s the various static alternatives, such as continuous creation, were still viable. Yet, like their medieval ancestors, some cosmologists found their imaginations equal to the task of going beyond the new Aristotelianism.

If the expanding galaxies were once compacted together, we may well imagine a rough analogy with thermodynamics. As we calculate backward, the overall temperature of the universe becomes greater and greater until we reach a kind of cataclysmic explosion. George Gamow first suggested that initial creation may have been this fiery explosion, a "Big Bang." In 1948 Ralph Alpher and Robert Herman calculated the radiation value as the universe cooled. Their reasoning went roughly like this. For some time after the Big Bang, say the first 500,000 years or so, the temperature was so hot that even atoms could not exist; rather, electrons and nuclei were in constant collision with photons in the given volume of the universe. (Before three minutes the nuclei would not exist and we would be in a universe of quantum elementary particles.) Now according to statistical mechanics there is a thermal equilibrium of both matter and radiation, and Planck's formula for black-body radiation gives us the distribution of energy with wavelength depending upon that temperature. So as the universe expands and the temperature cools, we can follow Planck's distribution for universal radiation to the present. In short, Alpher and Herman predicted a radiation background still bathing the universe from the time when radiation reacted with matter in equilibrium—like the afterglow of a heated iron. Their prediction was very small: microwave radiation at about 5° Kelvin. Few cosmologists believed them.

Then in 1964 an unexpected event occurred at the Bell Laboratories in New Jersey. Robert Wilson and Arno Penzias were using a new radio antenna built for satellite communications to survey radio sources outside the Milky Way. They discovered an unusual amount of background "noise"—that is, microwave radiation coming from all directions (although they first suspected pigeons nesting in the antenna horn). At the same time P. J. E. Peebles was recalculating the supposed background radiation from the Big Bang. The figure turned out to be about 3° Kelvin, close to the value Wilson and Penzias had found. The implication was clear: Humanity for the first time was listening to the rumblings of creation!

Thus the Big Bang, or some variation of the theory, became for most cosmologists the working model for the early universe. Yet problems still exist. Why is there matter rather than antimatter (as far as we can tell)? How did the galaxies form? Will the Big Bang end in a "Big Crunch"? Will there be another expansion, a "Big Bounce"? Working out a physics of creation is inherently difficult. And yet many physicists believe that in those early moments of creation may be found the clues, perhaps even the answers, to the grand unification envisioned by Einstein. And here our history becomes open, flowing into the present toward possible futures. Let us briefly glance up the road, aware that we now enter the theoretical present in which judgment must be suspended.

One branch of that road begins with stars themselves. Laplace had speculated that if a star was massive enough, even light could not escape its enormous gravity. In the 1930s with general relativity this speculation became an astrophysical theory. Imagine a star as it burns out. A number of things may happen, generally depending upon the size of the star: It may die in a cataclysmic explosion, a nova; it may swell and die, eventually becoming a black dwarf; or it may explode leaving its atoms crushed into neutrons, an immensely dense neutron star. (Pulsars, discovered in the 1960s, are thought to be neutron stars.) It is believed that all heavier elements, other than hydrogen and helium, are the debris of star death. Nevertheless, there is another possibility—total collapse.

As the star burns out, its internal thermal pressure can no longer support the star's own massive weight, and the star begins to shrink. Simple Newtonian calculation shows that this increasing contraction results in greater surface gravity as surface area decreases while mass remains constant. According to general relativity spacetime itself must become distorted. The collapsing star's gravity eventually becomes so great that even the exclusion principle is overpowered and atoms break apart, causing the star to become a heavy mass of neutrons. If the star is massive enough, contraction continues, and eventually the escape velocity at its surface reaches the speed of light. Therefore, light itself cannot escape, and the light cones of an event at the surface are tilted toward the star so that at a certain radius, called an *event horizon*, nothing can be known about the star—no information about events inside the horizon can escape, since nothing can surpass the speed of light. To an outside observer (fortunate enough not to be dragged in) the star has become absolutely black, no light escaping, a *black hole* in space. Yet within the black hole collapse still continues until . . . .

Time reaches zero and density becomes infinite! And thus the curvature of space must become infinite too—an infinite mathematical point of infinite sharpness called a *singularity*. Surely something is wrong with the theory, for when physical values become infinite, physics itself ends and its laws cease to apply. Yet in 1965 Roger Penrose demonstrated mathematically that General Relativity cannot avoid singularities (they had troubled Einstein), and—collapsing star or not—spacetime itself must become torn apart by a point of infinite sharpness, the singularity. Then in the 1970s Stephen Hawking produced a theory which predicted that black holes would radiate elementary particles—matter would come out of them. At Cambridge, Hawking applied quantum mechanics to the black hole's event horizon and demonstrated, using the uncertainty principle, that a black hole could emit particles, since its enormous gravitational field may supply the necessary energy for quantum interactions. This black-hole radiation is called "Hawking radiation," and Hawking also showed that the black hole could erode over time and even explode into high-energy gamma rays. Finally, Hawking and Penrose demonstrated that General Relativity led inexorably to a singularity at the beginning of the universe, a point at the very instant of creation which included all the matter of the universe, all energy, its spatial dimensions, and even time. The physical universe may have exploded into existence from the physically unthinkable, the singularity, and may end there as well.

In those first fiery moments of creation cosmology and the quantum meet. The tremendous energies of the early minutes of the fireball universe can be in part duplicated by accelerators, but there remains an era beyond which experiment cannot reach, where no earthly accelerator is able to duplicate the energies of creation. It is generally held that up to the first three minutes after the Big Bang no nuclei could form, and henceforth the era is one of elementary-particle interactions. Pushing back the cosmic clock, thus raising the energies involved, makes it possible to develop quantum field theories which predict ultimate unification of the four forces.

Quantum electrodynamics may serve as a starting point. The electromagnetic field is quantized as electrons surrounded by their ghostly virtual photons, whose exchange generates the force. The range of electromagnetism is infinite, hence the exchange particle must have zero mass—the photon. The only other force of infinite range is gravity. Nonetheless, we may imagine that it too can be characterized by a quantum exchange particle, also of zero mass. Let us call it the *graviton*. No graviton has been observed, and of course quantum gravity is quite different from Einstein's geometrical relativity. Then there are still the two nuclear forces to consider.

The success of quantum electrodynamics led physicists in the 1950s and 1960s to attack the strong force and find a field theory for strongly interacting particles, the hadrons. Unfortunately the problem was complicated by the various types of hadrons which came pouring out of accelerators. Most were unstable and decayed quickly, but classifying them was a problem. Besides interaction, classification is done by decay products, spin number, and charge. Besides the electric charge, hadrons were found to carry something called "strangeness." Finally, in 1961, Murray Gell-Mann of Caltech found a pattern to the hadrons, classifying them into families by a mathematical symmetry called the "eightfold way." What accounted for this symmetry? Gell-Mann postulated that the hadrons were really built from more fundamental particles he labeled *quarks* (from James Joyce's *Finnegans Wake*—"Three quarks for Muster Mark!"). The idea was that all hadrons could be constructed from different "flavors" (simply a label to distinguish them) of quarks and antiquarks. No quark could be separated from a hadron and none could be created in the laboratory. They are confined to hadrons rather like Thomson's old raisin in the pudding.

In the 1970s, then, the strong force was reduced theoretically to a quantum field theory of quark binding based upon particle exchange. The term "gluons" was adopted to designate all exchange particles, and the new quantum field theory of quark binding was called *quantum chromodynamics*, since it seems that quarks and antiquarks come with a set of three charges called "colors" (hence "chromo"), and exchange gluons bind to these color charges. The weak interaction was believed mediated by gluons called *vector bosons*. Thus each force can now be seen in terms of exchange quanta, and each at least theoretically may be described by a quantum field theory. The task is to unify these fields, and a new mathematical tool has been developed to accomplish this. It is called *gauge field theory*.

Modern gauge field theories derive from the mathematical concept of symmetry. Say we have two different atoms, and no matter how we rotate them there is no measurable effect upon the individual atoms themselves. Each is completely symmetric. Yet when joined to form a molecule, their symmetry is broken as the electrons of each shift, and a nonsymmetric molecule is formed. Heating the molecule, however, reduces it to its constituent atoms, which once again reveals their symmetry. A similar situation may be imagined for two points on a piece of rubber; no matter how we rotate the piece, the measured scale between the two points is unaffected. Stretching the rubber changes the scale or gauge. Gauge symmetry is an abstract transformation theory involving gauge fields which allows us to restore the invariance of the measurement. Roughly speaking, it permits us to "see" or restore the original symmetry even when it is spontaneously broken.

Using our atom-molecule analogy, the basic idea is that at low energies the four interactions are "frozen" and their underlying symmetry concealed. In 1967 Steven Weinberg and independently Abdus Salam produced a gauge field theory for the weak and electromagnetic interactions. The two interactions were seen to be exchanging the same family of gluons, which appear different once their symmetry is broken spontaneously at low energies. In the 1970s it was predicted that under symmetry all four gluons are massless; once symmetry is broken, however, three take on mass (the vector bosons—W^+, W^-, Z) and the fourth remains at zero—the photon. In 1983 in Rome a powerful accelerator produced what is believed to be the predicted W-particle or vector boson. Hence it now appears that two of the interactions have been unified— the electroweak.

Roughly the same basic concepts apply to the theory of strong and electroweak unification called the Grand Unification Theory. In this case the exchange particle is called the X-particle; however, the energies required to reveal this symmetry are much too high for the laboratory. Yet the theory does provide one means of testing; it predicts that protons are ultimately unstable and may decay. Experiments have been set up and physicists are waiting.

Creation may be ultimately seen as a series of successive symmetry breaking or freezing as energies dropped. For an amazingly brief moment after the Big Bang—called "Planck Time," 10^{-43} sec—the ultimate unity of forces, including gravity, may have existed in the form of one family of elementary particles and one force. The beautiful goal of simplicity and harmony, the quest begun long ago by the ancient Greeks, may lead to that first instant of creation as described by gauge theories. Yet opinions are still divided.

It has been a long road indeed, and while many questions remain unanswered, here our story must end. Ahead we hear the soft voices predicting the ultimate unity and simplicity—predictions we have heard many times before throughout our story. Perhaps we have now learned that is it an untiring faith in the human ability to understand the universe which makes such predictions possible. In this faith we take pride; from it we learn humility.

SUGGESTIONS FOR FURTHER READING

JAKI, STANLEY L. *The Milky Way: An Elusive Road for Science.* Devon, England: David and Charles, 1973.

MUNITZ, MILTON K. *Space, Time and Creation: Philosophical Aspects of Scientific Cosmology,* 2d ed. New York: Dover, 1981.

NORTH, J. D. *The Measure of the Universe: A History of Modern Cosmology.* Oxford: Clarendon Press, 1965.

SMITH, ROBERT W. *The Expanding Universe: Astronomy's "Great Debate" 1900–1931.* Cambridge: Cambridge University Press, 1982.

TREFIL, JAMES S. *The Moment of Creation: Big Bang Physics From Before the First Millisecond to the Present Universe.* New York: Charles Scribner's Sons, 1983.

WEINBERG, STEVEN. *Gravitation and Cosmology: Principles and Applications of the General Theory of Relativity.* New York: John Wiley & Sons, 1972.

————— . *The First Three Minutes: A Modern View of the Origin of the Universe.* New York: Bantam Books, 1977.

26

~~~~~~

The Secret of Life

Physics had passed through the fires of revolution; no longer could we mechanically visualize the macrocosmic world of relativity and no longer did causal determinism apply to the fleeting microworld of the quantum. Some physicists, philosophers, and popular writers were even ready to proclaim the end of strict objectivity and reductionism. The issues are still debated. Biology, on the other hand, followed a different course. Here there was no shattering revolution, rather a synthesis, from which many biologists emerged dedicated adherents of mechanism. The old nineteenth-century debate between vitalism and mechanism gave way in the twentieth century to the question of reductionism or holism. Can life be reduced ultimately to the categories of chemistry and physics, or, even though these methods prove successful, does not a complete understanding of life require a qualitative distinction between the properties of its parts and the emergent characters arising from their interactions? One question, however, did not change: what was the mechanism of evolution?

At the turn of the century the German Darwinist Ernst Haeckel published a work entitled *The Riddle of the Universe*—a "riddle" which Haeckel believed had been solved. All nature, wrote Haeckel, was governed by "eternal iron laws" of which natural selection was the foundation, physics and chemistry the motor, and the study of morphology or form of the species the methodological scaffolding. Darwinians, then, were interested in searching for common ancestors and building evolutionary family trees—phylogenies. The method they used was observation, deduction, and comparison, and their causes were ultimate causes. Haeckel's own law of recapitulation, for example, stated that the individual development (ontogeny) of the embryo recapitulates the adult stages of its evolutionary history (phylogeny). The embryo developed this way because of its evolutionary history—an ultimate cause.

Yet during the second half of the nineteenth century embryologists and physiologists had gone to the laboratory asking *how* did the embryo develop? Wilhelm Roux, who was influenced by Haeckel's ideas, found such explanations too vague and sought the chemical processes which operated on the cellular level. He called his method "developmental mechanism"; its method was to replace observation and deduction with experiment. Another German physiologist, Jacques Loeb, envisioned the goal of biology to be the reduction of all living phenomena to interacting molecules. He declared that the methods of physics and chemistry were the *only* methods by which to study the life sciences. Hence one argument against Darwinism in general was its methodology and mode of explanation.

Another argument involved inheritance and variation. Without a solid theory of inheritance many biologists turned to various forms of what may be called Neo-Lamarckism: direct environmental influences, inheritance of acquired characteristics, use and disuse. Other arguments revolved about the question whether variation was discontinuous—nature making huge jumps—or continuous and infinitesimal, as Darwin believed. Francis Galton's statistical approach to hereditary patterns in populations presupposed continuity and blending inheritance. But in 1894 William Bateson announced that natural selection, if true, could work only upon large and discontinuous variations. Ironically, the rediscovery of Mendel would initially confuse the problem, as the early geneticists were more prone to think in terms of genotypes than populations.

And there were good philosophical reasons for rejecting natural selection in favor of Neo-Lamarckism, owing mostly to the latter's implied progressionism. Teleology in general was a difficult thing to surrender. In 1907 the French philosopher Henri Bergson proposed the concept of an *elan vital,* a spiritual force injected into matter, driving it upward to higher states of complexity and consciousness. The French Jesuit Pierre Teilhard de Chardin developed during the Second World War a religious evolution of cosmic proportions. As matter becomes more complex, so does consciousness (all matter is to some degree conscious!), developing in a predetermined cosmic direction toward what Teilhard labeled the *Omega*—Christ. Teilhard's evolution claimed a wide allegiance even as late as the 1960s.

An interesting example of the influence of philosophy (or ideology) upon biology occurred in the Soviet Union in the 1930s. While Russia had been one of the focal points in the synthesis of Mendelian genetics and Darwinism, especially in the school of population genetics, the belief arose that the rejection of acquired characteristics contradicted the Marxist primacy of environmental conditioning. Thus, even after new theories of inheritance had removed the need for Lamarck, T. D. Lysenko with Stalin's support made Neo-Lamarckism the official creed of Russian biology.

Returning to the last decades of the nineteenth century, we find, then, a great many reasons for the so-called "eclipse of Darwinism." Again—biologists no longer doubted evolution, rather Darwin's mechanism. Yet in the last decade of the century August Weismann, a physician whose failing eyesight forced him to theoretical studies, proposed what proved to be a prophetic answer. It lay in the cell. Weismann ruled out soft inheritance; inheritance was a stable transmission of material particles found within the nucleus of the germ cells, and it was *biochemical*, the particles being of a specific molecular composition having definite material arrangements in the germplasm. The body was like a vehicle, determined by the germplasm yet carrying the immortal genetic material into the future. Weismann accounted for variation—upon which selection acted—by the recombination or rebuilding of the immortal parental particles in the gamete. What, then, were these material particles?

Observation had shown that while the body cells divided *after* the duplication of chromosomes in the nucleus—preserving their number— the sexual cells divided their number in half to be fused again in fertilization. Thus in 1895 E. B. Wilson wrote that "chromatin" must be the chemical compound of inheritance. Yet all of this was still merely speculation. Nevertheless, Weismann's Neo-Darwinism, as it was called, with its insistence upon hard inheritance, served to draw sharp lines between Lamarckism and selection, joining this issue to the battle over continuous or discontinuous variation. Then came further confusion—the "rediscovery" of Mendel.

Actually the rediscovery was more of a new realization of Mendel's significance. From March to May of 1900 three biologists working independently with plant heredity recognized the importance of regular segregation and independent assortment of parental characters as well as Mendel's use of statistics. Now Mendel's independent "traits" implied discontinuous variation and hard inheritance; if they were material particles, they also implied some sort of preformation! Thus one of the three biologists, Hugo de Vries, suggested that evolution proceeded by large mutations, instantaneous jumps. But Darwinians believed in continuous variation, blending, and soft inheritance. The biometricians— Galton's followers—dealt with the statistical analysis of populations, not individuals. Further, de Vries admitted selection as mainly a negative factor; mutations may be exterminated by it, yet selection could not account for the creation of new species. Many at first believed that Mendelism had actually killed natural selection!

In 1902 Bateson coined the word "allelomorphs" (alleles) to represent the pairs of unit characters contributed to the offspring by the parents. Then Wilhelm Johannsen drew the distinction between hereditary particles he called *genes* and the observed adult form. Inheritance did not consist of traits themselves but of genes, which were actually potentials for these observable characters; that is, the genotype was potential while the phenotype was the actual observed organism. Preformation was thus reformed: the germ cell did not actually carry a character, rather it held codes, genes, some of which were not expressed in the phenotype yet inherited. And in 1906 Bateson found that some groups were inherited together—linkage groups. But what was the actual material mechanism? How may we test the theory?

Chromosomes come in pairs, we know their exceptional behavior in germ cells, and in 1905 it was shown that sex itself was linked to the chromosomes (X and Y). While there are surely more genes than chromosomes, linkage groups do suggest that genes may be "parts" of chromosomes. Bateson would not accept this materialism! In the United States, however, T. H. Morgan began breeding the so-called fruit fly, *Drosophilia melanogaster*, in his laboratory at Columbia University. The flies told the story.

Morgan discovered a strange mutation, a white-eyed *male* fly (normally the eye was red) which was not a new species. Continual breeding and the following of the trait through generations indicated that the mutant white eye was sex-linked. A sex-linked gene and the inheritance of sex through specific chromosomes implied that genes were "located" on the chromosomes. The next logical question was: Could the inheritance of certain genes together occur because of their position on the chromosome? In 1915 Morgan and his students published their work, *The Mechanism of Mendelian Heredity*, which showed that this was indeed the case. In fact, they were able to draw chromosome maps showing the relative loci of genes; they discovered complicated processes of chromosome breakage, recombination (exchange of parts between pairs), and other rearrangements during inheritance. Mendelism through the work of Morgan became mechanistic.

And in the 1920s Mendelism finally became Darwinian. The phenotype was actually an intricate balance of genetic effects, some hidden, and thus variations would not be swamped. Mutation was but one source of variation; recombinations, inversions, and so forth provided a wealth of variable genetic material. All these sources of variation showed how slight and continuous changes could be preserved. The genotype, as Weismann had foreseen, was immortal, carrying a bank account of genetic variability which under certain conditions might be expressed. Hard inheritance without acquired characteristics could be made to agree with continuous variation without macromutation. Above all, Morgan's laboratory work demonstrated how Darwin could be made experimental, and in the 1940s ingenious experiments carried out by Theodosius Dobzhansky (studying adaptation under varying temperature conditions) actually showed evolution happening in the laboratory.

Still the picture was not complete, for the naturalists who initially opposed Mendelian genetics held in their hands important concepts which geneticists lacked—population thinking and statistical analysis of gene frequencies. In Russia, Sergei Chetverikov and his students came to view biological populations as statistical collections of genes—gene pools—thus shifting the emphasis from individuals to a fluid and dynamic stream of genes. In England R. A. Fisher published his *Genetical Theory of Natural Selection* (1930), which contained mathematical models showing how natural selection, acting upon a store of genetic variability, determined the probabilities of the inheritance of certain genes or groups of genes. J. B. S. Haldane expanded the model, adding new conditions; and in the United States Sewall Wright described evolution as direction shifts in the equilibrium or frequency of the gene pool.

Perhaps the most important work was done by Dobzhansky, his *Genetics and the Origin of the Species* of 1937. Here was finally the full combination of the population approach with the rigorous breeding and experimental methods of the geneticist, the integration of the naturalist approach with the laboratory. Here and in later works the gene-pool concept was no longer analogous to a bean bag; it became a dynamic and complex giant organism, a complex of genes interacting together and selected by a variety of biological factors. Recombinations and other forms of genetic variability became as important as mutation; complexes rather than atomic units were the stuff of selection.

The synthesis found its way into other disciplines as well. In paleontology George Gaylord Simpson showed that the fossil record was consistent with genetic mutation and variation. Ernst Mayr argued for the biological reality of the species defined by a host of factors which, added together, made species "reproductively isolated agregates of populations." Another ancient question had been answered.

The twentieth-century evolutionary synthesis was not a second revolution, it was the completion of the first. Both naturalists and Mendelians contributed indispensable concepts. Biological chance was tamed by mathematical probability. The living blueprint of nature had finally been read and, behold, its pages were bound together by Darwin.

If there was a twentieth-century revolution in biology, we must look for it upon another level. Back in 1868, amid debates over vitalism, Huxley had spoken of protoplasm as the physical basis of life, a foundation which could yet be resolved into the complex union and function of chemical molecules. To those who feared that the essence of life would be forever beyond its chemical basis, Huxley replied that this was like asking for the essence of "waterness" in H_2O. Weismann, too, wondered about the molecular constitution of the germplasm. The goal had been foreseen, but the nineteenth century lacked the prerequisite techniques and deeper chemical knowledge necessary to achieve it.

In 1897 Eduard Buchner found the substance zymase, an enzyme, in yeast cells. It was, thought Buchner, the "ferment" responsible for alcoholic fermentation, and indeed such enzymes might be responsible for all biochemical reactions, such as cell respiration and metabolism. At the turn of the century this was basically speculation and therefore questionable, yet in 1907 Loeb wrote that Buchner's discovery, exorcising the vital principle from fermentation, demonstrated that no problem should be considered beyond reach. Loeb, of course, was speaking in terms of the reductionist philosophy. Whether or not this was an acceptable *philosophy*, the basic method would indeed be to reduce the processes of life to the reactions, structures, and properties of chemical molecules.

In 1838 the name "protein" had been conferred upon albuminoid material. Chemists found that proteins could be broken down into amino acids, and Emil Fisher was convinced that proteins were constructed by amino acids in definite structures—Fisher had studied under Kekulé. In the 1920s techniques were developed which enabled chemists to produce protein crystals, the crystal being a definite latticework structure of the molecule. Some time earlier it had been suggested that X-rays with their short wavelengths could be passed through crystals and their atomic latticework be studied from diffraction patterns. This technique was developed by William H. Bragg and W. Lawrence Bragg (father and son) at Cambridge. Still other techniques (chromatography) helped chemists to establish the sequence of amino acids on protein molecules during the 1940s. And in the mid-1930s Linus Pauling produced a general theory of chemical bonding of proteins based upon the idea that the protein molecule chain folded back upon itself like a spring, the folding maintained by weak hydrogen bonds. The amino acid chain thus had the configuration of what is called an "alpha helix." Laboratory models of the alpha helix were built—another fruitful technique and rather reminiscent of nineteenth-century physics—and later X-ray crystallography validated Pauling's theory. Taken together, protein structures, the sequences of amino acids (some twenty of them), the establishment of enzymes as proteins, the discovery of coenzymes (many of them vitamins), and reductionist methods, led to an enormously complicated picture when these processes were recombined into the biochemical mechanics of the cell. They showed how matter and energy were transformed in living things, governed by the complex step-by-step action of enzymes. Added together, regular chemical processes, as Bernard had foreseen, gave rise to vitality.

Yet the structure and biochemistry of proteins still did not explain how molecules carried genetic information and how this information was replicated on the molecular level. In the early decades of the twentieth century it appeared likely that proteins with their amino acid sequences might be the candidate, and some biologists began to doubt the chromosome theory of inheritance. On the other hand, Morgan's group had demonstrated the genetic role of the chromosomes, although they had not investigated their chemical nature, and evolution made it clear that genes must somehow regulate the production of chemical substances in organisms. Since the nineteenth century it had been known that nucleic acids were found in chromosomes. Further studies of nucleic acids indicated that they consisted chemically of five nitrogenous bases—the purines (adenine and guanine) and the pyrimidines (thymine, cytosine, and uracil). Yet with all of the chemical work and interest in proteins, the molecular structure of chromosomes and their nucleic acids were largely ignored until the 1940s. Proteins with their greater numbers of amino acids seemed to be the logical carrier of the genetic code, but experimental genetics seemed to point to the chromosomes and in fact proved them to be the vehicles. How did they do it?

After 1930 the nuclean theory of heredity returned, and it was established that nucleic acids formed two large families of molecules: deoxyribonucleic acid, so labeled for its combination with phosphate (acid) and sugar (deoxyribose), and ribonucleic acid—for short, DNA and RNA. But genetics required of these molecules certain processes; they had to replicate themselves almost exactly in order to insure the stability of the genotype, and they also had to pass on instructions for the building of proteins and other chemical substances. Such activities were, perhaps, beyond the normal laws of organic chemistry, maybe even physics itself.

Some physicists felt that this might well be the case. In 1945 Schrödinger wrote that living matter might encompass physical and chemical laws as yet unknown (his book was called *What Is Life?*). Bohr speculated that the characteristics of living matter were quite different from those of the inorganic world and that biology might require a kind of complementarity approach between mechanism and vitalism like quantum physics. Schrödinger's book was especially influential. Many young physicists of the Second World War generation, feeling much of the excitement in physics to be past (and perhaps disillusioned by the atomic demon now loose), turned to biology. Might not the physical processes of genetics especially foster a new revolution in physics, as the great masters implied? Alas, such was not to be the case.

One of the young physicists who read Schrödinger's book was Francis Crick. After the war Crick went to the Cavendish Laboratories in order to work with the biologists and biochemists who were using the crystallography techniques. In 1951 he was joined there by J. D. Watson, a young American biologist interested in the molecular nature of the gene. By this time it was generally believed that DNA constituted the hereditary material, and the race was on to discover its structure and function. To find the structure and relate it to the function would, as Crick and Watson believed, provide the key to the puzzle of the genetic transfer of information—and, as Crick called it, "the secret of life."

Pauling's model building, X-ray analysis, knowledge of chemical bonding, and not least the proper mesh of personalities produced one of the most important biological discoveries of the twentieth century. By April of 1953 Crick and Watson had found the structure and glimpsed its genetic implications. The model of DNA they constructed was a "double helix," two spiral staircases of phosphate-sugar backbones twisted around each other, the stairs inside the outer backbones comprised of four paired nucleic bases linking the two chains together by weak hydrogen bonds (hence the double helix may unzip). Most important was the specific pairing sequence of the bases—adenine pairs with thymine and guanine pairs with cytosine. Thus the two backbones are complementary and, as Watson and Crick wrote in a follow-up paper, a long DNA molecule contains many possible permutations of base sequences, hence the base sequences are quite capable of carrying the genetic code. It is a kind of four-letter digital system.

The order of the bases on one backbone determines exactly the order of the bases complementary to it. Say the hydrogen bonds break. The chains unwind and each forms a "template" for the building of a new chain. Eventually there are two identical DNA double helixes, and this is how the genetic code is replicated on the chromosomes in the cell nucleus. In the 1960s it was also shown how the code was transcribed to amino acids in proteins. Copies of small parts of the DNA sequences are transcribed to a single strand of RNA. Carrying its genetic message, RNA becomes the template for translating the DNA four-letter code into the twenty-letter amino acid code. The molecular process is rather complex, since transfer or tRNA (what Crick called "adaptor molecules") is required to carry the amino acids to the template of the messenger RNA. Nevertheless, the process is mechanical, operating by *known* chemical laws. The old gene had become in effect a sequence of nucleotides, replicating themselves, building the organism, a typographical error here and there causing mutations. Evolution had finally reached the molecular level.

And no new laws were found, no new physics! In fact, the nature of the DNA code is the same everywhere, meaning that life on earth does have a common origin, as Darwin suspected. Further, the flow of genetic information is one-directional, DNA to DNA, and DNA to RNA to protein. There is no flow from protein to DNA and hence no inheritance of acquired characteristics. Molecular biology, as it is now called, became the great *mechanical* synthesis of the twentieth century (though of course beneath it lies the strange world of the quantum). While physicists questioned reductionism, many biologists became convinced adherents to it.

Like physics, however, biology too placed a bold new source of power in human hands. For in the 1970s biologists learned how to splice DNA and introduce it from one species to another. Portions of DNA from a donor organism are spliced to a virus, producting *recombinant DNA*, which is then allowed to infect recipient bacteria in which the virus replicates. Genetic engineering, like atomic physics, has placed into human hands enormous possibilities and, as with the modern alchemists, potential dangers.

Thus the story of the life sciences becomes an open history, in theory as well as practice, for there are still questions to be answered. One in particular haunts our thoughts: How did self-regulating life begin in those dim primeval ages? How did the DNA code itself evolve? And there is another, perhaps the most profound question of all: How does DNA give rise to the human brain, which is able to learn its own secrets? The next fortress to fall, Crick wrote, is the human brain (mind, according to Crick and many reductionists, being a function of the neurological system).

A brief glance at the science of psychology, its history, indicates a continual interplay between the physical sciences and conceptions of the mind. While Descartes had sundered the two, mind and body, the empiricists after Newton conceived the mind as a kind of mental mechanism governed by the same laws as the physical world, attraction and repulsion of ideas and sensations. Some in the eighteenth century emphasized the structure and complexity of matter as the key to mental activities, the human mind itself a product of physical activities activated by sensations.

Experimental psychology developed in Germany during the nineteenth century. Its foundations were the ideas formulated in mechanistic physiology, the theories of Helmholtz, du Bois-Reymond, and others who sought to reduce all phenomena to the laws of physiochemical matter and energy, studied by the familiar mathematical-physical methods. Its subject matter was basically the nervous system, response, and energy transformations in the organism. It was more of a philosophical position than a school or specific discipline.

The founder of nineteenth-century psychophysics as a discipline was Gustav Theodor Fechner. In 1860 Fechner made psychology into an "exact science" with his mathematical studies of the functional relations between magnitudes of stimuli and sensations. Most important, he showed how psychological experimentation was possible. In the later nineteenth century Wilhelm Wundt established the first laboratory dedicated to psychophysical experiment and founded a journal for experimental psychology. Psychology as an independent science had thus arrived, but it was not without problems. At Harvard, William James introduced the new German psychology, yet James felt that it was incomplete. The atomizing of experience as practiced in the laboratory neglected, James believed, the continuous flowing stream of consciousness. This consciousness is intentional, selective, and hence nonmechanical. It is beyond experiment, perhaps the ghost of free will in the deterministic machine.

Along with mechanistic physiology, evolutionary biology from Darwin to DNA profoundly influenced psychology. Closing the gap between humans and animals fostered new studies of animal behavior, as Darwin himself had done. In Russia Ivan Pavlov applied Loeb's mechanistic approach to the study of animal behavior and learning. His famous experiments on a dog demonstrated that through the repetition of stimuli a "conditioned response" is established, a new connection or reflex arc (he believed) in the cerebral cortex. Here was no mind-body dualism; for complex behavior and even learning could in theory be reduced to neuronal connections. All else was speculation.

In the second decade of the twentieth century the American J. B. Watson argued that for psychology to be a science it must completely sever its ties with philosophy and nebulous concepts such as consciousness. The mind is like a sealed box, its only observable data being behavior, and we can study only what goes in and what comes out. One pathway from this "behaviorism" has led to a strict determinism in the hands of B. F. Skinner. Human beings, according to Skinner, are complexes of possible behaviors whose frequencies of occurrence depend upon reinforcement. In fact, positive and negative reinforcements are the determinants not only of behavior but of values, feelings, and the entire spectrum of cultural mores. Consciousness, even the sealed box, are nonscientific ghosts. And with a scientific knowledge of value formation Skinner argued that we may be able to build a utopia (*Walden Two*).

Psychology developed in a somewhat different direction through the work of Sigmund Freud, his theory of the unconscious mind, psychological disorders, and later his psychobiological history of civilization. The unconscious mind (postulated by many before Freud) is the seat of repressed wishes and desires, mostly infantile and sexual in origin, the satisfaction of which would result in punishment or harm to the individual. Thus consciousness closes the gateway to motor activity and seals them off. In sleep, with its absence of conscious motor activity, the repressed wish finds an outlet in dreaming, but a disguised outlet, the manifest form of the dream which serves as a protective shield for the inappropriate latent reality of the desire. This dynamic theory of the mind was proposed by Freud in *The Interpretation of Dreams* (1900) and became the basis of psychoanalysis, which as Freud saw it was an independent science.

Yet Freud went on in later studies to give the development of this mental mechanism an evolutionary cast.[1] Repression was the origin of civilization, he wrote, and also the result of civilization. How did it arise? Are infantile erotic instincts and fears—the Odeipus complex, the castration complex, and so forth—purely psychological, the environmental result of personal experience?

In *The Descent of Man* Darwin had speculated that the early hominids lived in small bands and the most powerful male of the "primal horde" possessed all the females. The younger males, sons of the father, were thus denied access to the females and impelled to form their own bands. Now in *Totem and Taboo* Freud imagined that the brother clan banded together and committed the crime of Oedipus—killing (and eating) their father. Then, overcome by guilt, they deified the father and established the two great prohibitions against incest and parricide. Therefore, all religion (the deified father-god), morals, art (the totem), and laws derived from the crime of the sons.

How, then, does this cause modern repression? Freud's answer was peculiarly biological. The above primal events occurred many times and became sealed by repetition into the unconscious. They are, in short, Lamarckian acquired characteristics! In order for civilization to progress, the primal instinct of free sexual expression had to be repressed. But behind such repression stand the actual events of the father murder and accompanying guilt organically impressed into the unconscious, where they form the wild *id*. The personal history of the child, individual development, is a *recapitulation* of this psychosexual evolution and its acquired experience! Civilization in essence is the result of human evolutionary biology.

Later in *Civilization and Its Discontents* (1930) Freud conceived of an instinct inherent in life to return to the most simple state of existence before life itself. This is the death-wish, the instinct for destruction and a return to the inorganic. Civilization is, at heart, a struggle between Eros and the Death-Wish, a struggle not unlike that of the gods of order and the giants of chaos.

While most biologists reject the biological underpinnings of Freud's psychobiological history (recapitulation and Lamarckism), E. O. Wilson

---

[1]The case for the biological foundations of psychoanalysis in general is made by Frank J. Sulloway in *Freud, Biologist of the Mind* (New York: Basic Books, 1979).

in the 1970s formulated a new synthesis of evolutionary biology, human behavior, and the development of human culture. Called *sociobiology*, Wilson's theories propose a genetic basis, formed by natural selection, for such traits as values and morality, aggression, artistic talents, social cohesiveness, incest taboos—in short, a genetic basis for civilized culture. Wilson demonstrated that activities such as altruism would favor the passing of "altruistic determining genes" to future generations. Culture in general may be the result of natural selection reduced ultimately to its genetic basis in DNA, and the word "mind" becomes in effect a synonym for the genetically selected matrix of the brain. And with the possibilities, speculative or real, of genetic engineering . . . ?

We the survivors must live in the world science has built. The possibility of conscious intervention in genetic processes, like the power bestowed upon the human species by nuclear physics, requires a wise and careful contemplation of the ethical dimensions of scientific endeavor. Indeed, this task may well occur on a level coeval with internal scientific pursuits. And this brings us to the epilogue of our story.

SUGGESTIONS FOR FURTHER READING

ALLEN, GARLAND E. *Life Science in the Twentieth Century*. Cambridge: Cambridge University Press, 1978.

BORING, EDWIN G. *A History of Experimental Psychology*, 2d ed. Englewood Cliffs, N.J.: Prentice-Hall, 1950.

BOWLER, PETER J. *The Eclipse of Darwinism: Anti-Darwinian Evolutionary Theories in the Decades Around 1900*. Baltimore: The Johns Hopkins Press, 1983.

DUNN, L. C. *A Short History of Genetics: The Development of the Main Lines of Thought 1864–1939*. New York: McGraw-Hill, 1965.

FRUTON, JOSEPH S. *Molecules and Life: Historical Essays on the Interplay of Chemistry and Biology*. New York: John Wiley, 1972.

GRAHAM, LOREN R. *Between Science and Values*. New York: Columbia University Press, 1981.

LOWRY, RICHARD. *The Evolution of Psychological Theory: A Critical History of Concepts and Presuppositions*. New York: Aldine Publishing, 1982.

MAYR, ERNST, and PROVINE, WILLIAM. *The Evolutionary Synthesis: Perspectives on the Unification of Biology*. Cambridge, Mass.: Harvard University Press, 1980.

SULLOWAY, FRANK J. *Freud, Biologist of the Mind: Beyond the Psychoanalytic Legend*. New York: Basic Books, 1979.

WATSON, JAMES D. *The Double Helix: A Personal Account of the Discovery of the Structure of DNA*. New York: Atheneum, 1968.

Epilogue

"Human, All-too-Human"

Einstein wrote: "Man is, at one and the same time, a solitary being and a social being." The same might be said about science. At times it is a human, perhaps "all-too-human," endeavor to understand the material world, isolated and driven by its own internal dynamics; on the other hand, it often occupies a reciprocal role in the larger human, all-too-human world of values, politics, economics, and so on, interacting on many levels with the greater cultural milieu. It is not totally value-free, nor is it wholly value-laden; it combines elements of individual psychology along with inherited cultural presuppositions. Frequently these varied, often conflicting strivings, create a tension as traditional social values are brought into the domain of scientific criticism, or as external cultural developments influence and sometimes shape the material environment in which science operates or as science contributes to technological powers, human domination, and, alas, destruction of corporeal nature.

In the wake of the World Wars, most notably after the Second World War, there has been a steady if not unprecedented growth in most governments' financing and, at times, direction of scientific research and training. Here the emphasis comes to rest more and more upon science's utilitarian and practical applications through the mediation of technology. It is not difficult to see why: From communications, to the alleviation of many illnesses, to nuclear power, scientific developments have profoundly altered social, economic, and political life. Military

power, for example, is no longer concerned primarily with the possession of territory or the size of armies (although in terms of resources,
often gauged by their technical or industrial merits, these are still important); rather, military power becomes increasingly measured by
scientific and technological innovations.

While science has thus established itself as the preeminent force in
the dynamics of recent history, its very successes have in turn had a profound influence upon science itself. One of the great ironies is that many
of the discoveries from which technology springs had their origin in
purely theoretical exercises of the imagination, directed mainly by curiosity or unsolved esoteric problems which were of interest only to a
small, relatively isolated group. It seems almost superfluous to mention
nuclear power here, but we should also note that Planck's desperate
quantum of action has resulted in such things as lasers, masers, and especially solid-state physics, which has given us technological marvels all
the way from the transistor to the silicon semiconductors used in the
standard micro chip which has led to the computer age. The list can go
on, but the irony involved is simply that the vogue for a utilitarian and
practical problem-solving scientific research is in some ways opposed to
the more haphazard and "idle curiosity" of the past, a curiosity "idle" in
the utilitarian sense but vigorous in the intense desire to understand.

Scientists as a special interest group cannot help being attracted to
those avenues supported by governments or industries and offering
massively funded research facilities, economic rewards, and prestige.
While this development may result in advances, both theoretical and
practical, it also implies an ever-increasing exterior imposition of problems and goals. The latter may, in the long run, upset the delicate balance between accepted tradition and radical innovation, the kind of revolutionary change that comes from the human, all-too-human factors
such as those which drove Copernicus searching for harmony, or
Newton scouring alchemy for its hidden secrets, or Darwin speculating
on transmutation, or Einstein declaring the aether superfluous. Owing
to the immense economic and even political pressures brought to bear
upon the scientific community, a certain amount of freedom and open-
ness may be lost.

All this goes on at a time when no one can say with certainty that
our amazing and apparently successful picture of the universe is the
final word. Indeed, buried somewhere in that picture could lie a mysterious tiny fact, an unexplained anomaly, which holds the key to a completely new scientific vision; or an unforeseen bend in the road leading
to a higher ground of understanding upon which all the old facts are
burned away like mist beneath a midmorning sun. An ossified science,
frozen by its own success and following some predetermined course,
may not have the flexibility to take the revolutionary turn in the road or
even see it.

The influence of scientists upon political decision making has correspondingly increased. In technical matters of policy the political and economic decision becomes nearly synonymous with the scientific advice, which, since the layman really has no basis of judgment, becomes authoritative. This tendency has become even greater since the Second World War, as military power comes to depend more and more upon science. Scientists, then, are drawn into politics and diplomacy, where by the very nature of their position they command a great influence upon decision making. Although their advice is based upon expert knowledge, inevitably different groups of scientists will bring into that advice political and social ideologies. Their promotion of policies is based upon overtly technical expertise, yet imperceptibly their ideological positions tend to merge with this role as they assume more and more political power. Hence, because of their unrivaled authority, they are in the position to become a powerful new "priesthood," holding in their hands the future of society in general. In such a situation the ideal of an objective, value-free science, while never a totally accurate picture in the past, becomes a dangerous delusion.

Up to the rise of professionalism in the late nineteenth century, scientists mixed speculation on wider issues with their research. With professionalization and increasing specialization a discussion of values in scientific work came to be considered "unprofessional."[1] Ultimately the belief that knowledge itself is a value seemed to provide the fundamental moral usefulness of science, and questions dealing with ethical principles, values and such, were assumed to be sharply delineated from the pursuit of scientific knowledge per se. But this ideal distinction cannot easily be made, past or present, just as it becomes increasingly difficult to separate science and technology. The "usefulness" of science, itself a value, has come to be associated with much more; in popular terms science and technology has become a tautology as, perhaps, have science and truth itself. Many scientific disciplines, especially genetics, are by their very nature intimately connected with social values, and their knowledge in its social implications can hardly be considered value-free.

Nonetheless, many values which may appear irrational from the perspective of science have arisen from living social experience and conceal social as well as psychologically useful functions. And there exist other traditions, other sophisticated and profound ways of understanding the universe, often in conflict in Western rationalism, the loss of which to rational science could result in a certain spiritual impoverishment of humanity. It is also quite possible that many of these

---

[1]See Loren R. Graham, *Between Science and Values* (New York; Columbia University Press, 1981).

"nonscientific" systems, like Newton's alchemy, contain valuable hints or even jewels which may ultimately enrich science itself. The limits and problematic nature of Western rationalism have haunted many thinkers. There is, in fact, no guarantee other than from its own perspective that our modern science is superior to what we might label "nonscientific" attempts to understand the world. Philosopher Paul Feyerabend has called for more modesty from rationalists, and Feyerabend is prepared to admit that Western rationalism is but one of many myths and not necessarily the best. The apparent successes of technology, many of which are simply experience dependent, have tended to conceal such possibilities: The paradoxes of the quantum theory, for example, are of little interest to the engineer who uses its mathematical recipes in practical application. But to echo the ancient philosopher, wisdom begins in mystery, indeed a certain humility before the great wonder of nature.

In the last century Nietzsche exclaimed: "And even your atom, *messieurs* mechanists and physicists, how much error, how much rudimentary psychology, still remains in your atom!" And now, in the century of the quantum, we wonder whether objectivity—the invisible wall separating the observer from the observed—was ever really possible, whether every observation is not really a dialogue. In one sense and on one level our history has already spoken to the question; this ideal separation of human psychology from scientific contemplation of "facts" becomes a kind of myth when seen from the perspective of the history of science. The great "heroes" who discovered the "facts," often painted in the colors of Baconian induction, were still human beings, and their creations sprang from the total human matrix, from the extremely complex interplay of esthetics, values, religion, passions, all of which interacted with the physical world, shaping it, shaped by it. Science in the forge of creation is a fluid metal alloy of indefinite components, some visible and others not. But once the alloy has cooled and after it has been hammered into a finely crafted precision instrument, the textbook scientific achievement, these human components have melted and vanished from sight. Nonetheless they are there, hidden. Without them the instrument could not have been made. Separated by history, these enigmatic elements greet us with a surprising message and, perhaps, a profound whisper of wisdom: Science is a humanity, one of the humanities in the full sense of the term.

Is it not a wondrous thing that upon this infinitesimally tiny planet orbiting a very ordinary star in the suburbs of a galaxy of a billion stars, yet a very ordinary galaxy among thousands of millions, a species has evolved to the point of puzzling itself over the mystery of the universe? And more surprising still, that the universe itself may be, at least in part, comprehensible to this species? Einstein found the latter to be the most incomprehensible thing about nature. Perhaps it is really not. The delicately balanced physical laws of our universe set the very conditions, perhaps the *only* conditions, which made it inevitable that an intelligent species like us should arise capable of pondering the mystery. Physicists label this "the anthropic principle." Physicist John Wheeler has gone even further, taking his cue from the observer-participancy nature of the quantum: The universe produces beings capable of observing it, and such acts of observation, like collapsing the wave-function, impart tangible reality to the universe both past and present. We are not only the means by which the universe studies itself, we are the conscious eye, the decision-making lens through which the universe becomes what it is.

This may all be speculation, the free play of the imagination, metaphysics. But then if we have learned anything from the past, it is that such human, all-too-human traits are indispensable for science—these coupled with mystery and the curiosity it inspires. Nevertheless there are other important elements, and one in particular is faith; not a fanatical faith in answers, rather, a faith that we have the ability to understand, that our questions have answers, that we are even asking fruitful questions, and that there is something tangible and real in nature which corresponds to our conceptual schemes. And balancing this faith there must also be that peculiar human ability to doubt, to step back from apparent success, from the "facts," doubting everything, and then to transform that doubt into constructive action without being overwhelmed by it. From such alchemical opposites, fused in the human soul, springs the adventure we call science. That such a delicate and fragile venture has continued to the present day may be the most astonishing fact of all.

Index